The
Physiology of Tropical Orchids in Relation to the Industry

The

Physiology of Tropical Orchids in Relation to the Industry

C. S. Hew & J. W. H. Yong

Department of Botany
National University of Singapore

World Scientific
Singapore • New Jersey • London • Hong Kong

Published by

World Scientific Publishing Co. Pte. Ltd.

P O Box 128, Farrer Road, Singapore 912805

USA office: Suite 1B, 1060 Main Street, River Edge, NJ 07661

UK office: 57 Shelton Street, Covent Garden, London WC2H 9HE

Library of Congress Cataloging-in-Publication Data

Hew, Choy Sin.

The physiology of tropical orchids in relation to the industry / C. S. Hew & J. W. H. Yong.

p. cm.

Includes bibliographical references and index.

ISBN 9810228554. -- ISBN 9810231938 (pbk.)

1. Orchid culture -- Asia, Southeastern. 2. Orchids -- Asia, Southeastern -- Physiology. 3. Orchid culture -- Tropics. 4. Orchids -- Tropics -- Physiology. I. Yong, J. W. H. (Jean W. H.) II. Title.

SB409.5.A785H48 1997

635.9'344--dc21 97-18872

CIP

British Library Cataloguing-in-Publication Data

A catalogue record for this book is available from the British Library.

Printed in Singapore by Uto-Print

Foreword

I take great pleasure in writing the foreword to this book, *The Physiology of Tropical Orchids in Relation to the Industry*, which relates to a thriving industry.

Cut-flower orchid production and potted orchid cultivation have been a mainstay agro-industry in South East Asia and indeed, throughout the world. In order to sustain and nurture the growth of the industry, new and improved agro-technology is needed.

The scientific disciplines that contribute to improving orchid production technology have been developed to such sophisticated and specialised levels that the trial-and-error approach generally adopted by orchid hobbyists and commercial growers can no longer be depended upon to meet the demands of a global cut-flower market. Scientific studies on orchid biology are paving the way for the orchid industry. If orchid researchers, hobbyists and commercial growers can be provided with convenient access to more recent research findings, clearly these would greatly enhance their efforts in meeting the challenge of improving the production technology.

There are very few orchid books in the world that deal specifically with the scientific aspects of orchid biology and cultivation. In South East Asia, there is not, as yet, an organised source of tropical orchid literature suited for the study of orchid biology and the direct application of this knowledge to serve the industry. The contribution of Professor Hew Choy Sin and Mr Jean Yong to tropical orchid biology and industry is therefore both valuable and timely. This book is written in response to the growing demand for an orchid physiology book with a tropical perspective both in Singapore and her neighbouring South East Asian countries. This pioneering book aims at defining the status of our present knowledge of orchid physiology, with an emphasis on tropical orchids, and considers how existing knowledge can be put to greater and more practical use. The authors have identified the gaps in our knowledge and discussed how these gaps can best be filled through additional research. *The Physiology of Tropical Orchids in Relation to the Industry* will be an important and useful source of information for university students, orchid researchers and commercial orchid growers.

I congratulate the authors for sharing their expertise.

Professor Leo Tan
Director
National Institute of Education
Nanyang Technological University

President
Singapore National Academy of Science

Preface

The fundamental aim underlying the writing of this book is the desire to provide a comprehensive and exclusive text of tropical orchid physiology relevant to commercial growers, research workers and graduate students. Over the past few decades, the orchid industry is growing at a steady pace in the South East Asian and East Asian regions, and it is becoming an essential export item in some Asian countries. To maintain this progress, there is an urgent need for a comprehensive book that is relevant to the region to guide orchid growers in improving their cultivation and management skills, and to guide new students in understanding orchid physiology.

There are scientific books written on orchids that are very good, in our opinion, such as *The Orchids: A Scientific Survey*, *Orchids: Scientific Studies*, *Fundamentals of Orchid Biology* and the book series *Orchid Biology: Reviews and Perspectives*. We hope that this book would complement the existing scientific literature available to improve orchid cultivation and to set new research agenda especially in the tropics.

The bulk of the text is based on the research effort of past graduate students, research associates and visiting scientists working with Professor C. S. Hew in Nanyang University and later, in the National University of Singapore. The duration of orchid research spans 26 years, first started in 1970, and is still been actively pursued till today. To fill the relevant gaps in information and for comparison purposes, relevant publications from other research groups are also included. This inevitably includes some discussion of the temperate orchids. The idea of this book was conceptualized when we were making a computer database of publications related to orchid physiology in 1995. We decided to take a step further and to produce an integrated and unifying theme of tropical orchid physiology with a clearly written factual text and illustration. The present cultural technology has given growers and hobbyists the opportunity to grow orchids anywhere in the world. As such, the strict demarcation of whether an orchid is a tropical or a temperate one is no longer possible. We proposed that the term "Tropical orchids" be perceived in a broad sense.

There are nine chapters in this book. Each chapter is designed to provide a comprehensive, up-to-date information on an aspect of orchid physiology. References in the text are reduced to include only the leading authorities in the appropriate fields. Whilst it is recognised that the study of biological science follows no set pattern, the content of different chapters is written using a similar approach. Unlike the earlier chapters, Chap. 9 is a unique chapter where it deals with the problems and recent advances in orchid tissue culture. This chapter looks at the problems created by growing orchids in an artificial environment and offers practical solutions and new research directions to improve *in vitro* orchid growth.

C. S. Hew & J. W. H. Yong
Singapore, March 1996

Contents

Foreword v
Preface vii
Acknowledgements xiv

1. The Relevance of Orchid Physiology to the Industry 1
1.1. Introduction 1
1.2. Orchid Cultivation and Industry 2
1.3. How Basic Orchid Physiology Can Help the Industry 5
1.4. Concluding Remarks 8

2. A Brief Introduction to Orchid Morphology and Nomenclature 11
2.1. Introduction 11
2.2. Growth Habit 11
2.3. Orchid Plant Parts 13
Pseudobulbs 13
Flowers 15
Seeds 22
Leaves 22
Roots 23
2.4. Growth Cycle of Orchids Under Greenhouse Conditions 30
2.5. Nomenclature 30
Species 30
Hybrid 33
2.6. Summary 33

3. Photosynthesis 37
3.1. Introduction 37
3.2. Photosynthetic Pathways 37

3.3. What is $\delta^{13}C$ Value? 41
3.4. Patterns of CO_2 Fixation in Orchids 45
Thin-leaved orchids 45
Thick-leaved orchids 49
3.5. Photosynthetic Characteristics of Non-Foliar Green Organs 52
Aerial roots 54
Stems 61
Pseudobulbs 62
Flowers and fruit capsules 64
Varying $\delta^{13}C$ values in non-foliar green organs 66
3.6. Factors Affecting Photosynthesis 68
Effects of light 68
Effects of age 69
Effects of water stress 75
Effects of temperature 77
Effects of sink demands 81
Effects of pollutants 82
Effects of virus infection 84
Effects of elevated carbon dioxide 85
3.7. Concluding Remarks 86
3.8. Summary 87

4. Respiration **93**
4.1. Introduction 93
4.2. Respiratory Processes 93
4.3. Respiration in Plant Parts 96
Protocorms and Seedlings 96
Leaves 99
Flowers 101
Roots 106
4.4. Respiratory Drift During Flower Development 109
4.5. Photorespiration 118
4.6. Other Oxidases in Relation to Orchid Respiration 120
4.7. Concluding Remarks 122
4.8. Summary 123

5. Mineral Nutrition	**129**
5.1. Introduction	129
5.2. Mineral Requirements and Tissue Analysis	129
5.3. Fertiliser Application Practices	136
Effects of organic fertilisers on orchid growth	138
Effects of mulching on orchid growth	139
Effects of inorganic fertilisers on orchid growth	143
5.4. Foliar Application and Root Absorption	149
5.5. Ion Uptake	152
Ion uptake by orchid tissues	152
Ion uptake by orchid roots	153
5.6. Concluding Remarks	161
5.7. Summary	161
6. Control of Flowering	**168**
6.1. Introduction	168
6.2. Differentiation of Flower Bud	168
6.3. Factors Affecting Flower Induction	170
Juvenility in orchids	172
Response to low temperature	172
Photoperiodic response	177
Hormonal control	177
6.4. Seasonality in Flowering	179
6.5. Application of Flower Induction at the Commercial Level	183
6.6. Bud Drop	188
6.7. Controlling Orchid Flower Production	189
6.8. Concluding Remarks	192
6.9. Summary	193
7. Partitioning of Assimilates	**198**
7.1. Introduction	198
7.2. The Source–Sink Concept of Phloem Translocation	198
Sources and sinks	199
Phloem loading	200

Along the path 201
Phloem unloading 201
7.3. Patterns of Assimilate Movement in Most Higher Plants 202
7.4. Patterns of Assimilate Movement in Tropical Orchids 204
Assimilate partitioning in the sympodial orchids 205
Assimilate partitioning in the monopodial orchids 220
7.5. Import of Assimilates by Mature Orchid Leaves 226
7.6. The Role of Non-Foliar Green Organs in Assimilate Partitioning 228
7.7. Improving the Harvestable Yield of Orchids 228
7.8. Concluding Remarks 239
7.9. Summary 240

8. Flower Senescence and Postharvest Physiology **245**
8.1. Introduction 245
8.2. Senescence in Plants 245
8.3. Growth and Development of Orchid Flower and Inflorescence 247
8.4. Flower Senescence in Orchids 254
Post-pollinated phenomena 254
Ethylene and senescence 256
8.5. Postharvest Handling of Cut-Flowers 267
Preharvest conditions 269
Extension of vase-life 270
Formulation of various solutions 271
Bud opening 276
8.6. Storage and Transport 276
Low-temperature storage 277
Hypobaric storage/controlled storage 277
8.7. Concluding Remarks 280
8.8. Summary 280

9. Recent Advances in Orchid Tissue Culture **288**
9.1. Introduction 288
9.2. Factors Affecting Orchid Growth *in Vitro* 289
Sugar 290
Carbon dioxide 292
Ethylene 293
Nitrogen sources 296
Light 297
Other factors 299
9.3. Improving Orchid Cultures 300
Gas-permeable culture system 300
Alternative supporting media 306
Carbon dioxide enrichment 308
Development of a flow system 310
9.4. *In-Vitro* Flowering 312
9.5. Thin-Section Culture 313
9.6. Synthetic Seeds 314
9.7. Concluding Remarks 315
9.8. Summary 317

Subject Index 323
Plant Index 329

Acknowledgements

We thank Mrs. Hew Yik Suan and Miss Gan Kim Suan for their help in preparing and editing the manuscript. The technical support of Mr. Ong Tang Kwee over the years is greatly appreciated. We are grateful to the following for their help in many ways: Multico Orchids Private Limited, Lee Foundation, Professor M. Tanaka, Dr. Hugh Tan and Dr. S. C. Wong. We are grateful to the publishers and journals for allowing us to reproduce their illustration and acknowledgement is given beside the illustration. The strong institutional support provided for orchid research by Nanyang University, and later, the National University of Singapore, is acknowledged.

Chapter 1

The Relevance of Orchid Physiology to the Industry

1.1. Introduction

Layman and scientists alike have always been fascinated by the beauty and mystery of orchids. The appreciation of orchid beauty has a very long history in both the Western and Eastern cultures. Much of this is attributed to the diverse form and structure of orchids and the large number of species in the orchid family. Arditti (1992) has given an excellent historical account of orchids in Asia, Africa, Europe, New Guinea and Australia. Suffice to say, the beauty and appreciation of orchids are subjective to the beholder. Some like them small while others like them to be showy. In oriental literature, *lan* (which means orchid in Chinese), for example, is often personified as a man of virtue who strives for self-discipline, champions his principles and does not succumb to poverty and distress.

Confucius wrote:

> "*Lan* that grows in deep forests never withholds its fragrances even when no one appreciates it."

These very ethereal qualities of *lan* have been much appreciated in the Orient since some 2,500 years ago.

1.2. Orchid Cultivation and Industry

Orchid cultivation has come a long way. Over the years, it has evolved from a hobbyists' market into a highly commercial market. Large-scale cultivation of orchid cut-flowers and potted orchids is now the trend. In the past, orchid growers and hobbyists relied solely on the collection of orchid species from the wild because the technique of breeding and selection (either by conventional or genetic manipulation) is not available. Mass cultivation becomes possible with the breakthrough in orchid seed germination. This laid the foundation for intensive breeding and selection of new orchid hybrids. The discovery and development of an asymbiotic method to germinate orchid seeds in 1921 by Lewis Knudson. This has also paved the way for the development of tissue culture technique for mass clonal propagation of orchids. The availability of asymbiotic germination and tissue culture has made large-scale orchid cultivation economically feasible. Today, orchids such as *Cymbidium*, *Dendrobium*, *Phalaenopsis* and *Oncidium* are marketed globally and the orchid industry has contributed substantially to the economy of many ASEAN (Association of the South East Asian Nations) countries (Hew, 1994; Laws, 1995).

The market potential for both orchid cut-flowers and potted orchids is very favourable (Laws, 1995). This is evident from the world market demand of planting materials for orchids grown for cut-flowers and potted plants (Table 1.1). In the year 2000, the total demand is estimated to be 1,598 million units of plant stock. Based on the Japanese flower auction sale figures for 1993, orchid cut-flowers accounted for 32% of the total market share, amounting to US$ 53.7 million, and all the orchid cut-flowers are imported from Thailand, Singapore, Malaysia and the Philippines (Fig. 1.1). Japan is now the major market for ASEAN orchid cut-flowers, replacing Germany, and the import of orchid cut-flowers into Japan has been increasing steadily from 1985 to 1995. In 1993, orchid cut-flowers formed about 7% of the US$ 3 billion cut-flower market in Japan (Fig. 1.2). The Japanese market for potted orchids was estimated to be at US$ 261 million in 1993 (Fig. 1.3). The status and future development of the orchid industry in ASEAN have been reviewed recently and the prospects for ASEAN orchid growers are indeed bright (Hew, 1994).

Table 1.1. World demand for orchid planting material.

	Plant stock turnover (million units)			Total estimated sales in 5 years (Millions of US$)
	1995	2000	Change in percentage	
Planting materials for cut-flower production	66	109	Increased by 11%	170
Planting material for potted plants	1220	1489	Increased by 4%	1891
Total	1286	1598		2061

Note: Sales values are based on blooming size plants priced at US$ 1.50 per plant.

Source: Unpublished market estimate of world orchid (tropical, sub-tropical and temperate) demand, provided by Multico Orchids Private Limited, Singapore.

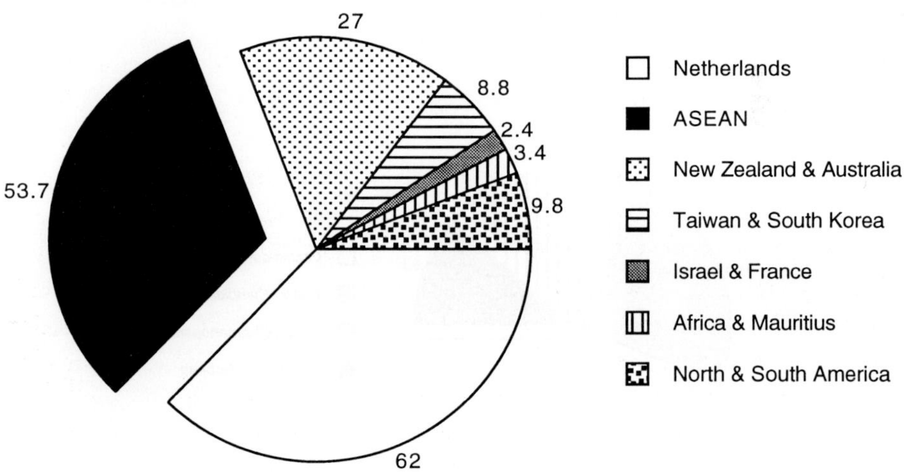

Fig. 1.1. Japanese flower imports in 1993.
Note: Figures are quoted in millions of United States dollar.

Redrawn from Suda (1995).

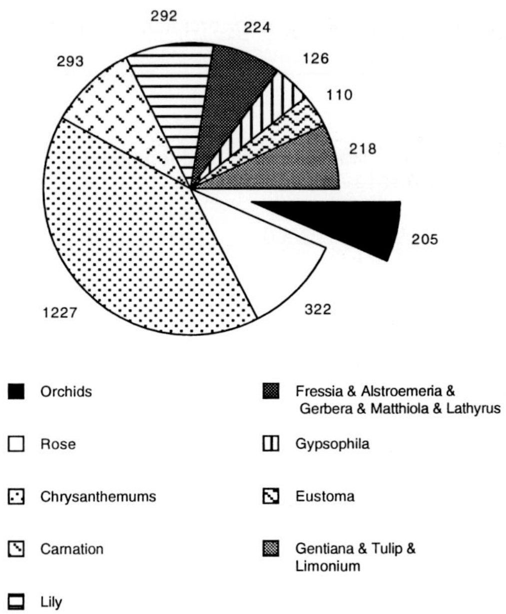

Fig. 1.2. Japanese cut-flower auction sales in 1993.
Note: Figures are quoted in millions of United States dollar.

Redrawn from Suda (1995).

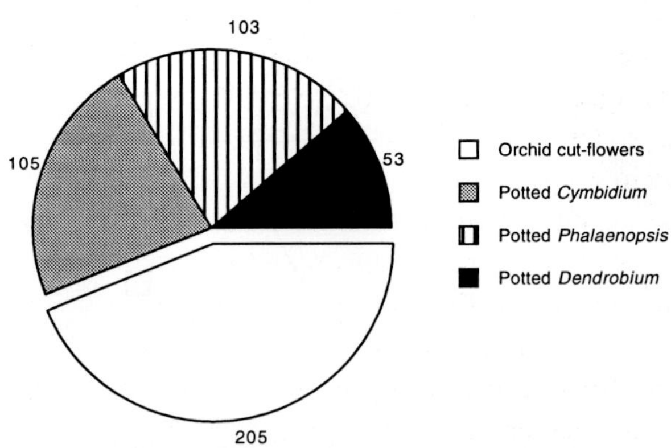

Fig. 1.3. Japanese auction sales for orchid cut-flowers and potted orchids in 1993.
Note: Figures are quoted in millions of United States dollar.

Redrawn from Suda (1995).

There are three major factors that contribute significantly to the success of the orchid industry:

1. Excellent environmental conditions that favour low production cost.
2. High production technology that results in high productivity and good product quality.
3. Good marketing and distribution leading to market advantages.

Being in the tropics, ASEAN countries are endowed with a climatic condition well-suited for large-scale orchid cultivation. Hence, it is not surprising that considerable efforts have been made to upgrade technology pertaining to commercial orchid cultivation. A good understanding of orchid physiology is the key step to improving orchid cultivation.

1.3. How Basic Orchid Physiology Can Help the Industry

The physiological basis of crop yield has been dealt with in great details for most agricultural crops (Evans, 1975). Physiological processes that determine crop yield are canopy structure, photosynthesis (pathways and rates), crop respiration, photorespiration, water relations, mineral nutrition, partitioning of assimilates and storage capacity. A thorough understanding of all these processes is essential to improve crop yield. In the following chapters, we would like to use this similar approach to improve orchid cultivation by studying the various physiological processes affecting orchid growth. We have resolved the orchid cut-flower production cycle into a series of processes and examine the relevance of orchid physiology in each process (Fig. 1.4). The resolution of the orchid cut-flower production cycle into discrete processes is a logical approach to identify any possible limiting factor. We believe that this approach is an effective way to optimise orchid cultivation for cut-flower production and to a lesser extent for potted orchids.

In starting an orchid farm, an important consideration is to ensure a steady supply of good quality planting materials. Obtaining planting material through conventional vegetative propagation method is a slow and costly affair. Today, the supply of uniform clonal planting material comes mainly from tissue culture. This demand for micropropagated orchids also explains the recent rapid increase in the number of commercial orchid tissue culture laboratories operating in ASEAN countries.

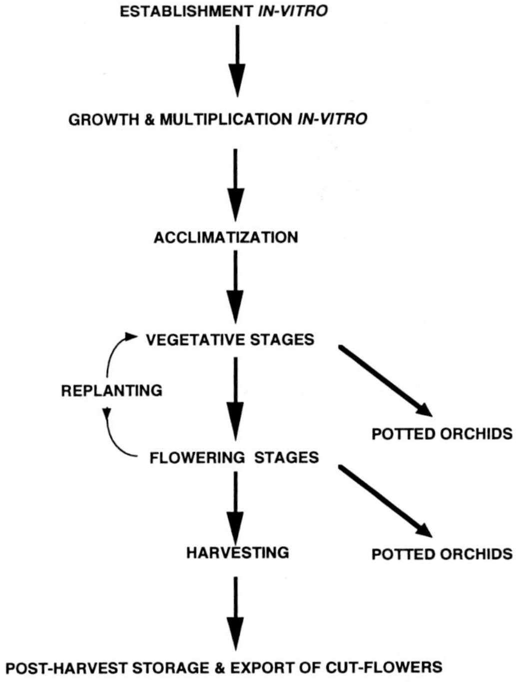

Fig. 1.4. Key production processes of the orchid industry.

Rapid and large-scale clonal propagation of orchids is made possible by using the batch tissue culture procedure. To date, more than 43 orchid genera have been mericloned successfully using different plant parts including leaves, roots, flower stalks, axillary buds and apical meristem (Arditti and Ernst, 1993). Clonal propagation of orchids using batch tissue culture has been the mainstay throughout the world since 1960. There are, however, problems associated with the batch tissue culture approach. In batch culture, the explant is cultured on a defined liquid or solid medium. Given an appropriate culture medium, the explant proliferates and then differentiates. Batch culture is essentially a closed system and the *in-vitro* conditions will change with time and may not be optimal for cell growth. Since the tissues are grown in a fixed volume of medium, there is a continual depletion of nutrients and accumulation of toxic materials. To optimise cell growth, it is important to maintain all factors at optimal conditions. In batch culture, this is only possible by very frequent subculturing. Subculturing involves considerable time and effort and will certainly cause a major increase in production cost. In recent years, there have been considerable improvements made in this area. The improved cultural methodology is essentially based on a better understanding of basic plant physiology.

Generally, orchid seedlings that are grown in flasks are first transferred to a community pot, then to thumb pots, after that to a 8 cm (in diameter) pot, and finally to a 15 cm (in diameter) pot. The duration for each transfer is about six months. It is surprising that few scientific studies have been made on the growth and survival rate of plantlets during and after the transfer from culture flask to community pots in the greenhouse. In fact, high plantlet mortality rates have often been experienced with some orchid hybrids. The hardening or acclimatisation of plantlets in flasks and community pot certainly deserves more research. The development of new approaches such as the photo-autotrophic culture system with CO_2 enrichment represents a significant contribution to improve the growth and acclimatisation of orchid plantlets under *in vitro* culture and during transplanting.

In the tropics, it may take more than two years for the orchid plantlets to reach the flowering stage. Orchids, particularly those with an epiphytic origin, are notoriously slow-growing plants. The slow growth of epiphytic orchids may be attributed to its mode of carbon acquisition. Incidentally, most economically important orchids for cut-flower production in the tropics are epiphytic in origin with Crassulacean Acid Metabolism (CAM). In their natural habitat, epiphytes usually meet with a greater degree of environmental stress (e.g., the supply of water and minerals). An understanding of how these orchids cope physiologically with the environmental stress will certainly improve the cultivation of orchids. If a commercial orchid grower wants to optimise orchid growth and flowering, he or she needs to have an understanding of the structure and physiology of orchids. Some basic physiological processes that are relevant to orchid cultivation include photosynthesis, respiration, mineral nutrition, control of flowering and partitioning of assimilates. For example, the grower may want to know the light requirement of an orchid, type of fertiliser to use, method of fertiliser application (either through leaves or roots), or the possible use of plant hormones to induce flowering. Such information can only be obtained from physiological experiments conducted on orchids.

Flower production is a major concern of an orchid farm. As in the other flower crops, the number of spray produced by an orchid varies from time to time. Flower production depends on the genetic make-up of the orchid hybrids and how well they are grown. To achieve maximum yield, proper agronomic practices must be observed. Equally important is the control of flowering to meet market demand. For example, in Japan and Taiwan, large-scale cultivation of *Phalaenopsis* and *Cymbidium* is made possible by the success in controlling flowering. Therefore, the ability to control flowering in tropical orchids using physiological tools is indeed crucial.

The importance of proper postharvest handling of cut-flowers has often been overlooked in the ASEAN orchid cut-flower industry. The management of any floricultural production requires adequate postharvest technology to ensure good marketable quality for the product. The apparent lack of proper postharvest management in many ASEAN orchid farms is attributed to the little information available for postharvest physiology of orchid flowers. This has made it difficult to formulate appropriate postharvest technology and management of orchid cut-flowers, an issue that has been repeatedly raised for discussion in the ASEAN Orchid Congresses.

1.4. Concluding Remarks

It is envisaged that growing tropical orchids for cut-flower production and potted plants will benefit from the recent advances in plant physiology and biotechnology. For the orchid industry, producing an improved hybrid, through conventional breeding or genetic engineering, is only the beginning.

Optimisation of the production processes and ensuring a quality product for the market is equally important. To achieve this goal, a good basic understanding of orchid physiology is essential to solve key physiological issues (Fig. 1.5).

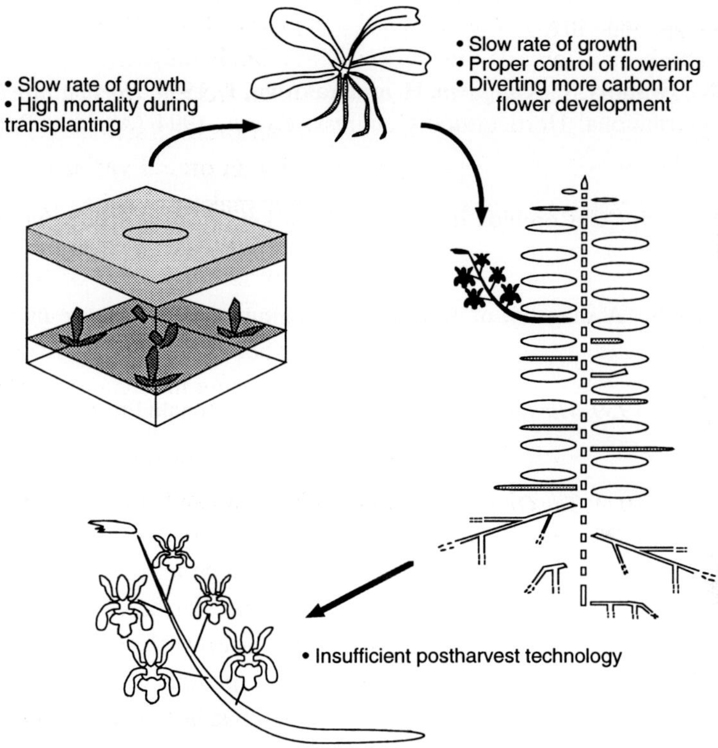

Fig. 1.5. Some key physiological issues affecting the orchid industry.

General References

Arditti, J., 1992, *Fundamentals of Orchid Biology* (John Wiley and Sons, New York), 691 pp.

Arditti, J. and Ernst, R., 1993, *Micropropagation of Orchids* (John Wiley and Sons Inc., New York), 640 pp.

Evans, L. T., 1975, "The physiological basis of crop yield," in *Crop Physiology: Some Case Histories*, ed. L. T. Evans (Cambridge University Press, London), pp. 327–550.

Hew, C. S., 1994, "Orchid cut-flower production in ASEAN countries," in *Orchid Biology: Reviews and Perspectives, Vol. VI*, ed. J. Arditti (John Wiley and Son Inc., New York), pp. 363–401.

Konishi, K., Iwahori, S., Kitagawa, H. and Yakuwa, T., 1994, *Horticulture in Japan.* XXIVth International Horticultural Congress, Kyoto, 1994 (Asakura Publishing, Tokyo), 180 pp.

Laws, N., 1995, "Cut orchids in the world market," *FloraCulture International* **5 (12)**: 12–15.

Suda, S., 1995, "A snapshot of Japanese horticulture," *FloraCulture International* **5 (2)**: 16–19.

Withner, C. L., 1959, *The Orchids: A Scientific Survey* (Ronald Press Co., New York), 648 pp.

Withner, C. L., 1974, *The Orchids: Scientific Studies* (Wiley-Interscience, New York), 608 pp.

Chapter 2

A Brief Introduction to Orchid Morphology and Nomenclature

2.1. Introduction

Few plants can create such an aura of mystique and grandeur as orchids. Their intricate appearance has enthralled many people. The orchid family is probably the largest in the plant kingdom, having about 750 different genera with at least 25,000 native species and more than 30,000 cultivated hybrids — the result of interbreeding — and more are being registered and added to the ever growing list of hybrids. Orchids as a plant family is systematically placed with the Monocotyledons (flowering plants with one seed-leaf or cotyledon).

A good basic understanding of the different plant parts within an orchid and the usage of appropriate orchid nomenclature is important for anyone involved in orchid research and business. In this chapter, many of the examples used for illustration, description and naming are based on economically important orchids.

2.2. Growth Habit

Orchid shoots can grow in two basic ways: sympodial (Fig. 2.1) and monopodial (Fig. 2.2). In sympodial orchids, the growth of the shoot is limited. For flowering shoots, it terminates in a flower or inflorescence, so that continued growth is possible only by the formation of a laterally located axillary bud. In the case of

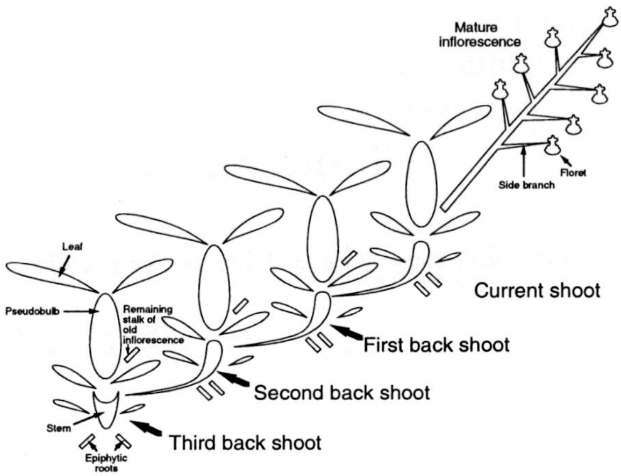

Fig. 2.1. Diagrammatic representation of the growth habit of a sympodial orchid *Oncidium* Goldiana.

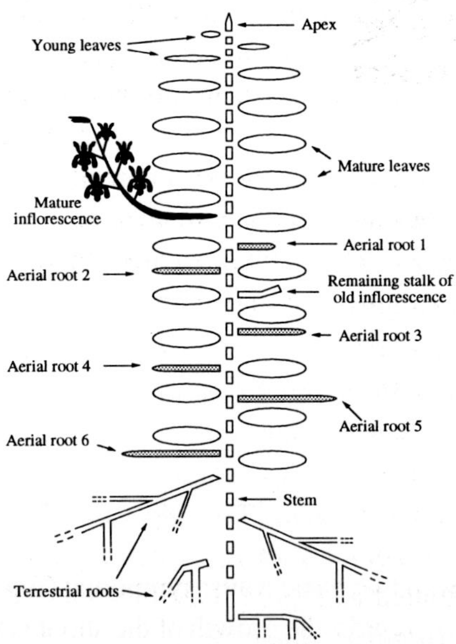

Fig. 2.2. Diagrammatic representation of the growth habit of a monopodial orchid *Aranda* Noorah Alsagoff.

non-flowering shoots, new axillary shoot arises from the laterally located bud. Growth is continuous and theoretically unlimited at the apex for the monopodial orchids, e.g., *Vanda*, *Aranda* and *Mokara*.

2.3. Orchid Plant Parts

Pseudobulbs

Most epiphytic orchids possess a prominent, enlarged bulbous structure at the base of their leaves, termed a pseudobulb (Dressler, 1981). The term 'pseudobulb' is first used by John Lindley in 1837 (Curtis, 1943). In general, the pseudobulb is the enlarged portion of the stem from which all leaves and inflorescences arise. Pseudobulbs can be classified, regardless of shape, to be of homoblastic (many internodes) or heteroblastic (single internode) type on basis of the number of internodes forming the pseudobulb (Fig. 2.3). The pseudobulb of *Dendrobium crumenatum* (Pigeon orchid) is an example of a homoblastic pseudobulb while the pseudobulb of *Oncidium* Goldiana is of the heteroblastic type.

Numerous studies on pseudobulbs of several orchids have revealed the absence of stomata. However, openings in the tissue do occur at the base of ant-inhabited pseudobulbs. The role of the pseudobulb as a water and food storage organ is well-recognised.

Withner and coworkers (1974) reported that although considerable differences can be seen in the external features of pseudobulbs, little variation occurs in the internal tissue arrangements for the different orchid species. Pseudobulbs have a unique structure where the entire organ is covered with thick cuticle and is lacking in stomata. The epidermis of the pseudobulb consists of two, three or four layers of thick-walled parenchyma cells. The groundmass is not sharply differentiated and there is no discernible cortex. The vascular bundles are scattered irregularly throughout the groundmass. Two major cell types have been reported in the parenchymatous groundmass of mature pseudobulbs for several orchid species. They are small 'assimilatory' cells

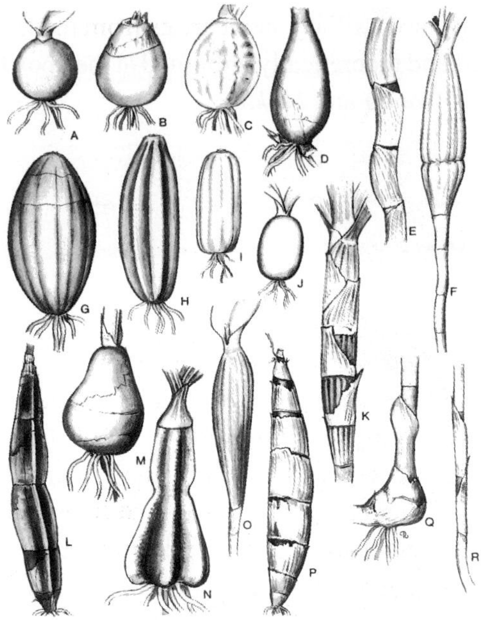

Fig. 2.3. Pseudobulb shapes in orchids.

Note: (A) Globose or round [*Sophronitis*]; (B) Ovoid (*Neomoorea*]; (C) Ovoid-compressed [*Laelia*]; (D) Oblong- or ovate-elongate [*Encyclia*]; (E) Jointed [*Dendrobium*]; (F) Unguiculate [*Myrmecophila*]; (G) Elliptic [*Grammatophyllum*]; (H) Elliptic-elongate, sulcate or furrowed [*Gongora*]; (I) Oblong-sulcate or furrowed [*Pholidota*]; (J) Oblong-cylindrical [*Bulbophyllum*]; (K) Cylindrical [*Ansellia*]; (L) Four-sided [*Dendrobium*]; (M) Pyriform [*Encyclia*]; (N) Constricted or hour-glass shaped [*Calanthe*]; (O) Obovoid or club-shaped [*Cattleya*]; (P) Fusiform or spindle-shaped [*Catasetum*]; (Q) Swollen base [*Cattleya*]; (R) Stem-like or reed-like [*Isochilus*].

Reproduced from Sheehan & Sheehan (1994), courtesy of Timber Press, Inc.

that are living and containing predominantly chloroplasts or starch grains; and larger dead cells that are irregularly shaped with pleated walls (Fig. 2.4). Compared to the outer portion of the pseudobulb, the central portion is of a lighter shade of green. This is attributed to the distribution of living cells: Living cells nearer to the epidermis are rich in chloroplasts but lacking in starch grains while those nearer to the centre of the pseudobulb are rich in starch grains and lacking in chloroplasts. Based on the anatomical studies, a

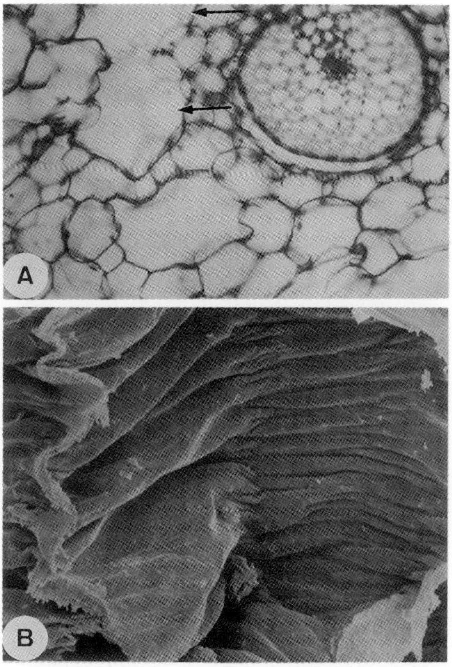

Fig. 2.4. Living assimilatory cells and water-storage cells in pseudobulbs of *Stanhopea*. Note: (A) *S. wardii*, cross section of pseudobulb showing collateral vascular bundle; living assimilatory cells and dead water-storage cells [arrows] [250 X]. (B) *S. grandiflora*, scanning electron micrograph of pleated cell wall of pseudobulb water-storage cell [920 X].

Reproduced from Stern & Morris (1992), courtesy of Lindleyana.

possible storage function for starch was suggested for the smaller living cells while the larger dead cells may be used for water storage.

Flowers

For most orchids, the inflorescence consists of an axis that bears individual flowers along its length. The axis is divided into two regions: The peduncle (or stalk) is the axis region from the stem or base of pseudobulb to the point of insertion for the lowermost flower; rachis, the remaining part of the axis containing the flowers. Each flower is subtended by a modified leaf (bract) which is connected to the axis. Generally, the oldest flower is found nearer to the base of the axis and the flowers are progressively younger along the axis towards the tip of the inflorescence.

Orchid flowers are zygomorphic (symmetrical about a single plane) in nature (Fig. 2.5). The size of the flowers can range from minute types to those up to

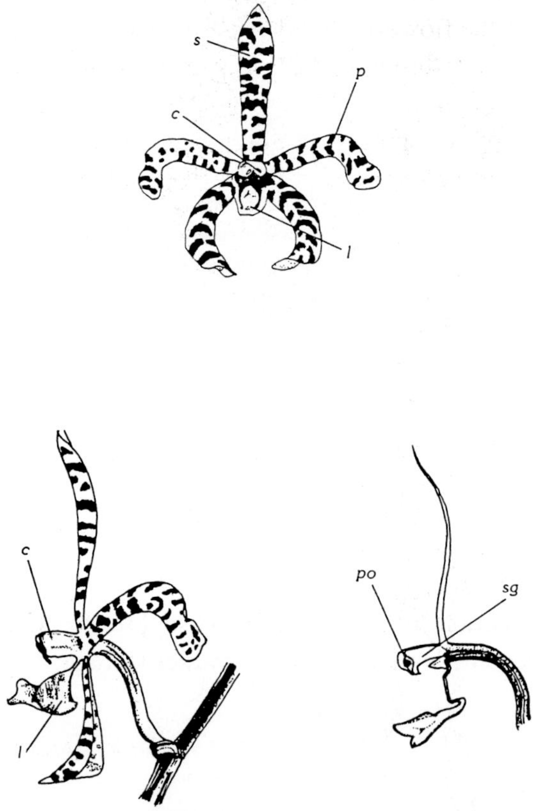

Fig. 2.5. Flower structure of *Arachnis* Maggie Oei.
Note: Explanation of symbols: *c*, column; *l*, lip; *p*, petal; *po*, pollinium; *s*, sepal; *sg*, stigma.

Redrawn from Teo (1979).

20 cm wide. Even within a genus, their size, shape and colour vary considerably although all orchids have the same basic structure.

Each orchid flower has three sepals (the outermost segments of a flower) and three petals (Fig. 2.5). All of these are coloured, unlike many non-orchid flowers where the sepals are green and leaf-like. The uppermost sepal is symmetrical and often larger than the other two lateral sepals. The petals on either side of the flower are usually equal in size and shape, whereas the bottom one is formed into the shape of a lip and known as the labellum. The labellum in many orchids is modified to form a spur (a cone-like structure that protrudes towards the back of the flower) where nectar is produced (Fig. 2.6).

Many orchid flowers turn upside down during its development and this is termed resupination (Arditti, 1992). For example, the process of resupination can be followed easily by tracing the location of the spur on flowers of different ages along the axis of a *Dendrobium* inflorescence (Fig. 2.7). As the flowers

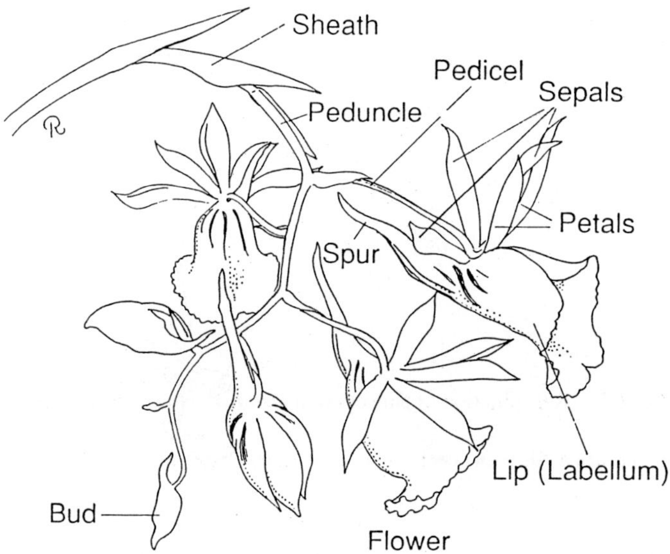

Fig. 2.6. The orchid inflorescence and its parts.

Reproduced from Sheehan & Sheehan (1994), courtesy of Timber Press, Inc.

Fig. 2.7. Resupination of flowers of a *Dendrobium* inflorescence.

Reproduced from Sheehan & Sheehan (1994), courtesy of Timber Press, Inc.

open, the buds twist so that the spur is positioned lowermost. Alternatively, we can look at the ovary of each flower to decide whether resupination has taken place.

The column is unique to orchids. It is a coalescence of both the male and female reproductive organs (Fig. 2.8). The anther cap lies at the tip of the column, enclosing the pollinarium and the rostellum that lies beneath the pollinarium. Generally, the pollinarium consists of pollinia (masses of pollen), viscidium (a sticky disc) and stipe (thin strip of tissue that connects the pollinia to the viscidium). Beneath the rostellum lies the stigma that is a cavity filled with sticky fluid. The stigma is connected to the ovary by the column that allows the growth of pollen tubes towards the ovules during fertilisation. The ovary (inferior type) containing the ovules is below the point of insertion for

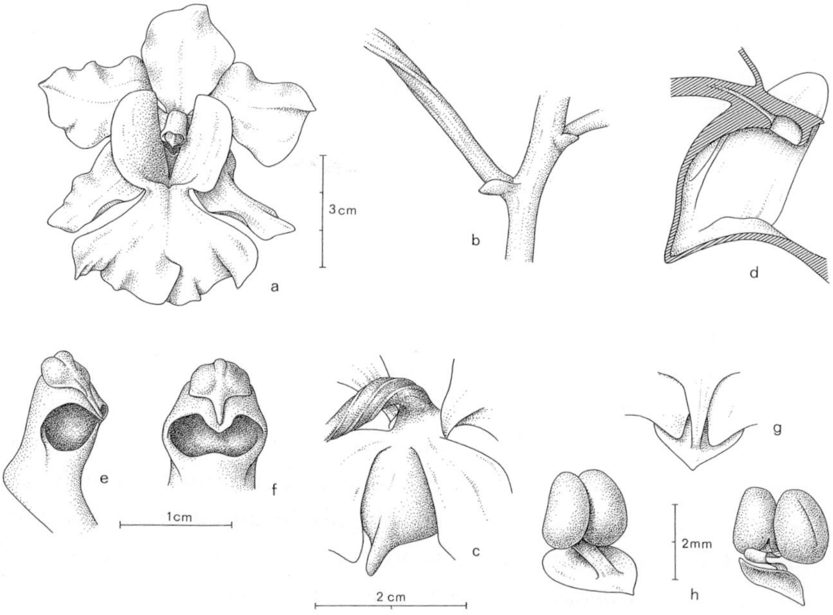

Fig. 2.8. Flower structure of *Vanda* Miss Joaquim.

Note: (a) Front of flower; (b) Base of ovary, showing twist (giving rise to resupination), and bract; (c) Base of flower from behind, showing junction of lateral sepals and lip; (d) Longitudinal section of flower (anther removed); (e) & (f) Two views of the column; (g) Tip of rostellum, showing viscidium; (h) Two views of pollinia with viscidium and stipes, after bending of stipes.

Reproduced from Seidenfaden & Wood (1992), courtesy of Olsen and Olsen, Fredensborg, Denmark.

the sepals and petals. A simplified outline of an orchid 'half-flower' is shown in Fig. 2.9.

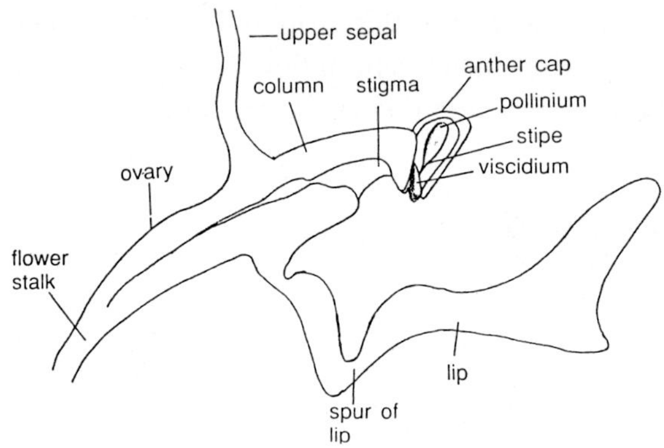

Fig. 2.9. A simplified outline of an orchid flower.

Redrawn from Tan & Hew (1995).

Stomata can be found on the various parts of the orchid flower such as the column, pollen cap and petals (Fig. 2.10). Generally, there are fewer stomata in petals than in the column (Table 2.1). In petals, stomata are found either on the upper surface (e.g., *Vanda* Miss Joaquim), lower surface (e.g., *Dendrobium superbum*) or on both (e.g., *Oncidium* Norman Gaunt). Stomata in the petals may be scattered (e.g., *Vanda suavis*) or highly localised (e.g., *Dendrobium superbum*). For some orchids, there is no stomata on either side of the petals (e.g., *Angraecum giryamae*). The occurrence of stomata in the pollen cap (which is small in area and easily dislodged) makes it an ideal material for studying stomata in orchid flowers. Almost all the stomata observed in the petals, column and pollen cap of tropical orchids are either closed or partially opened (Fig. 2.11). This implies that the orchid flower stomata are probably vestigial and practically non-functional.

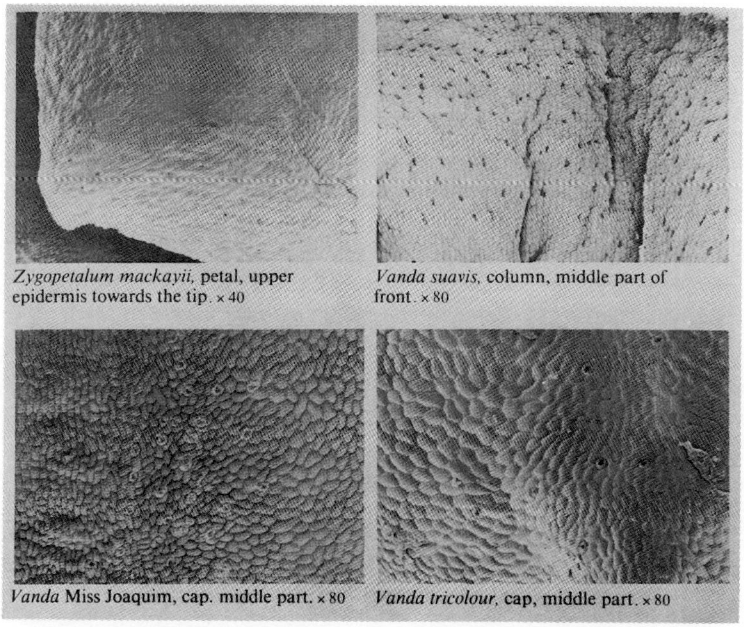

Zygopetalum mackayii, petal, upper epidermis towards the tip. × 40

Vanda suavis, column, middle part of front. × 80

Vanda Miss Joaquim, cap. middle part. × 80

Vanda tricolour, cap, middle part. × 80

Fig. 2.10. The distribution of stomata in some orchid flowers.

Reproduced from Hew & Veltkamp (1985), courtesy of the Malayan Orchid Review.

Table 2.1. Distribution of stomata in some tropical orchid flowers.

Orchid	Sepal		Petal		Labellum	Column
	Lower epidermis	Upper epidermis	Lower epidermis	Upper epidermis	Lower epidermis	
Thin-leaved orchids						
Arundina graminifolia	71	50	150	—	—	Scanty
Oncidium Goldiana	582	392	433	358	—	291
Thicked leaves orchids						
Arachnis Maggie Oei	131	88	74	102	507	1,133
Aranda Wendy Scott	67	124	45	47	22	950
Vanda Tan Chay Yan	55	64	46	96	23	1,170

Adapted from Hew, Lee & Wong (1980).

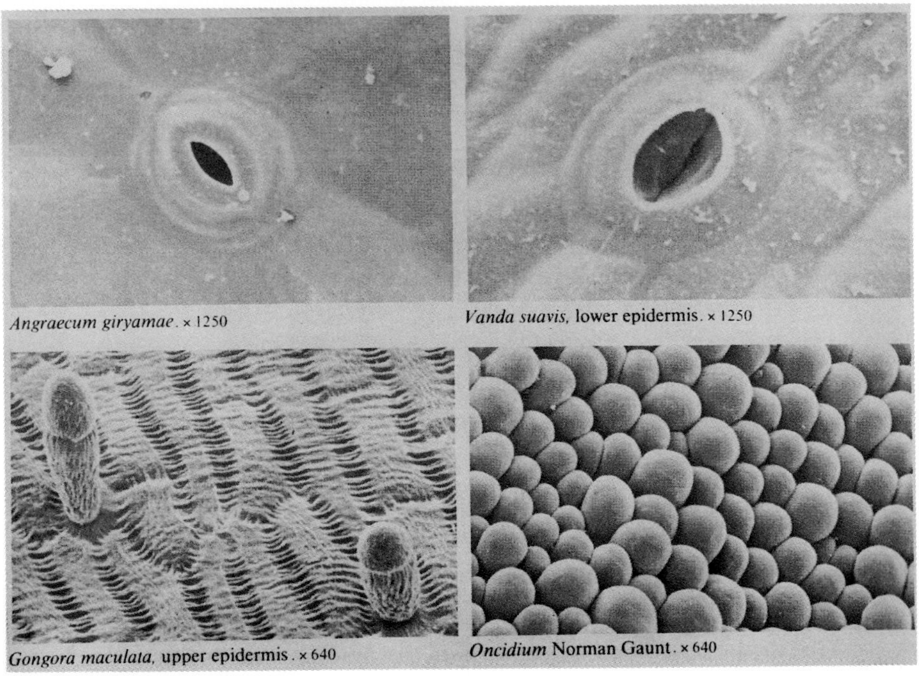

Fig. 2.11. The surface contour of some orchid flower petals.

Reproduced from Hew & Veltkamp (1985), courtesy of the Malayan Orchid Review.

Seeds

After pollination, the ovary develops into a fruit capsule containing millions of seeds. The time required for development into the fruit capsules varies for different orchids. The orchid seed consists of a mass of undifferentiated mass of cells enclosed by a seed coat (Fig. 2.12).

Leaves

Leaves of orchids are variable in shapes, sizes and thickness. Information on anatomy and morphology of orchid leaves are important for both horticultural

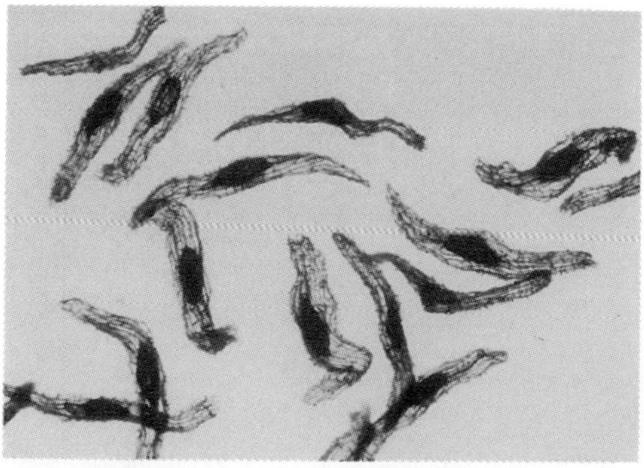

Fig. 2.12. Seeds of *Spathoglottis plicata.*

By courtesy of Dr. Hugh Tan, The National University of Singapore, Singapore.

and scientific practices. Generally, orchid leaves can be divided into two types based on leaf thickness: Thin-leaved or thick-leaved. Figure 2.13 shows the cross-section of an orchid leaf with the following structures: Cuticle, upper epidermis, mesophyll layer, vascular bundles and lower epidermis.

Both thin- and thick-leaved orchids lack stomata on the upper epidermis. Economically important thin-leaved orchids include *Oncidium* Goldiana, *Spathoglottis plicata* and *Cymbidium sinense*. Thin-leaved orchids have higher density of stomata on the lower epidermis in comparison to thick-leaved orchids (Table 2.2). Thick-leaved orchids include *Dendrobium*, *Aranda* and *Mokara*. Figure 2.14 shows the distribution of stomata on the abaxial (lower) side of a *Mokara* leaf. Interestingly, thin and thick-leaved orchids are associated with C_3 and CAM mode of photosynthesis respectively (see Chap. 3 on Photosynthesis).

Roots

The morphology of orchid roots is dependent on its habitat, either terrestrial or epiphytic. Aerial roots of epiphytic orchids are often exposed and free hanging, or sometimes appressed to a supporting structure. Conversely, roots of terrestrial orchids are usually hidden in the soil.

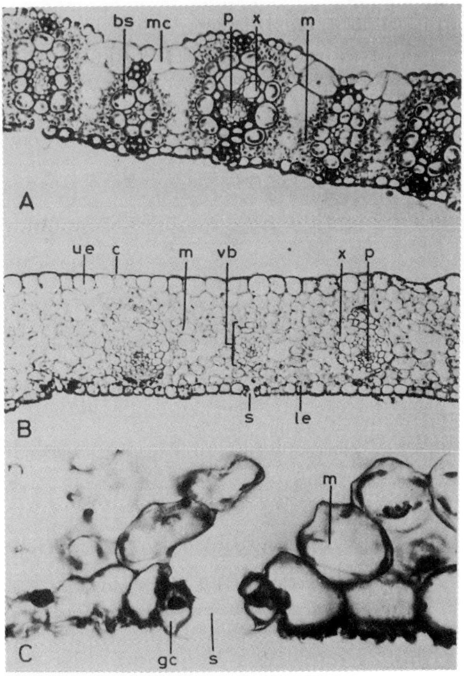

Fig. 2.13. Leaf cross section of *Arundina graminifolia.*
Note: (A) Leaf cross section of *Imperata cylindrica*, a known C_4 plant [for comparison, 320 X]; (B) *Arundina graminifolia*, leaf cross section [160 X]; (C) *Arundina graminifolia*, leaf cross section showing stoma [1,000 X]. Explanation of symbols: bs, bundle sheath; c, cuticle; gc, guard cell; le, lower epidermis; m, mesophyll; mc, motor cell; p, phloem; s, stoma; ue, upper epidermis; vb, vascular bundle; x, xylem.

Adapted from Wong (1974).

Epiphytic orchids

The great majority of economically important orchids for cut-flowers and potted plants are epiphytic in origin; e.g., *Vanda, Aranda, Dendrobium* and *Oncidium.* Aerial roots of epiphytic orchids are characterised by a green tip (sometimes reddish, as in the case for some dendrobiums) whilst the remainder part of the root is covered with velaman. Roots are produced at the basal joints in sympodial orchids. In contrast, root production for the monopodial orchids is at regular

Table 2.2. Leaf characteristics of some tropical orchids.

Orchid	Leaf thickness (mm)	No. of cell layers in the mesophyll	Cuticle thickness (μm)		Stomatal density stomata (cm^{-2})	
			Lower epidermis	Upper epidermis	Lower epidermis	Upper epidermis
Thin-leaved orchids						
Arundina graminifolia	0.3	11–12	2	2	15,100–18,000	none
Oncidium Goldiana	0.5	10–12	3	3	6,500–7,500	none
Spathoglottis plicata	0.3	5	2	2	14,000	none
Thicked leaves orchids						
Aranda Deborah	1.6	18–21	11	14	3,000	none
Aranda Wendy Scott	1.5	16–18	14	14	3,000–3,300	none
Arachnis Maggie Oei	1.2	12–15	9	9	4,000	none
Dendrobium Caesar	1.5	15	6	6	3,800	none

Adapted from Hew, Lee & Wong (1980) and Avadhani, Goh, Rao & Arditti (1982).

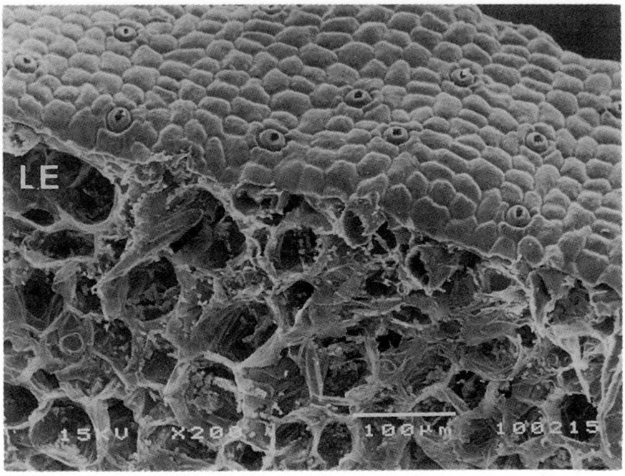

Fig. 2.14. Scanning electron microscopy of stomata on a *Mokara* Yellow leaf.
Note: Stomata are present on the abaxial surface of the leaf. Explanation of symbol: LE, lower epidermis.

intervals near the nodal region along the stem axis and up to three roots may be produced at each node. For example, aerial roots of *Aranda* Deborah may be produced at successive nodes, but the occurrence of roots along two adjacent nodes is rare. There is generally no distinct pattern for the occurrence of roots along the monopodial stem axis although roots are usually present on alternate nodes or every third node.

Terrestrial orchids

For terrestrial orchids, the various species and hybrids of *Cymbidium* and *Spathoglottis* are important as potted plants. Roots of terrestrial orchids are frequently ground-dwelling, thick and fleshy with a probable storage function. Sometimes, these roots may appear tuber-like. While the tuber-like roots are observed in numerous temperate orchid genera (e.g., *Acres*), they are uncommon in the tropical orchids except for a few genera (e.g., *Habenaria*). Roots of most terrestrial orchids contain a fungus that usually infects the orchid at the seed stage. This mycorrhizal fungus is known to provide carbohydrate and mineral nutrients to both young and adult orchids.

Generally, orchid roots can be divided into several distinct layers: Velamen, cortex (exodermis and endodermis) and stele (Fig. 2.15). A unique feature of the aerial root is the presence of velamen, which covers the whole root except the tip (Fig. 2.16). Lying beneath the velamen and exodermis is the chloroplast-

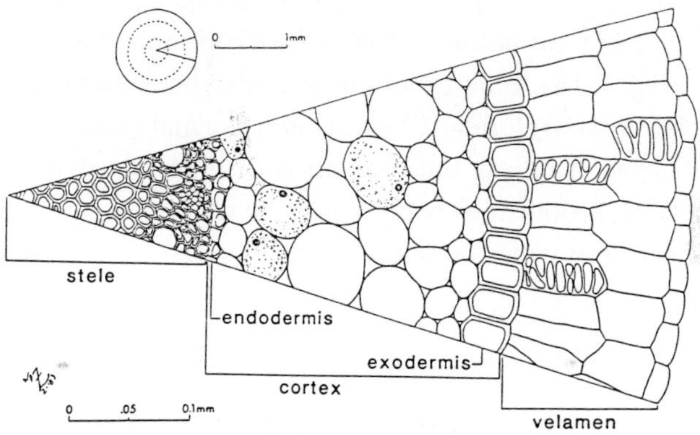

Fig. 2.15. Transection of an orchid root.
Note: The figure is drawn from a free-hand section of a root of *Restrepiella ophiocephala*.

Reproduced from Pridgeon (1987), courtesy of Cornell University Press.

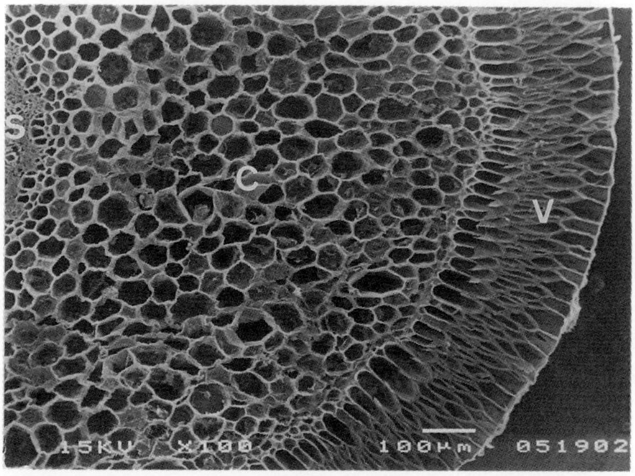

Fig. 2.16. Scanning electron microscopy of an orchid aerial root of *Arachnis* Maggie Oei.
Note: Explanation of symbols: V, velamen; C, cortex; S, stele.

containing cortex. A highly specialised layer of cells, the exodermis, lies between the cortex and the velamen. The exodermis consists of two components: Small and dense cytoplasmic passage cells that are evenly interspersed among the larger, elongated and more vacuolated cells with thick walls.

Root hair formation has been observed under certain circumstances. For example, fine root hairs are produced on the *Vanda* aerial root under certain conditions (Fig. 2.17). Sometimes, roots of micropropagated plantlets produce fine root hair (Fig. 2.18). Fine root hairs can also be found in the roots of the terrestrial orchid *Spathoglottis plicata*. Under normal conditions, aerial roots do not usually branch unless the root tip is of a certain distance away. The production of lateral roots does occur when the root tip is damaged (Fig. 2.19) or when submerged in water for more than 24 hours.

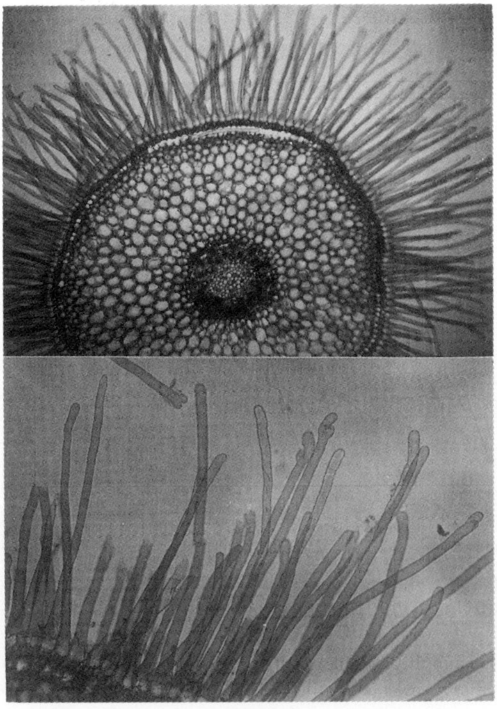

Fig. 2.17. Root hairs in the aerial root of *Vanda* Miss Joaquim.

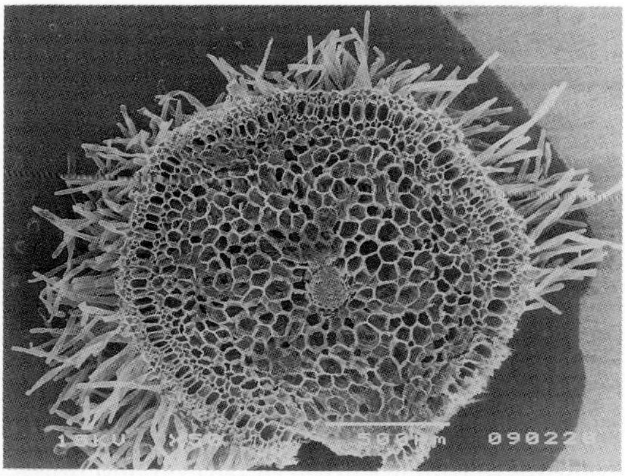

Fig. 2.18. Scanning electron microscopy of root hairs in the aerial root of *Mokara* Yellow.

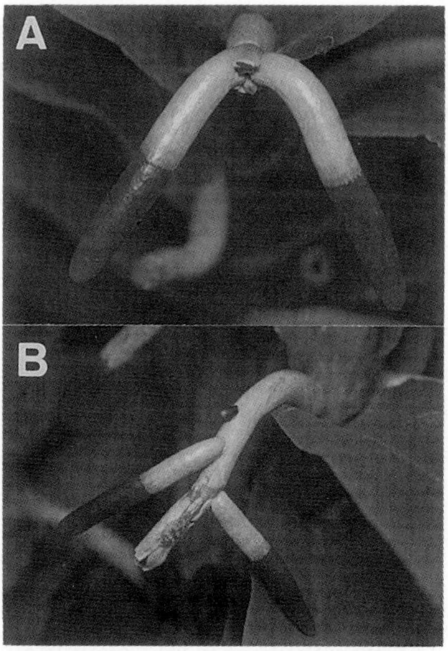

Fig. 2.19. The development of lateral roots in aerial roots of *Aranda* Noorah Alsagoff after decapitation.

Note: (A) The development of lateral roots from the cut end and (B) from various positions behind the cut end.

2.4. Growth Cycle of Orchids Under Greenhouse Conditions

The growth cycle of an orchid is important to both scientists and commercial growers. For the scientists, a proper understanding of the different growth stages would ensure that experiments are carried out with plants of the appropriate growth stage under certain environmental conditions. The clonal nature of many sympodial orchids makes choosing and standardisation of plant materials for experiments difficult. A good experimental set-up requires careful observation and selection of plant materials. For example, the different number of connected shoots of *Dendrobium* must be an important consideration for any experiments relating to translocation of carbon and nutrients. This ensures that experiments are reproducible and allows other scientists to understand and participate in future related research. Figure 2.20 gives an example of how a systematic approach can be used to standardise orchids used as an experimental material.

For the commercial orchid growers, predictability and reliability of flower production are important requisites of a good farm. The growth cycle of an orchid allows the growers to predict the probable harvest time and to adopt sound farm management practice to modulate flower supply. To illustrate, the growth cycle of an economically important orchid cut-flower is shown in Fig. 2.21 as an example.

2.5. Nomenclature

Species

The name (or specific epithet) of a species is always italicised (or underlined in some books) but never capitalised. For example, let us use the name *Eulophia graminea.* The generic name is *Eulophia* and *graminea* is the specific epithet. The species name (or binomial) should also be followed by the person(s) who described the plant. Take the example of *Eulophia graminea* Lindl., it is implied

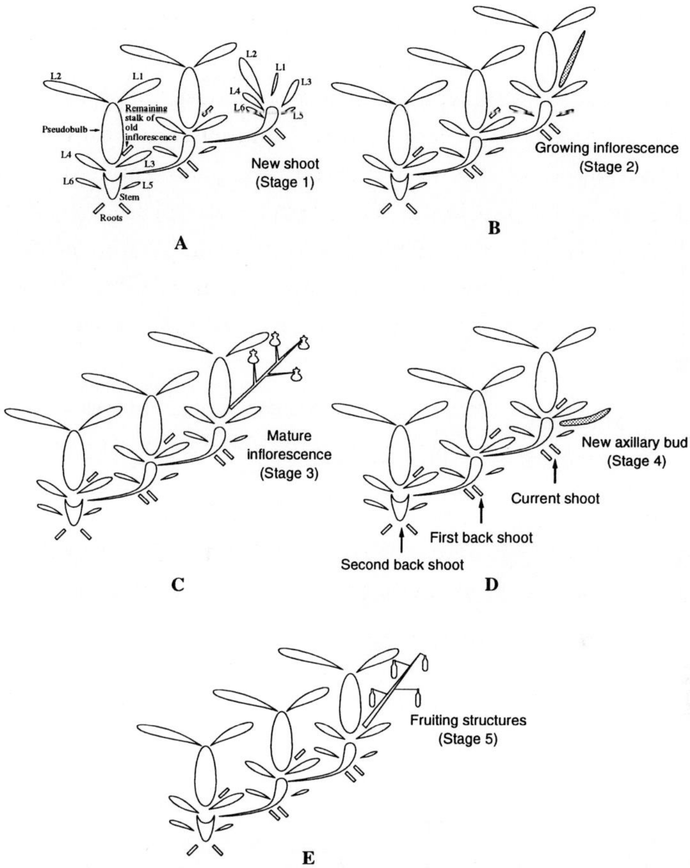

Fig. 2.20. Diagrammatic representations of *Oncidium* Goldiana with current shoots at growth stage 1, 2, 3, 4 or 5 connected to two back shoots.

Note: (A) Current shoot at stage 1 connected to two back shoots; (B) Current shoot at stage 2 connected to two back shoots; (C) Current shoot at stage 3 connected to two back shoots; (D) Current shoot at stage 4 connected to two back shoots; (E) Current shoot at stage 5 connected to two back shoots.

Redrawn from Yong (1995).

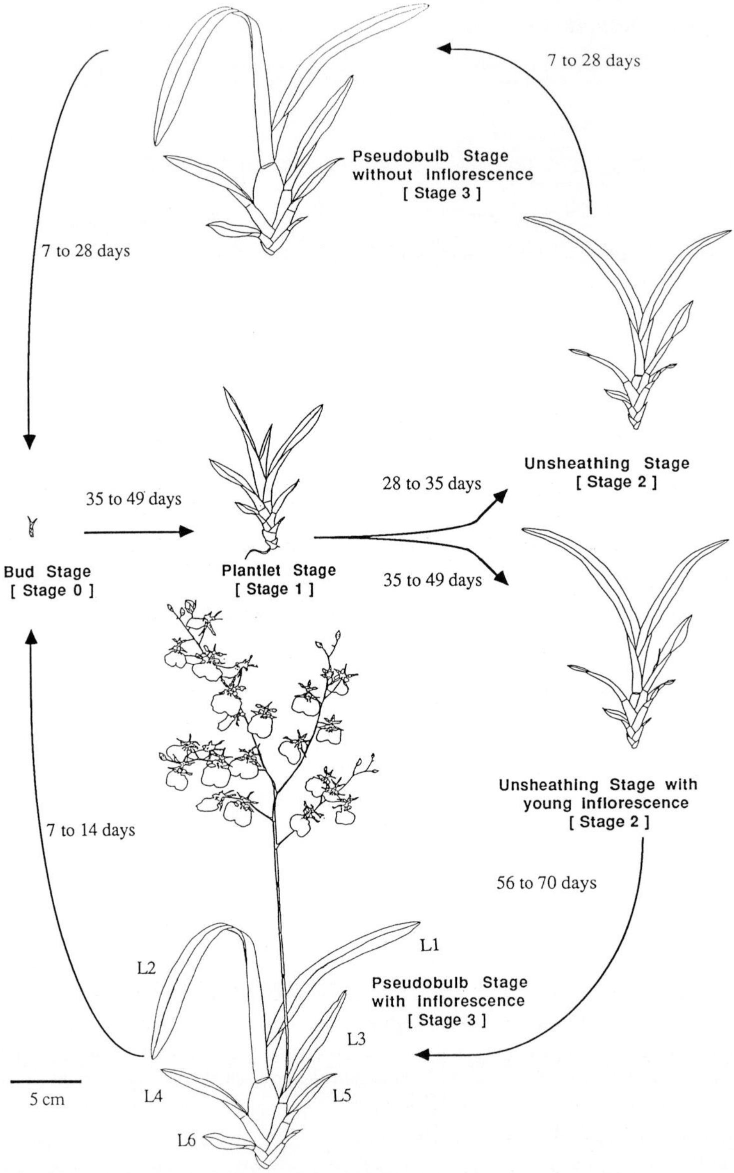

Fig. 2.21. The growth cycle of *Oncidium* Goldiana under tropical greenhouse conditions in Singapore.

Redrawn from Hew & Yong (1994).

that John Lindley is the first person who described the species *Eulophia graminea.*

Hybrid

The name of a hybrid consists of a generic name and a grex epithet, following the rules laid down in *Handbook of Orchid Registration and Nomenclature* (Cribb *et al.*, 1985). For example, let us use the name *Vanda* Miss Joaquim. This hybrid is produced by crossing two species of the same genus: *Vanda hookerana* × *Vanda teres*. The generic name is *Vanda* and the grex epithet is 'Miss Joaquim', a fancy name. The fancy name is in normal print and not written in Latin. The grex name refers to all the progeny of a particular cross. The grex epithet is usually named after a person, flower colour and even places. Hybrid names must be officially registered with the International Registration Authority (Royal Horticultural Society in London) to be valid. Names of bigeneric hybrids are derived from the parent genera. For example, *Aranda* is an artificial hybrid generic name with an obvious combination of *Ara*chnis and Va*nda*. For trigeneric hybrids, the hybrid name should consist of the three parent genera or a new name. For example, the artificial genus *Mokara* is derived from the combination of *Arachnis* × *Ascocentrum* × *Vanda*.

2.6. Summary

1. Orchids can be divided into two groups by its growth habit: Monopodial and sympodial. These subgroups can be further divided on the basis of leaf thickness: Thick or thin-leaved orchid. For example, *Oncidium* Goldiana is a sympodial thin-leaved orchid hybrid whereas *Mokara* White is a monopodial thick-leaved orchid hybrid.
2. Most economically important tropical orchids for cut-flowers and potted plants are epiphytic in origin although they can be planted on the ground or in pots. There are a few terrestrial orchids that are used as potted plants (e.g., *Spathoglottis plicata*).

General References

Arditti, J., 1992, *Fundamentals of Orchid Biology* (John Wiley and Sons, New York), 691 pp.

Avadhani, P. N., Goh, C. J., Rao, A. N. and Arditti, J., 1982, "Carbon fixation in orchids," in *Orchid Biology: Reviews and Perspectives, Vol. II*, ed. J. Arditti (Cornell University Press, Ithaca, New York), pp. 173–193.

Cribb, P. J., Greatwood, J. and Hunt, P. F., 1985, *Handbook of Orchid Registration and Nomenclature*, Third edition (International Orchid Commission, London), 143 pp.

Dressler, R. L., 1981, *The Orchids: Natural History and Classification* (Harvard University Press, Cambridge, Massachusetts), 332 pp.

Pridgeon, A. M., 1987, "The velamen and exodermis of orchid roots," in *Orchid Biology: Reviews and Perspectives, Vol. IV*, ed. J. Arditti (Cornell University Press, Ithaca, New York), pp. 139–192.

Rasmussen, H., 1987, "Orchid stomata — Structure, differentiation, function and phylogeny," in *Orchid Biology: Reviews and Perspectives, Vol. IV*, ed. J. Arditti (Cornell University Press, Ithaca, New York), pp. 105–138.

Seidenfaden, G. and Wood, J. J., 1992, *The Orchids of Peninsular Malaysia and Singapore* (Olsen and Olsen, Fredensborg, Denmark), 779 pp.

Sheehan, T. and Sheehan, M., 1994, *An Illustrated Survey of Orchid Genera* (Timber Press Inc., Oregon, USA), 421 pp.

Sinclair, R., 1990, "Water relations in orchids," in *Orchid Biology: Reviews and Perspectives, Vol. V*, ed. J. Arditti (Timber Press, Portland, Oregon), pp. 63–119.

Tan, H. T. W. and Hew, C. S., 1995, *A Guide to the Orchids of Singapore*, Revised edition (Singapore Science Centre, Singapore), 160 pp.

Withner, C. L., Nelson, P. K. and Wejksnora, P. J., 1974, "The anatomy of orchids," in *The Orchids: Scientific Studies*, ed. C. L. Withner (Wiley-Interscience, New York), pp. 267–348.

References

Ando, T. and Ogawa, M., 1987, "Photosynthesis of leaf blades in *Laelia anceps* Lindl. is influenced by irradiation of pseudobulb," *Photosynthetica* **21**: 588–590.

Chiang, S. H. T., 1970, "Development of the root of *Dendrobium kwashotense* Hay, with special reference to the cellular structure of its exodermis and velamen," *Taiwania* **15**: 1–16.

Chiang, Y. L. and Chen, Y. R., 1968, "Observations on *Pleione formosana* Hayata," *Taiwania* **14**: 271–301.

Curtis, C. H., 1943, "Pseudobulbs," *Orchid Review* **51**: 137.

Goh, C. J., 1983, "Aerial root production in *Aranda* orchids," *Annals of Botany* **51**: 145–147.

Hew, C. S., Lee, G. L. and Wong, S. C., 1980, "Occurrence of non-functional stomata in the flowers of tropical orchids," *Annals of Botany* **46**: 195–201.

Hew, C. S. and Veltkemp, C. J., 1985, "Orchid floral stomata under the scanning electron microscope," *Malayan Orchid Review* **19**: 26–32.

Hew, C. S. and Yong, J. W. H., 1994, "Growth and photosynthesis of *Oncidium* Goldiana," *Journal of Horticultural Science* **69**: 809–819.

Rasmussen, H., 1986, "The vegetative architecture of orchids," *Lindleyana* **1**: 42–50.

Stern, W. L. and Morris, M. W., 1992, "Vegetative anatomy of *Stanhopea* (Orchidaceae) with special reference to pseudobulb water-storage cells," *Lindleyana* **7**: 34–53.

Tanaka, M., Yamada, S. and Goi, M., 1986, "Morphological observation on vegetative growth and flower bud formation in *Oncidium* Boissiense," *Scientia Horticulturae* **28**: 133–146.

Teo, C. K. H., 1979, *Orchids for Tropical Gardens* (FEP International Sdn. Bhd., Malaysia), 137 pp.

Wong, S. C., 1974, "A study of photosynthesis and photorespiration in some thin-leaved orchid species," M.Sc. Dissertation, Department of Biology, Nanyang University, Singapore, 148 pp.

Yong, J. W. H., 1995, "Photoassimilate partitioning in the sympodial thin-leaved orchid *Oncidium* Goldiana," M.Sc. Dissertation, Department of Botany, The National University of Singapore, 132 pp.

Chapter 3

Photosynthesis

3.1. Introduction

During photosynthesis, carbon dioxide is fixed and reduced to carbohydrate. Green plants can be divided into three groups with respect to their patterns and biochemistry of CO_2 fixation. The first group of plants has been generally referred to as C_3 plants. This group of plants that includes spinach, pea and sunflower, assimilates carbon dioxide primarily through Calvin's cycle. The second group of plants that includes maize, sugarcane and sorghum, is known as C_4 plants. These plants fix CO_2 through the C_4 pathway. The third group of plants are those with Crassulacean Acid Metabolism (CAM). Some common examples of CAM plants include cactus, pineapple and bromeliads. The carboxylation and decarboxylation events that drive the CO_2 concentrating mechanism of C_4 and CAM plants are similar, but they operate on different anatomical, physiological and biochemical principles.

This chapter will provide a brief introduction to the three photosynthetic pathways, photosynthetic characteristics of orchid leaves and non-foliar green organs, and the factors which affect photosynthesis in orchids.

3.2. Photosynthetic Pathways

In C_3 plants, the fixation of carbon dioxide is mediated by RUBPC (ribulose bisphosphate carboxylase) and a three-carbon compound, phosphoglycerate, is the first stable photosynthetic product. These intermediates are reduced eventually to carbohydrate using the photochemically generated ATP and NADPH. The cycle is completed by the regeneration of a five-carbon acceptor molecule (Fig. 3.1).

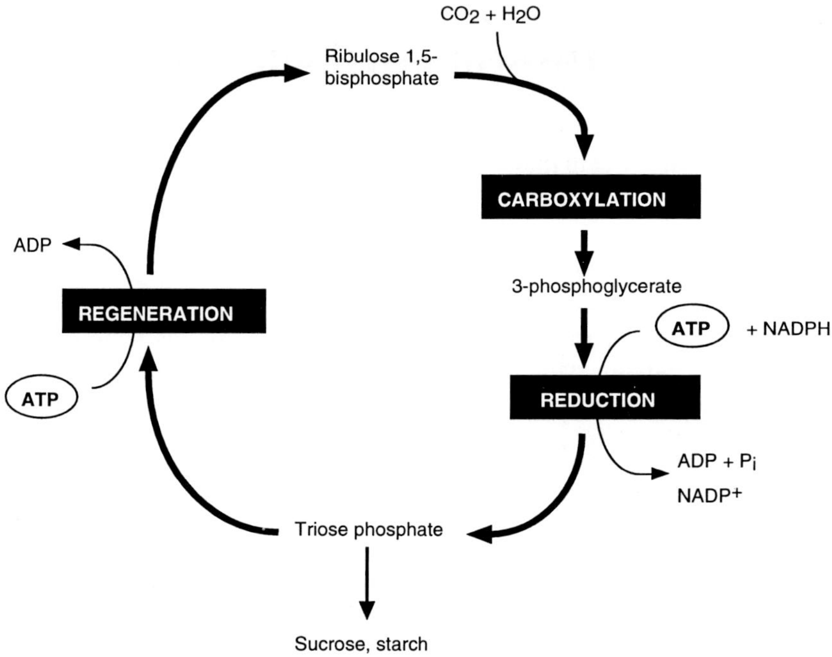

Fig. 3.1. The C_3 photosynthetic carbon reduction cycle.
Note: The cycle proceeds in three stages: (1) carboxylation, during which CO_2 is covalently linked to a carbon skeleton; (2) reduction, during which carbohydrate is formed at the expense of the photochemically derived ATP and reducing equivalents, NADPH; and (3) regeneration, during which the CO_2-acceptor molecule, ribulose 1,5-bisphosphate is re-formed.

Redrawn from Taiz and Zeigler (1991).

Plants exhibiting 'Hatch–Slack–Kortschak' pathway of carbon fixation or C_4 plants are usually characterised by the following feature: Kranz anatomy (leaf anatomy with chloroplasts showing size and structural dimorphism), chlorophyll a/b ratio of 4, low CO_2 compensation point (0–5 ppm) and $\delta^{13}C$

values of −9‰ to −14‰. The apparent absence or low activity of photorespiration is due to the suppression of oxygenase activity by high partial pressures of CO_2 present in the bundle sheath cells. The C_4 carboxylation acts as a CO_2 concentrating device for the C_3 cycle. The distinguishing biochemical feature of C_4 plants is the first carboxylation of CO_2 which is carried out by PEPC (phosphoenolpyruvate carboxylase). The CO_2 acceptor is the three-carbon compound phosphoenolpyruvate (PEP), and the product is the four-carbon compound oxaloacetate (OAA), which is readily converted to malate or aspartate. The fate of OAA is of the same general pattern in all C_4 plants, but varies in detail for both malate and aspartate formers (Edwards and Walker, 1983). In C_4 plants, the aspartate or malate formed is transported to the bundle

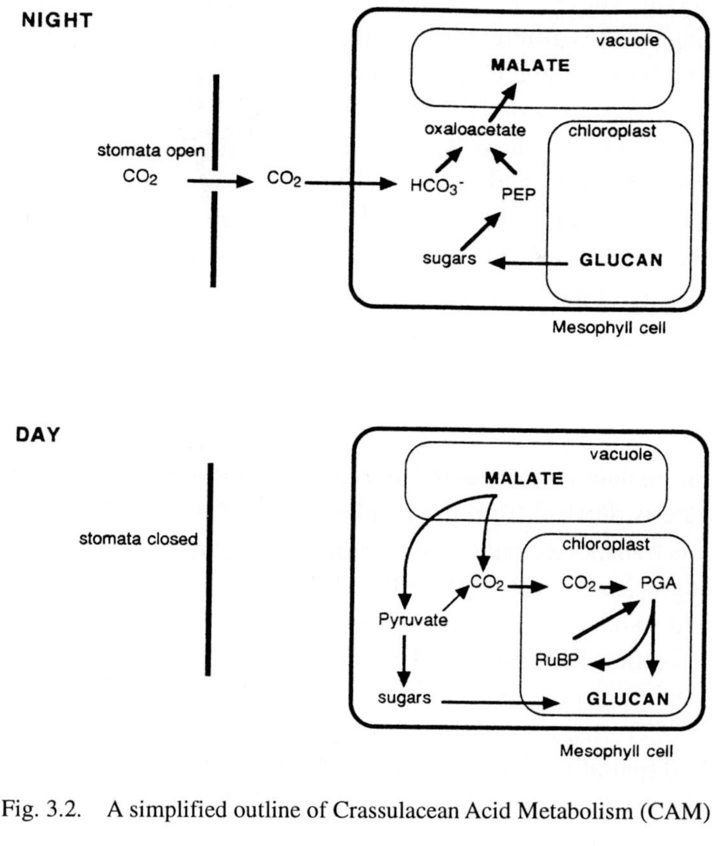

Fig. 3.2. A simplified outline of Crassulacean Acid Metabolism (CAM).

sheath cells where it is decarboxylated and the CO_2 released is then fixed by RUBPC. There are at least three variants of C_4 pathway.

The C_3 plants can be separated from the C_4 plants by their respiratory response to illumination. C_3 plants have high CO_2 compensation point (30–70 ppm) and have sizable photorespiration. The C_4 plants have low CO_2 compensation point (0–10 ppm). Photorespiration is suppressed by high CO_2 concentration in bundle sheath cells resulting from the remarkable CO_2 concentrating mechanism through PEPC (phosphoenolpyruvate carboxylase) in C_4 plant.

Unlike the C_3 and C_4 plants that assimilate CO_2 in light and evolve CO_2 in dark, CAM plants fix CO_2 mainly in the dark (Fig. 3.2). They exhibit diurnal fluctuation of titratable acidity. Also, their stomata are closed in the day and opened at night. These features have resulted in a diurnal gas exchange pattern in CAM plants that is different from that of C_3 or C_4 plants. The CAM pathway integrates both the C_3 and C_4 pathways over a diel cycle. In CAM plants, the initial carboxylation occurred through RUBPC during the light period and PEPC during the dark period. The $\delta^{13}C$ value of the CAM plant is determined by the relative contribution of carbon from either pathways, which is known to be dependent on leaf age, tissue type and environmental conditions (Kluge and Ting, 1978). The diurnal CO_2 exchange patterns of CAM plants can be divided into four phases: Phase I (nocturnal fixation of atmospheric CO_2 into malic acid using PEPC), phase II (beginning of the light phase that is associated with rapid uptake of CO_2), phase III (active decarboxylation of malate to release CO_2 internally) and phase IV (late light period of CO_2 uptake using RUBPC) (Fig. 3.3).

At night, malate is formed and stored in the vacuoles of leaves. Phosphoenolpyruvate is derived from the breakdown of starch or glucan. In the day, the malate is transported out of the vacuole and decarboxylated and the CO_2 is fixed through Calvin cycle (Phase III). The pyruvate formed is subsequently converted to starch or glucan. There are, therefore, similarities in the pathway of carbon fixation between CAM and C_4 plants. However, in CAM plants, there is temporal separation between the initial CO_2 fixation (through PEPC) and the final CO_2 fixation (through Calvin cycle). In C_4 plants, the two fixations are separated spatially in the mesophyll and vascular bundle sheath chloroplasts respectively.

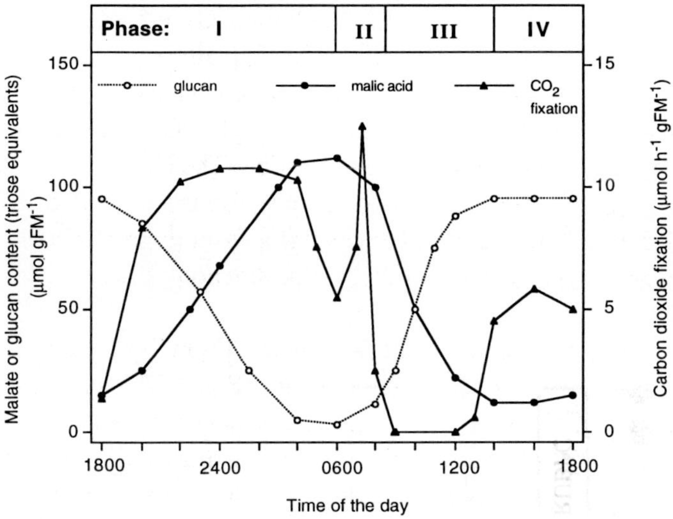

Fig. 3.3. Generalized schematic representation of malic acid and glucan levels, and rates of net carbon dioxide fixation in air in CAM plants.

Note: Levels of malic acid and glucan and rates of net carbon dioxide fixation in air are used to identify the four phases of CAM. Some salient characteristics of each phase in CAM plants (ME-type) is as follows: Phase I = Acidification using PEPC, with net carbon dioxide fixation; Phase II = transition from using PEPC to RUBPC; Phase III = Deacidification, carbon dioxide refixation using RUBPC; Phase IV = Transition from using RUBPC to PEPC.

Adapted from Osmond (1978).

A comparison between the various features of the three major groups of higher plant is given in Table 3.1. Based on the distribution of CO_2 fixation pathways in the various taxonomic groups, it has been suggested that the CAM and C_4 pathways are recent addenda to the more primitive Calvin cycle.

3.3. What is $\delta^{13}C$ Value?

Recent evidence shows that during photosynthesis, green plants preferentially take up the lighter of two naturally occurring isotopes of carbon (^{12}C and ^{13}C)

Table 3.1. Some characteristics distinguishing C_3, C_4 and CAM plants.

Characteristics	C_3	C_4	CAM
First stable product	C_3 compound (phosphoglycerate)	C_4 compound (aspartate and malate)	C_4 and C_3 compounds (night and day respectively)
Initial CO_2-fixing enzyme	RUBPC	PEPC	PEPC and RUBPC (night and day respectively)
Leaf chlorophyll a to b ratio	2.8 ± 0.4	3.9 ± 0.6	2.5 to 3.0
Theoretical energy requirement for net CO_2 fixation (CO_2: ATP : NADPH)	1 : 3 : 2	1 : 5 : 2	1 : 6.5 : 2
Leaf anatomy in cross section	Diffuse distribution of organelles in mesophyll or palisade cells with similar or lower organelle concentrations in bundle sheath cells if present	A definite layer of bundle sheath cells surrounding the vascular tissue which contains a high concentration of organelles: layer(s) of mesophyll cells surrounding the bundle sheath cells	Spongy appearance. Mesophyll cells have large vacuoles with the organelles evenly distributed in the thin cytoplasm. Generally lack a definite layer of palisade cells
Chloroplasts	similar in all tissues	dimorphic	similar in all tissues
Leaf isotopic ratio ($\delta^{13}C$)	−22‰ to −34‰	−11‰ to −19‰	−13‰ to −34‰

(Continued)

Table 3.1. (*Continued*)

Characteristics	C_3	C_4	CAM
Response to net photosynthesis to increasing light intensity at temperature optimum	Saturation reached at about 1/4 to 1/3 full sunlight	Either proportional to or only tending to saturate at full sunlight	Uncertain, but apparently saturation is well below full sunlight
Optimum day temperatures for net CO_2 fixation	15°C to 25°C	30°C to 47°C	35°C
Maximum rate of net photosynthesis (mg CO_2 $m^{-2}s^{-1}$)	0.4 to 1.1	1.1 to 2.9	< 0.4
CO_2 compensation point (ppm of CO_2)	35 to 70	0 to 5	0 to 5 in dark; 0 to 200 with daily rhythm
Leaf photorespiration detection:			
(a) exchange measurements	Present	Difficult to detect	Difficult to detect
(b) glycolate oxidation	Present	Present	Present
Photosynthesis sensitive to changing O_2 concentration from about 1% to 21%	Yes	No	Yes
Transpiration ratio (g of water/g of dry mass)	450 to 950	250 to 350	50 to 55

Adapted from Black (1973) and Bidwell (1979).

(Farquhar *et al.*, 1989). As a consequence, the ratio of these two carbon isotopes in plant tissue can be used to indicate the possible mechanism involved in the derivation of the carbon. The $^{13}C/^{12}C$ ratio is measured by mass spectrometry. The carbon isotope discrimination ratio is expressed conventionally as $\delta^{13}C$ value relative to a standard. The standard is limestone from the Peedee formation, South Carolina (PDB).

$$\delta^{13}C \text{ (parts per thousand or ‰)} = ([R_{sample}/R_{standard}] - 1) \times 1000$$

where R represents the $^{13}C/^{12}C$ ratio.

Since most samples are more deficient in ^{13}C than the standard, the scale is all on the negative side. The extent of isotope discrimination by plants is

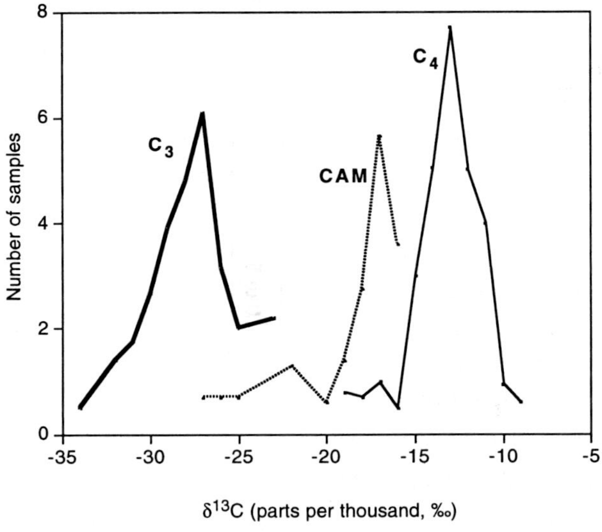

Fig. 3.4. The ^{13}C composition of C_3, C_4 and CAM species expressed as $\delta^{13}C$ in parts per thousand (‰).

Note: Unpolluted air has a $\delta^{13}C$ of -7‰, indicating that air has less ^{13}C than the standard prepared from a fossil carbonate. The average $\delta^{13}C$ for C_3 and C_4 plants is -27‰ and -11‰, respectively. Hence C_4 plants have a higher $\delta^{13}C$ composition than C_3 plants. CAM plants show a variable isotope composition because of the nature of their carbon metabolism pathway.

Adapted from Lerman (1975).

variable. However, a close correlation existed between $^{13}C/^{12}C$ ratio in plant tissue and the carbon pathway of photosynthesis. In fact, it has been suggested that the $\delta^{13}C$ value of plant tissue could be used to trace the evolutionary development of carbon pathway during geological times.

Angiosperms can be divided into three major groups (i.e., C_3, C_4 and CAM) on the basis of the $\delta^{13}C$ value (Fig. 3.4). Lerman (1975) has reported $\delta^{13}C$ values of $-17‰$, $-27‰$ and $-10‰$ for CAM, C_3 and C_4 respectively. However, CAM plants have a more variable carbon isotope composition than C_3 or C_4 plants. These plants usually show $\delta^{13}C$ values between the extremes of C_3 and C_4 plants.

3.4. Patterns of CO_2 Fixation in Orchids

Thin-leaved orchids

Current evidence suggests that thin-leaved orchids fix CO_2 through the C_3 pathway or Calvin's cycle. The photosynthetic light response curves of some thin-leaved orchids, such as *Arundina graminifolia* and *Oncidium* Goldiana are presented in Fig 3.5. Some physiological characteristics for this pathway of carbon fixation in the thin-leaved orchids include: $\delta^{13}C$ values (ca. $-27‰$), relatively high CO_2 compensation point (45–55 ppm) in gas exchange studies (Table 3.2), chlorophyll a/b ratio of 2 and prominent post-illumination outburst of CO_2 in gas exchange studies. Conclusive evidence for C_3 pathway of carbon fixation in thin-leaved orchids is shown using ^{14}C feeding experiments where the three-carbon compound phosphoglycerate is the initial product after short-term $^{14}CO_2$ fixation.

There are published reports that orchids may exhibit C_4 pathway of carbon fixation. Malate was detected as an early product of photosynthesis in young leaves of *Arundina graminifolia* and this has led to the suggestion that young leaves of *Arundina graminifolia* may photosynthesise in part through the C_4 pathway in contrast to the mature leaves (Table 3.3). Hocking and Anderson (1986) reported that leaf extracts of *Cymbidium canaliculatum* and *Cymbidium*

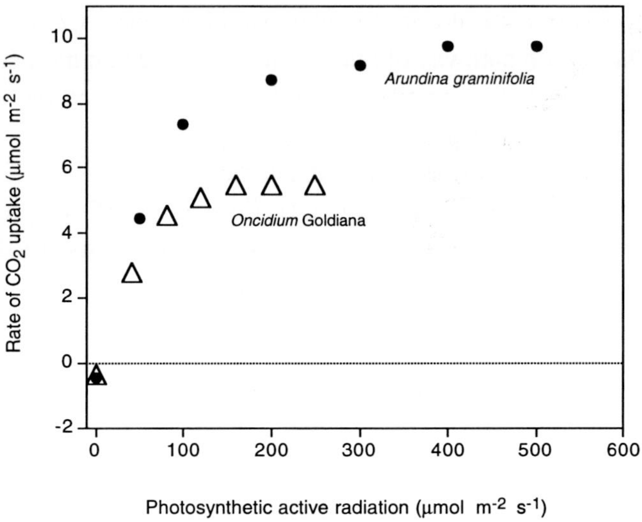

Fig. 3.5. The photosynthetic light response curves of leaves of two thin-leaved orchids.
Note: Fully expanded leaves were used for measurement.

Redrawn using data from Wong & Hew (1973) and Hew & Yong (1994).

Table 3.2. Carbon dioxide compensation point of some thin-leaved orchids.

Orchid	CO_2 compensation point (ppm)
Arundina graminifolia	55
Coelogyne mayeriana	50
Coelogyne zochusseni	50
Eulophia keithii	50
Oncidium flexuosum	55
Oncidium spacelatum	56
Oncidium Goldiana	53–55
Paphiopedilum barbatum	55
Spathoglottis plicata	48–50
Tainia penangiana	58

Adapted from Wong & Hew (1973) and Hew & Yong (1994).

Table 3.3. Percentage distribution of radioactivity following $^{14}CO_2$ fixation in two thin-leaved orchids.

Orchid species	Leaf age	Period of fixation (s)	Total ^{14}C fixed (cpm gFM^{-1})	% of the total activity in PGA	% of the total activity in Malate
Bromheadia	Young	5	19×10^4	34.5	0
finlaysonia		180	186×10^4	18.4	5.6
	Mature	5	35×10^4	13.0	2.3
		180	361×10^4	20.0	2.2
Arundina	Young	5	16×10^4	8.9	24.6
graminifolia		180	162×10^4	10.6	14.5
	Mature	5	23×10^4	37.8	8.4
		180	275×10^4	13.5	3.5

Adapted from Avadhani & Goh (1974).

madidum contain substantial pyruvate phosphate dikinase (PPD, EC 2.7.9.1) activity similar to most C_4 plants (Table 3.4). PPD is usually absent or occurs in very low activities in leaves of C_3 and CAM plants. The synthesis of PEP through the action of PPD is regarded as an essential adjunct to the C_4 mechanism. The results of Hocking and Anderson (1986) seem to suggest that the two *Cymbidium* orchids may fix CO_2 through C_4 photosynthesis.

On the contrary, recent studies on *Arundina graminifolia* have shown that both young and mature leaves of this orchid fixed carbon through C_3 photosynthesis. Supporting evidences for the operation of C_3 pathway include: Phosphoglycerate (PGA) as the early product of short term $^{14}CO_2$ fixation, substantial glycolic acid oxidase activity, glycolic acid accumulation in the presence of α-hydroxylsulfonate, low PPD activities and prominent post-illumination CO_2 outburst in gas exchange studies (Tables 3.5, 3.6). It is important to ascertain that the C_4 acid (malate) reported by Avadhani and Goh (1974) is due to the photosynthetic reactions implicit in the term C_4 photosynthesis but not from β-carboxylation. Moreover, the sole evidence of labelling of C_4 acids such as malate and aspartate as early products of short-term $^{14}CO_2$ fixation is not sufficient to define a plant as a C_4 plant. For a complete analysis,

Table 3.4. Pyruvate phosphate dikinase activity in some orchids.

Orchid	PPD (μmole mg Protein^{-1} min^{-1})	Photosynthetic pathway
Cattleya × Mary Jane	12.1	CAM
Coelogyne massangeana	0.4	C_3
Cymbidium canliculatum	80.5	CAM
Cymbidium madidum	42	C_3
Cymbidium suave	3.8	C_3
Zea mays (Maize, a known C_4 plant)	191.2	C_4
Saccharum officinarum (Sugar cane, a known C_4 plant)	55.7	C_4

Adapted from Hocking & Anderson (1986).

Table 3.5. Glycolic acid accumulation and glycolic acid oxidase activities in thin-leaved orchids.

Orchid species		Glycolic acid accumulation (μmole gFM^{-1})		Glycolic acid oxidase (n mole glyoxylate mg Protein^{-1} min^{-1})
		Water	α-HPMS	
Arundina graminifolia	Young	6.2 ± 0.03	18.7 ± 0.03	14.9 ± 1
	Mature	7.6 ± 0.1	26.7 ± 0.4	—
Cymbidium sinense	Young	8.1 ± 0.05	12.6 ± 0.2	37.5 ± 0.9
	Mature	7.4 ± 0.08	14.9 ± 0.3	24.5 ± 0.2
Saccharum officinarum (Sugar cane, a known C_4 plant)	Young	—	—	7.4 ± 0.8

Note: Leaf sections were either treated with water or 10 mM α-hydroxylsulfonate (α-HPMS) and illuminated with 200 μmol m^{-2}s^{-1} for one hour.

Adapted from Hew, Ye & Pan (1989).

Table 3.6. Activities of pyruvate phosphate dikinase in two thin-leaved orchids.

Plant species		PPD (n mole AMP mg $Protein^{-1}$ min^{-1})
Arundina graminifolia	Young leaves	4.2
Cymbidium sinense	Young leaves	3.2
	Mature leaves	3.3
Saccharum officinarum (Sugar cane, a known C_4 plant)	Young leaves	45.3

Adapted from Hew, Ye & Pan (1989).

a pulse-chase study is needed to demonstrate the transfer of label from carbon-4 of C_4 acids to carbon-1 of PGA (Edwards and Walker, 1983).

Hocking and Anderson (1986) have also expressed reservation over their own findings of C_4 photosynthesis in *Cymbidium* orchids. Uncertainty exists whether PPD activity can be used to establish the mechanism of CO_2 assimilation in orchids. The high PPD activity found in leaves of *C. canaliculatum* is not typical of CAM plants (e.g., *Kalanchoe daigremontiana*) studied elsewhere. In an earlier paper published in 1983, Winter and coworkers (1983) have proposed that *C. canaliculatum* and *C. madidum* are CAM and C_3 plants respectively, based on $\delta^{13}C$ values. In conclusion, direct evidence supporting the occurrence of C_4 photosynthesis in orchids is lacking and awaits further experimentation.

Thick-leaved orchids

The gas exchange of thick-leaved orchids is different from that of C_3 and C_4 plants (Fig. 3.6). It exhibits the four typical phases of gas exchanges as in other CAM plants. For example, in *Aranda* Wendy Scott leaf, no net gas exchange is observed from 9 am to 12 noon. CO_2 uptake begins after mid-day

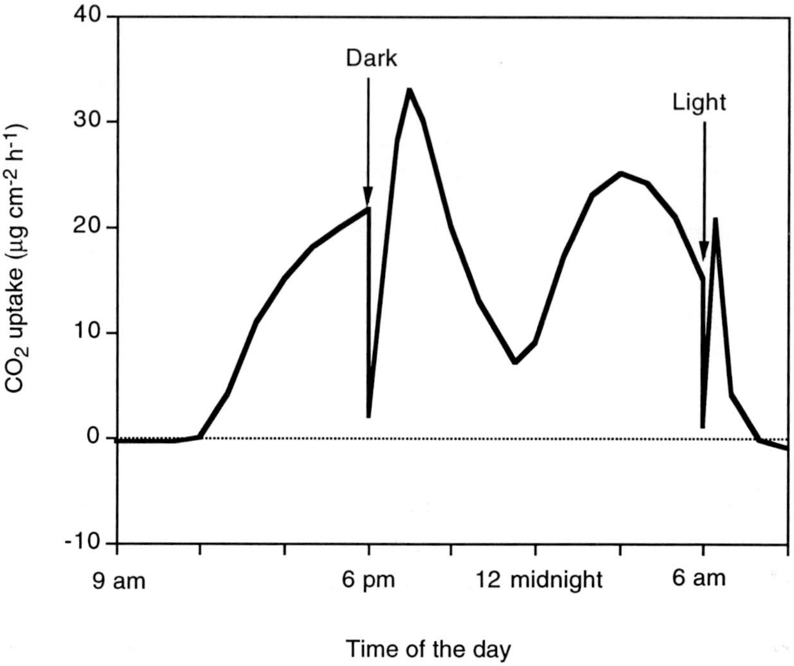

Fig. 3.6. Diurnal carbon dioxide gas exchange of an *Aranda* leaf.

Redrawn from Hew (1976).

and the rate increases with time and reaches a value of 21 µg CO_2 $cm^{-2}h^{-1}$ at 6 pm. Immediately after the light is turned off, there is a sharp dip in CO_2 uptake that is followed by a rapid CO_2 uptake. A peak of value 33 µg CO_2 $cm^{-2}h^{-1}$ is observed at about 7 pm and a second peak at 3 am. When the light is turned on at 6 am, there is a sharp dip followed by CO_2 uptake. The rate begins to decline rapidly and the leaf releases CO_2.

Thick-leaved orchids have features that are characteristic of CAM plants. This includes leaf and cell succulence, diurnal fluctuation in titratable acidity and nocturnal CO_2 fixation and inverted stomatal physiology. Titratable acidity fluctuation in certain tropical orchids is given in Table 3.7.

Table 3.7. Titratable acidity fluctuation in some orchids.

Orchids	Titratable acidity (μeq gFM^{-1})	
	9.30 am	5 pm
Thick-leaved orchids		
Leaves of mature plants		
Dendrobium taurinum	176.0	4.9
Dendrobium crumenatum	136.4	10.0
Vanda dearei	121.3	14.3
Vanda Ruby Prince	95.3	7.5
Protocorms (0.5 mm to 1 mm)		
Dendrobium taurinum	22.0	15.0
Dendrobium crumenatum	26.2	0.8
Vanda dearei	9.1	0.8
Thin-leaved orchids		
Leaves of mature plants		
Spathoglottis plicata	13.7	16.2
Arundina graminifolia	4.6	4.8
Protocorms (1 mm to 3 mm)		
Spathoglottis plicata	16.4	10.5
Arundina graminifolia	12.9	14.5

Adapted from Hew & Khoo (1980).

Table 3.8 gives the $\delta^{13}C$ value for a number of thin- and thick-leaved orchids. Leaf thickness is positively correlated to $\delta^{13}C$ value. Thin-leaved orchids (e.g., *Spathoglottis plicata*, *Arundina graminifolia*, *Coelogyne rochussenii*, *Coelogyne mayerianna* and *Oncidium flexuosum*) have $\delta^{13}C$ values of −23‰ to −24‰ while thick-leaved orchids (e.g., *Dendrobium taurinum*, *Cattleya* Bow Bell, *Aranthera* James Storie, *Aranda* Wendy Scott and *Arachnis* Maggie Oei) have $\delta^{13}C$ values ranging between −15‰ and −16‰.

Table 3.8. $\delta^{13}C$ values and leaf thickness of some orchids.

Orchid species or hybrid	$\delta^{13}C$ values (‰)	Leaf thickness (mm)
Thick-leaved orchids		
Arachnis Maggie Oei	−15.4	1.5
Aranda Wendy Scott	−15.1	1.5
Aranthera James Storie	−14.9	1.5
Cattleya Bow Bells	−16.2	2.5
Cymbidium canliculatum	−18.7	1.67
	−16.7 (Pseudobulbs)	—
Dendrobium taurinum	−15.5	1.5
Thin-leaved orchids		
Spathoglottis plicata	−27.3	0.3
Arundina graminifolia	−28.1	0.3
Coelogyne rochussenii	−28.0	0.2
Coelogyne mayerianna	−27.5	0.4
Oncidium flexuosum	−22.0	0.4
Cymbidium madidum	−27.0	0.65
Cymbidium suave	−27.0	0.59
Shootless orchids		
Chiloschita phyllorhiza	−14.8 (roots)	—
Taeniophyllum malianum	−15.8 (roots)	—

Adapted from Neales & Hew (1975), and Winter, Wallace, Socker & Roksandic (1983).

3.5. Photosynthetic Characteristics of Non-Foliar Green Organs

Leaves are the main sources of assimilates for growth, especially in leafy orchids. There are numerous non-foliar green organs in leafy orchids such as pseudobulbs, flowers, fruit capsules and roots that can potentially contribute to the overall carbon balances (Table 3.9). Recent evidences indicate that the sole contribution of carbon from non-foliar sources in most leafy orchids is not sufficient for growth and that the major portion of photoassimilates obtained from regenerative photosynthesis in these organs is utilised within the organs and not exported to other sink organs. This is unlike the shootless orchids

Table 3.9. Carbon fixation in non-foliar green organs of some orchids.

Plant organ	Species/hybrid	Physiological observation
Fruit capsules	*Laeliocattleya* hybrid	Demonstrated gas exchange
	Encyclia tampensis	Weak CAM
	Oncidium Goldiana	Fixed $^{14}CO_2$
Flowers	*Arachnis* Maggie Oei	Weak CAM
	Aranda Deborah	Weak CAM
	Cymbidium hybrid	Fixed $^{14}CO_2$
	Dendrobium Mary Mak	Weak CAM
	Oncidium Goldiana	Non-CAM
		Fixed $^{14}CO_2$
		High PEPC/RUBPC ratio
	Phalaenopsis hybrid	Non-CAM
Flower stalks	*Phalaenopsis* hybrid	Weak CAM
Pseudobulbs	*Laelia anceps*	Regulates CAM activity in leaves
	Oncidium Goldiana	No gas exchange in light except with the removal of cuticle
Roots	<u>I: Leafy orchids</u>	
	Arachnis Maggie Oei	No net photosynthesis
	Aranda Wendy Scott	No net photosynthesis
		High PEPC activity
	Aranda Deborah	No net photosynthesis
	Cattleya hybrid	No net photosynthesis
		Fixed $^{14}CO_2$
	Encyclia tampensis	No net photosynthesis
	Epidendrum sp.	Fixed $^{14}CO_2$
	Kingidium taeniale	No net photosynthesis
	Phalaenopsis hybrid	Fixed $^{14}CO_2$
	Rangaeris amaniensis	No net photosynthesis
	Saccolabium bicuspidatus	No net photosynthesis
	Vanda paraishi	No net photosynthesis
	Vanda suavis	Well developed chloroplasts
	Vanda paraishi	Fixed $^{14}CO_2$
	Oncidium Goldiana	Fixed $^{14}CO_2$

(*Continued*)

Table 3.9. (*Continued*)

Plant organ	Species/hybrid	Physiological observation
	II: Leafless orchid	
	Campylocentrum tyrridion	Net photosynthesis observed
	Campylocentrum pachyrrbizum	Net photosynthesis observed
	Chilochista usneoides	Net photosynthesis observed
	Polyradicion lindenii	Net photosynthesis observed
	Sarcocbilus segawai	Net photosynthesis observed
Stems	*Epidendrum xanthium*	Fixed $^{14}CO_2$
	Phalaenopsis hybrid	Fixed $^{14}CO_2$
	Vanda suavis	Fixed $^{14}CO_2$

Adapted from Hew (1995).

where the roots form more than half of the biomass of the orchid and the non-foliar organs (in this case, roots) are the only source available for photo-assimilates acquisition.

Distinction has been made between regenerative and net photosynthesis. Fixation of CO_2 by non-foliar organs is primarily regenerative. Nitrogen investment is high in leaf that shows net photosynthesis. For non-foliar organs involved in regenerative photosynthesis, nitrogen investment is low but high in water use efficiency. This phenomenon could be adequately explained by the relative cost effectiveness of investing scare resources in an epiphytic habitat.

Aerial roots

The photosynthetic efficiency of aerial roots in leafy orchid has attracted considerable attention. Although the gas exchange pattern of aerial roots in leafy orchid is different from that of the leaf (Fig. 3.7), it exhibits acidity fluctuation similar to the leaf (Fig. 3.8). Aerial root will lose its chlorophyll and become branched when it penetrates into the mulch. Interestingly, this terrestrial form of aerial roots does not show fluctuation in titratable acidity.

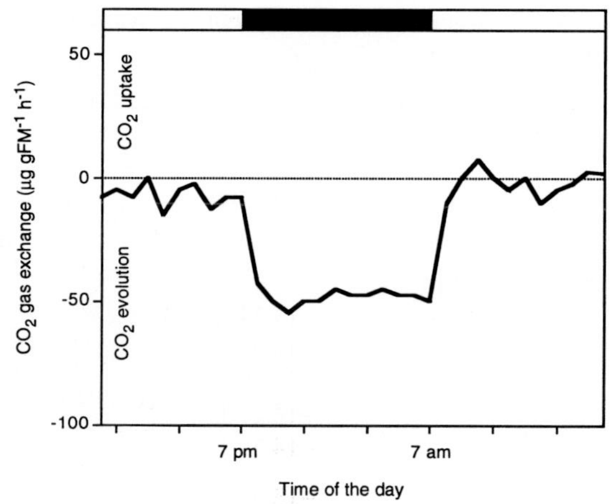

Fig. 3.7. Diurnal carbon dioxide exchange in detached aerial roots of *Arachnis* Maggie Oei.
Note: Roots were detached and placed in vials containing a known amount of water. Three roots were used for each determination.

Redrawn from Hew, Ng, Wong, Yeoh & Ho (1984).

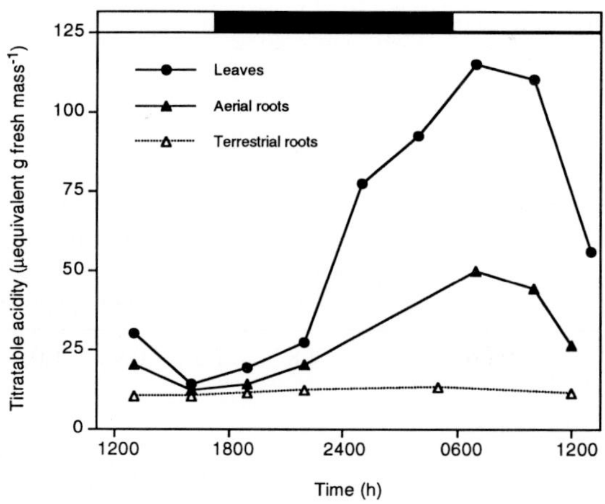

Fig. 3.8. Diurnal fluctuation in titratable acidity levels of leaves, aerial roots and terrestrial roots of *Arachnis* Maggie Oei.

Adapted from Hew, Ng, Wong, Yeoh & Ho (1984).

Table 3.10. $\delta^{13}C$ values of *Arachnis* Maggie Oei aerial roots at various distances from the root tip.

Plant material	$\delta^{13}C$ values (‰)	
	Cortex	Velemen
Aerial root (distance from the root tip)		
0–1 cm	−13.34 ± 0.23	—
1–2 cm	−13.68 ± 0.35	−13.90 ± 0.19
3–4 cm	−14.18 ± 0.15	−14.13 ± 0.11
5–6 cm	−14.22 ± 0.10	−14.35 ± 0.17
7–8 cm	−14.55 ± 0.14	−14.75 ± 0.29
Mean	−13.99 ± 0.48	−14.28 ± 0.36
Leaf	−14.54 ± 0.18	—

Note: mean ± SD.

Adapted from Hew, Ng, Wong, Yeoh & Ho (1984).

Table 3.11. Comparison of PEP carboxylase and RUBP carboxylase activities in *Arachnis* Maggie Oei aerial roots and leaves at different time of the day.

Plant material	Chlorophyll content (mg g FM^{-1})	Enzyme activity (μmol HCO_3^- [mg Chl]$^{-1}$ min^{-1}) PEPC		RUBPC		Ratio of PEPC: RUBPC	
		7 am	3 pm	7 am	3 pm	7 am	3 pm
Aerial root (0–2 cm) from the tip	0.06	4.8	7.5	0.9	1.0	5 : 1	8 : 1
Aerial root (12–14 cm) from the tip	0.03	3.7	8.0	0.8	1.2	5 : 1	7 : 1
Leaf 3 (young)	0.15	2.5	20.9	1.3	2.2	2 : 1	10 : 1
Leaf 9 (mature)	0.20	4.4	22.2	1.1	1.6	4 : 1	14 : 1

Adapted from Hew, Ng, Wong, Yeoh & Ho (1984).

Aerial roots and leaves of a CAM orchid have similar $\delta^{13}C$ values (Table 3.10). Although aerial roots of the leafy orchid exhibit dark acidification and have $\delta^{13}C$ value typical of CAM plants, there is no net dark CO_2 uptake. Instead, the aerial roots fix CO_2 in the light and evolve CO_2 in darkness. Nevertheless, CO_2 fixation in both light and dark could be demonstrated by feeding the aerial roots with $^{14}CO_2$. Apparently, orchid roots are capable of

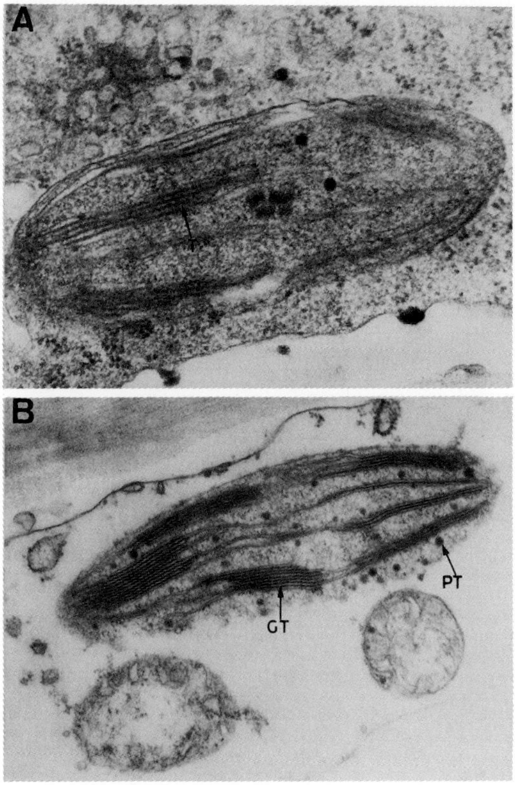

Fig. 3.9. Transmission electron microscopy of chloroplasts isolated from the cortex of root tip and mature root segment of *Vanda suavis*.
Note: (A) Chloroplast from cortex of root tip [30000 X]; (B) Granal chloroplast from cortex of mature root segments [20000 X]. Explanation of symbols: GT, grana thylakoid; PT, plastoglobulus.

Adapted from Ho, Yeoh & Hew (1983).

considerable CO_2 fixation in the dark. It is unlikely that the low dark CO_2 fixation is limited by PEPC levels. Aerial roots contain as much as one half of the activity of PEPC as in the leaves (Table 3.11). The PEPC activity in orchid roots is low at the start of the light period and becomes higher in the late afternoon. The K_m (Michelis–Menton constant, a measure of enzyme kinetics) for PEP is the same for PEPC in roots and leaves. The occurrence of grana-typed chloroplasts in the cortical layer of aerial roots (Fig. 3.9) is consistent with the view that aerial roots of leafy epiphytic orchids have well-developed photosynthetic apparatus. Hill's reaction and O_2 evolution have also been demonstrated in isolated root chloroplasts.

Evidently, the CO_2 fixation in darkness by aerial roots is masked by the high respiration rate (Fig. 3.10) (See Chap. 4 on RESPIRATION). The seemingly high CO_2 partial pressures arising from respiration favours CO_2 fixation within the roots. In a way, the CO_2 fixation pattern in aerial roots of a leafy orchid is not unexpected. The CAM mode of carbon fixation in aerial roots is associated with drought tolerance. The behaviour of aerial roots is

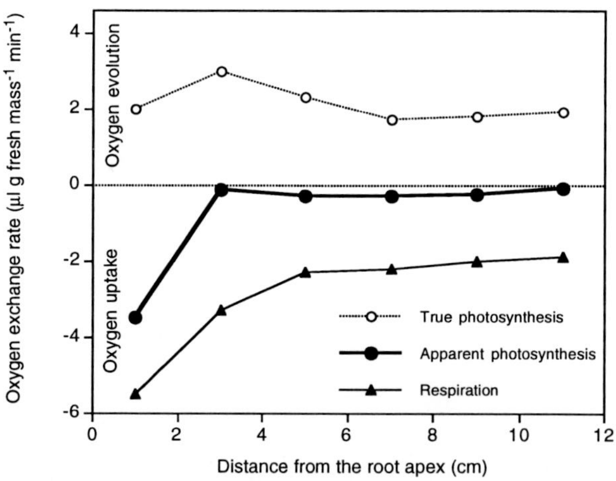

Fig. 3.10. Photosynthesis and respiration in aerial roots of *Aranda* Wendy Scott at various distances from the root tip.

Redrawn from Hew (1987).

similar to cactus plants conserving carbon by refixing respired CO_2 when the water potential of tissue mandates that the stomata remain closed for weeks during the dry season.

Another possible explanation to account for the zero net photosynthesis of aerial roots is the velamen. When the velamen is dry, its surface scatters light, thus reducing the proportion of incident light available for photosynthesis by the chloroplasts located in the cortex. Furthermore, a water saturated velamen may impede gas exchange. It seems that the rate of CO_2 fixation by aerial roots is affected by the velamen when it is either dry or wet (Fig. 3.11). It therefore appears that aerial roots of leafy epiphytic orchids are not able to provide sufficient carbon to maintain themselves. Based on the CO_2 gas exchange pattern of aerial roots, it was estimated that a *Cattleya* root of at least 21 cm long under continuous irradiance is necessary to offset the energy used

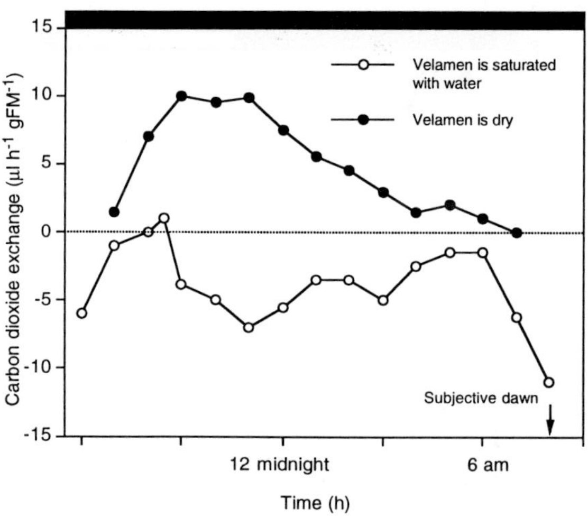

Fig. 3.11. Effect of saturating the velamen with moisture on the progress of carbon dioxide uptake in darkness for the shootless orchid *Chiloschista usneoides*.
Note: Gas exchange was followed at 25°C until it was established that a normal carbon dioxide exchange pattern was developing. At the time indicated, the orchid was sprayed with distilled water until the velamen was saturated. Positive values (above zero) indicate net carbon gain.

Redrawn from Cockburn, Goh & Avadhani (1985).

in respiration. The same seems to hold true for aerial roots of the other orchids studied so far.

Perhaps what is important here is the ability of roots to recycle or refix at least part of the respiratory CO_2. This would provide a substantial portion of the total carbon and energy requirement for the continuous production and growth of aerial roots for anchorage, water storage and acquisition of minerals. The ability to economise all resources with great efficiency is closely tied to the remarkable success of the orchid as an epiphyte.

Roots may form more than half of the biomass of an orchid plant. Thus, in terms of carbon budget, it would be of interest to know how much the roots are dependent on the leaves for nutritional support. The situation in roots of shootless orchid species is unique where the roots become the sole organ for photosynthesis. Net CO_2 gas exchange and typical acidity fluctuation are observed in roots of shootless orchids (Figs. 3.12, 3.13). Photosynthetic carbon

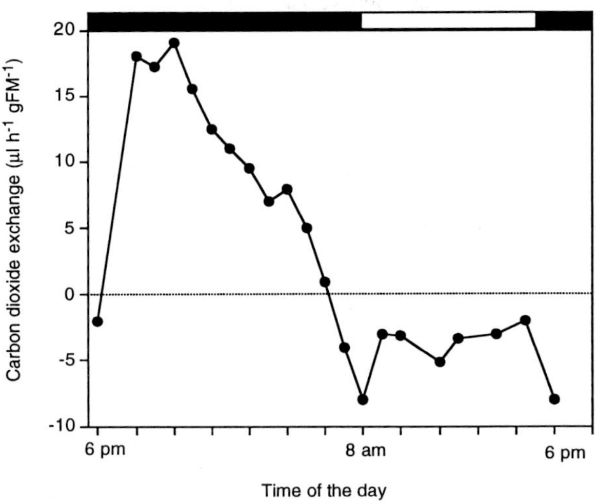

Fig. 3.12. Carbon dioxide exchange in darkness and in light for the roots of *Chilochista usneoides*.

Note: Following incubation in darkness at 25°C for 15 h, the orchid was illuminated with 300, 600 and 900 $\mu mol\ m^{-2}s^{-1}$ of photosynthetically active radiation. Finally, the plant was returned to darkness. Positive values (above zero) indicate net carbon gain.

Adapted from Cockburn, Goh & Avadhani (1985).

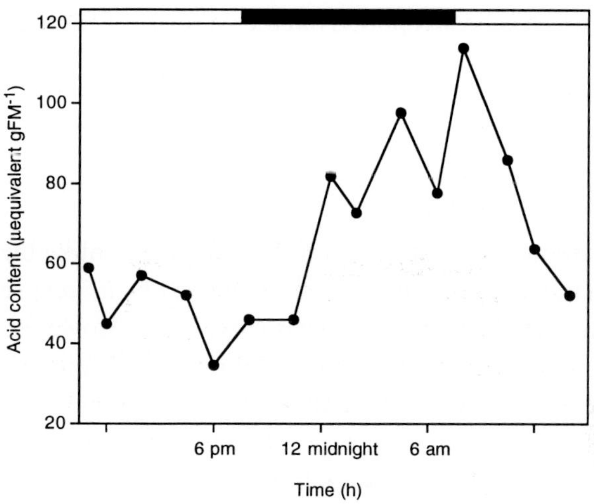

Fig. 3.13. Titratable acid content in the roots of the shootless orchid *Chiloschista usneoides*.

Redrawn from Cockburn, Goh & Avadhani (1985).

assimilation by these roots involves the synthesis and accumulation of malic acid from CO_2 in the darkness. The malic acid accumulated during darkness is utilised in the light. The $\delta^{13}C$ values of two shootless orchid species (*Chiloschista phyllorhiza* and *Taeniophyllum malianum*) are −14.5‰ and −15.8‰ respectively. Unlike the leaves, the roots do not possess stomata or any means to regulate the CO_2 diffusion between the internal gas phase of the plant and the atmosphere. The absence of stomatal control in root CAM activity of shootless orchid is unique. This may represent an addendum to the presently recognised mechanisms (C_3, C_4 and CAM) by which plants acquire atmospheric CO_2 and the term 'Astomatal CAM' for this variant of photosynthetic carbon metabolism has been proposed.

Stems

Stems of monopodial orchid are green and can clearly contribute positively to the total carbon gain of the orchid. The $^{14}CO_2$ fixation by stems of *Cattleya* and *Phalaenopsis* has been reported but the pathway of carbon fixation remains unclear.

Pseudobulbs

Pseudobulbs are modified stems with thick cuticle. Unlike the stems, there is no stomata on the pseudobulbs of orchids. Intact *Oncidium* pseudobulbs show no gas exchange in light or in darkness (Fig. 3.14). However, CO_2 evolution can be detected in darkness after the partial removal of cuticle (2 cm by 2 cm) from each side of the pseudobulb. Using the same pseudobulb, there is a gradual decrease in CO_2 evolution when exposed to light indicating that there is some degree of CO_2 fixation by the pseudobulb tissue. However, there is no net CO_2 gain by the pseudobulb tissue of *Oncidium*.

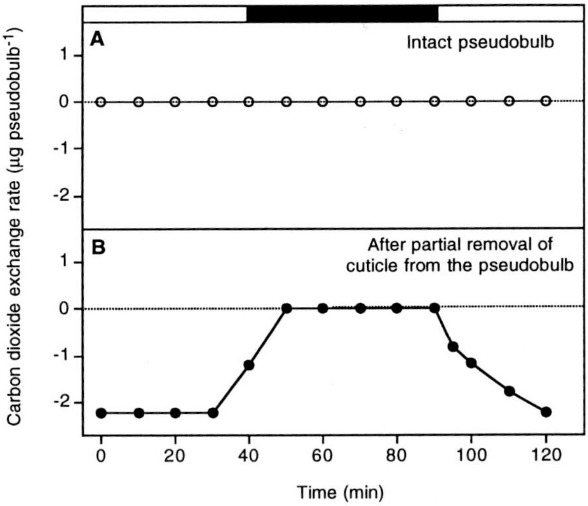

Fig. 3.14. Gas exchange patterns in pseudobulbs of *Oncidium Goldiana*.
Note: (A) Intact pseudobulbs; (B) pseudobulbs after the partial removal of cuticle. Uniform illumination of 150 µmol $m^{-2}s^{-1}$ was provided for both sides of the pseudobulb. In (B), 2 cm^3 of cuticle was removed from each side of the pseudobulb. Each reading is a mean of three replicates. Positive values (above zero) indicate net carbon gain.

Redrawn from Hew & Yong (1994).

No significant diurnal fluctuation in titratable acidity is observed in the pseudobulbs of the C_3 orchid, for example *Oncidium* Goldiana (Table 3.12). The chlorophyll content (expressed in terms of per gram fresh mass) in *Oncidium* pseudobulbs is only 4–6% when compared to the leaves. In addition, these tissues contain substantial RUBPC and PEPC activity (Table 3.13). It

Table 3.12. Some physiological parameters of *Oncidium* Goldiana pseudobulbs.

Water content (%)	94.4 ± 0.2
Total chlorophyll content (mg gFM^{-1})	0.071 ± 0.001
Chlorophyll a/b ratio	1.86 ± 0.04
Titratable acidity (μeq g FM^{-1})	9.2 ± 1.2 (9 am) 10.1 ± 0.6 (4 pm)

Note: mean ± SE.

Adapted from Hew & Yong (1994).

Table 3.13. Total chlorophyll content and the ratio of Phosphoenolpyruvate carboxylase and Ribulose bisphosphate carboxylase activity in different plant parts of *Oncidum* Goldiana.

Plant part	PEPC/RUBPC ratio	Chlorophyll (mg g DM^{-1})
Leaf L2	0.3	10.50 ± 0.82
Leaf L4	0.3	10.98 ± 0.51
Pseudobulbs	0.4	2.38 ± 0.21
Peduncle (flower stalk)	0.5	1.68 ± 0.10
Buds	3.9	1.09 ± 0.07
Florets	1.8	0.59 ± 0.03
Fruit capsules	0.6	1.21 ± 0.16
Epiphytic roots	5.0	1.39 ± 0.08

Note: n = 3 or 4, ± SE.

Adapted from Hew, Ng, Gouk, Yong & Wong (1996).

appears that pseudobulb photosynthesis is involved primarily in the refixation of respiratory CO_2 produced by the underlying massive parenchymatous tissues. Evidently, the development of water conservation feature in the pseudobulb, with an impermeable layer of cuticle and the absence of stomata, is at the expense of CO_2 diffusion. At present, the importance of pseudobulbs of C_3 orchid to leaf photosynthesis remains to be established.

For CAM pseudobulbs, an exposure of light to the pseudobulb is thought to be necessary for the daily net CO_2 uptake by leaves (Fig. 3.15). It is suggested that the organic acids fixed in the leaves move into the pseudobulb during the night for storage. During the day, the CAM pseudobulbs act as a CO_2 releasing organ for carbon fixation. While this speculation awaits further study, this observation implies that pseudobulbs of CAM orchids may function actively in the regulation of CAM photosynthesis.

Flowers and fruit capsules

Stomata in orchid flowers are generally non-functional and it is unlikely that gas exchange in orchid flowers is under the same diurnal stomatal control as in the leaves. Green *Cymbidium* flowers are able to photosynthesise and more ^{14}C is fixed in light than in darkness. However, their rates of CO_2 fixation are comparatively lower than other organs such as roots, stems and leaves. Fluctuations in titratable acidity have been observed in flowers of CAM orchid (e.g., *Arachnis*, *Dendrobium* and *Vanda*) but not in the C_3 orchid *Oncidium* Goldiana. However, there is a report that flowers of *Phalaenopsis* (a known CAM orchid) do not exhibit acidity fluctuation. On the other hand, flower stalks of *Phalaenopsis* do show weak CAM activity. Flowers of the C_3 orchid *Oncidium* Goldiana have high PEPC/RUBPC ratio, indicating that it may fix CO_2 primarily through β-carboxylation (Table 3.13).

Fruit capsules are formed from flowers after fertilisation. Fruit capsules of the CAM orchid *Encyclia tampensis* exhibit CAM-like activity, which decreases with fruit development. The decrease in CAM activity is attributed to the increasing resistance to CO_2 diffusion as the fruit capsules mature (Fig. 3.16).

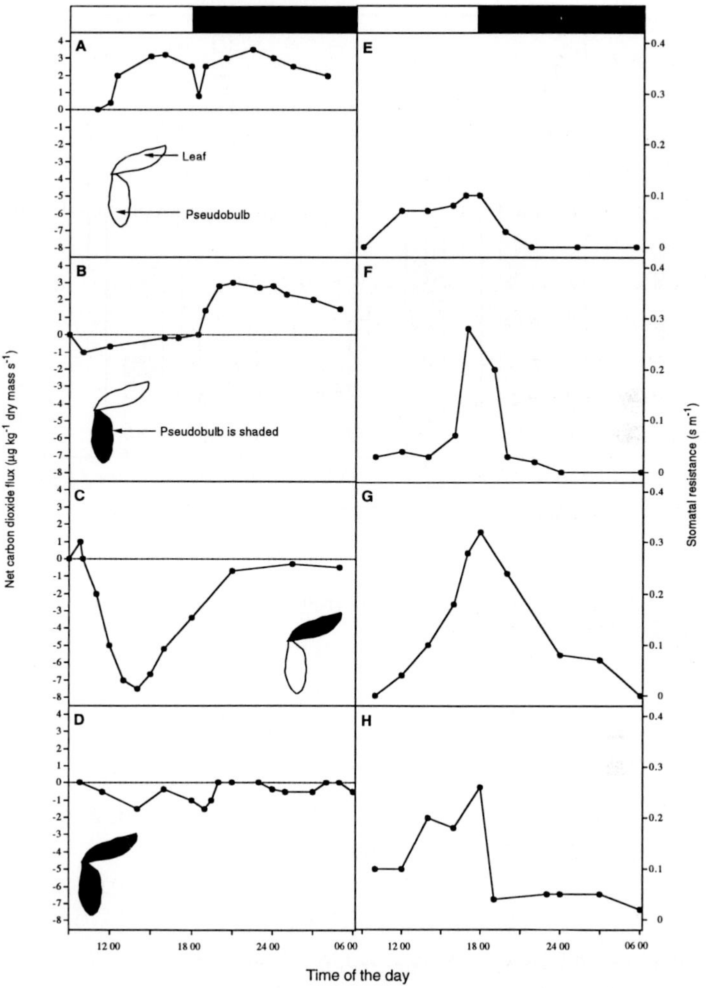

Fig. 3.15. The effect of light on leaves and pseudobulbs of *Laelia anceps* on the rate of carbon dioxide exchange and stomatal resistance of leaf.

Note: (A) & (E): Both the leaf and pseudobulb are kept in the light. (B) & (F): The leaf is placed in the light while the pseudobulb is kept in darkness. (C) & (G): The leaf is placed in darkness while the pseudobulb is kept in the light. (D) & (H): Both the leaf and the pseudobulb are kept in darkness from 09 00 h to 18 00 h.

Adapted from Ando & Ogawa (1987).

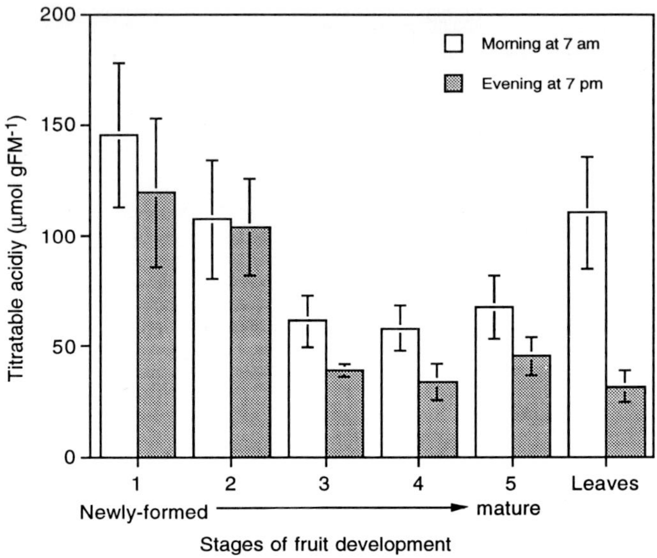

Fig. 3.16. Titratrable acidity changes during fruit capsule development of the CAM orchid *Encyclia tampensis*.
Note: mean, ± SD.

Redrawn from Benzing & Pockman (1989).

Similarly, fruit capsules of the C_3 orchid *Oncidium* Goldiana are able to fix CO_2 but the pathway of fixation is not known.

Varying $\delta^{13}C$ values in non-foliar green organs

The contribution of regenerative photosynthesis in non-foliar green organs of orchids is reflected in the $\delta^{13}C$ values obtained. For example, varying $\delta^{13}C$ values for the different plant parts of *Oncidium* Goldiana has been reported (Fig. 3.17). As discussed earlier, the $\delta^{13}C$ value is a measure of the relative abundance of ^{13}C in a given plant material. The degree of depletion varies, depending on the mode of carbon assimilation operating in the plant.

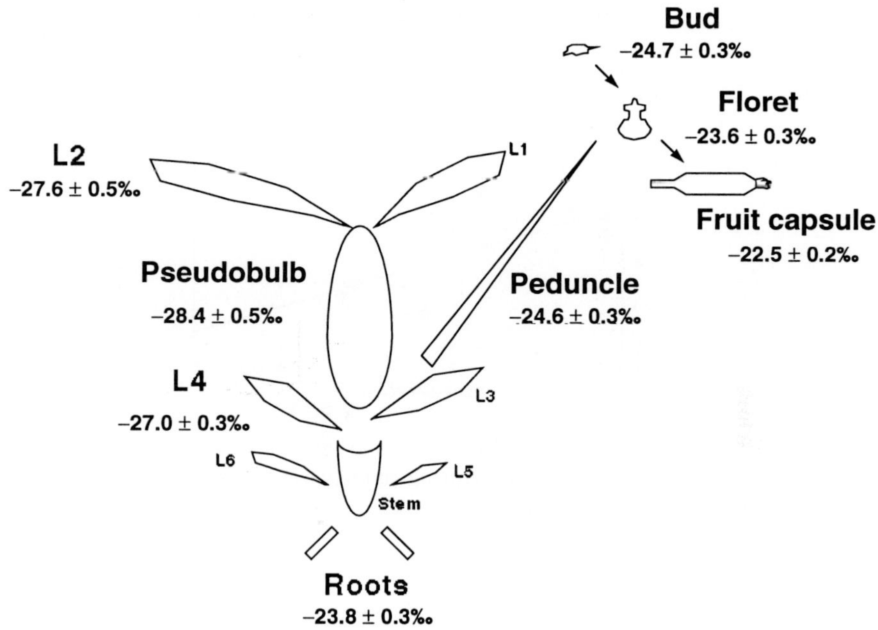

Fig. 3.17. $\delta^{13}C$ values in the different plant parts of *Oncidium Goldiana*.

Redrawn from Yong (1995).

Discrimination against ^{13}C is most pronounced in C_3 plants that utilised RUBPC as the initial carboxylase while CO_2 fixation through PEPC shows less discrimination against ^{13}C during the uptake of atmospheric CO_2. Since inflorescences, epiphytic roots and fruiting structures have two sources of carbon (import and regenerative photosynthesis), it is likely that regenerative photosynthesis within non-foliar green tissues modifies the proportion of ^{13}C inside the tissues that use predominantly imported carbon from the leaves. The enrichment of ^{13}C in these organs is due to a low RUBPC/PEPC ratio. This postulation is supported by a significant correlation ($p < 0.05$) between RUBPC/PEPC ratio within these tissues and its corresponding $\delta^{13}C$ values (Fig. 3.18).

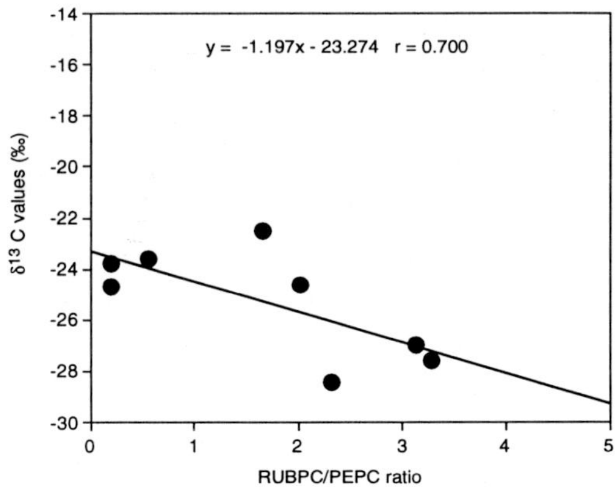

Fig. 3.18. The relationship between $\delta^{13}C$ values and the ratios of ribulose bisphosphate carboxylase (RUBPC) and phosphoenolpyruvate carboxylase (PEPC) activities within the different plant parts of *Oncidium Goldiana*.

Adapted from Hew, Ng, Gouk, Yong & Wong (1996).

3.6. Factors Affecting Photosynthesis

Effects of light

Photosynthesis of C_3 orchids saturates at different light intensities depending on whether it is sun loving or shade loving. The sun-loving thin-leaved orchids, *Arundina graminifolia* and *Spathoglottis plicata*, saturate at light intensities beyond 200 μmol $m^{-2}s^{-1}$ whereas the shade orchids, *Oncidium* Goldiana and *Cymbidium sinense*, saturate at light intensity of 80–100 μmol $m^{-2}s^{-1}$ and 150 μmol $m^{-2}s^{-1}$ respectively (Fig. 3.5).

The light compensation point is defined as the light intensity at which photosynthetic rate equals the rate of respiration. For example, the light compensation point for two shade-loving orchids, *Oncidium* Goldiana and *Cymbidium sinense* , is around 5–8 μmol $m^{-2}s^{-1}$.

Like in other CAM plants, the light intensity in the day does affect dark CO_2 fixation in thick-leaved orchids during the night (Phase I). For *Arachnis*, CO_2 fixation at night is markedly enhanced with an increase of light intensity in the day. For *Phalaenopsis*, a shade loving CAM orchid, the day- and night-time CO_2 fixation increase with increasing light intensity in the day up to 130 μmol $m^{-2}s^{-1}$.

At present, we have no accurate information pertaining to the light requirement of commercially important thick-leaved orchids under cultivation. In practice, thick-leaved orchids like *Arachnis* and *Aranda* are grown under full sun while *Dendrobium*, *Vanda* and *Mokara* are cultivated under partial shade. It is not known whether the present conditions used by commercial growers are the optimal light requirement for these orchids. In the ASEAN region, orchids for cut flowers are grown in the open field either under full sunlight or in partial shade. There is no information about photosynthesis in relation to the productivity of an orchid community. Light interception, light requirement and photosynthetic efficiency of an orchid crop stand are areas that deserve more investigation.

Effects of age

Photosynthesis of the C_3 orchid, *Oncidium* Goldiana, has been studied at four different stages of development: Bud stage (youngest), plantlet stage, unsheathing stage and pseudobulb stage (oldest). Leaf photosynthesis changes as the leaves age. For example, the quantum yield is highest at stage 1 (0.08) and lowest at stage 4 (0.047).

Similarly, the capacity for CAM appears to change with leaf age. For example, the effects of leaf age on CAM in relation to changes in fresh mass, dry mass, protein content, chlorophyll content and leaf area are presented in Table 3.14. The fifth *Aranda* leaf is considered fully expanded in terms of area, chlorophyll, protein, fresh mass and dry mass while titratable acidity is highest in the tenth leaf. Lowest CAM activity is found in the first leaf that is actively growing. When the leaf ages (e.g., fifteenth leaf), the titratable acidity decreases significantly. For *Aranda*, it appears that CAM capacity reaches a maximum only after a leaf has attained maturation. Similarly, changes in CAM activity with leaf age are also observed in *Arachnis* (Fig. 3.19) and *Phalaenopsis* (Fig. 3.20).

Table 3.14. Leaf characteristics of *Aranda* Wendy Scott.

Leaf position*	Fresh mass (g)	Dry mass (g)	Protein content (mg g FM^{-1})	Chlorophyll content (mg g FM^{-1})	Nocturnal acidity increases (μeq gFM^{-1})	Leaf area (cm^2)
L 1 (young, expanding leaf)	0.93 ± 0.28	0.16 ± 0.02	33.5 ± 6.33	0.20 ± 0.07	61.7	14.4 ± 1.5
L 5	3.26 ± 0.21	0.56 ± 0.01	26.79 ± 6.08	0.25 ± 0.04	142.3	30.8 ± 1.4
L 10	3.31 ± 0.57	0.56 ± 0.09	29.38 ± 1.19	0.29 ± 0.05	151.4	29.9 ± 3.9
L 15 (oldest leaf)	3.50 ± 0.35	0.59 ± 0.26	21.80 ± 3.71	0.24 ± 0.04	127.1	31.4 ± 2.0

*Note: Counting down from the apex.

Adapted from Hew (1978).

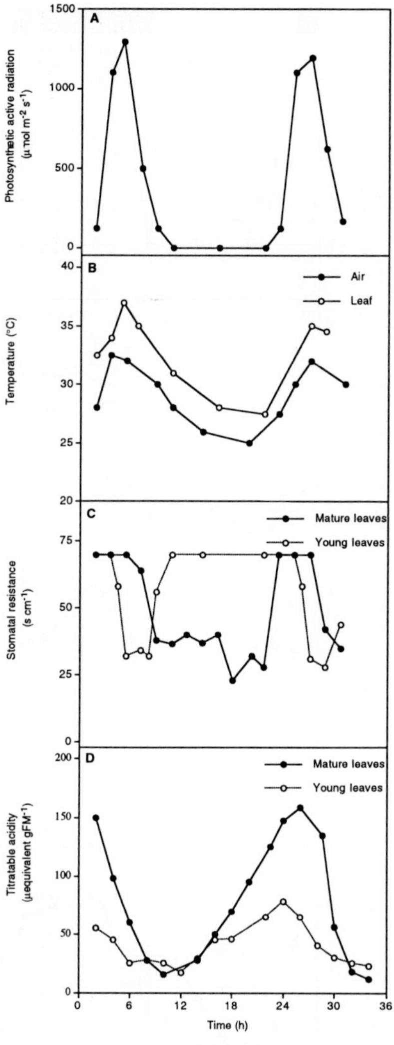

Fig. 3.19. Photosynthetic characteristics of young and mature leaves of *Arachnis* Maggie Oei in a normal day–night cycle.

Note: (A) Photosynthetic active radiation received by the leaves. (B) Atmospheric and leaf temperatures. (C) Stomatal resistance in young and mature leaves. (D) Titratable acidity content in young and mature leaves.

Adapted from Goh, Avadhani, Loh, Hanegraaf & Arditti (1977).

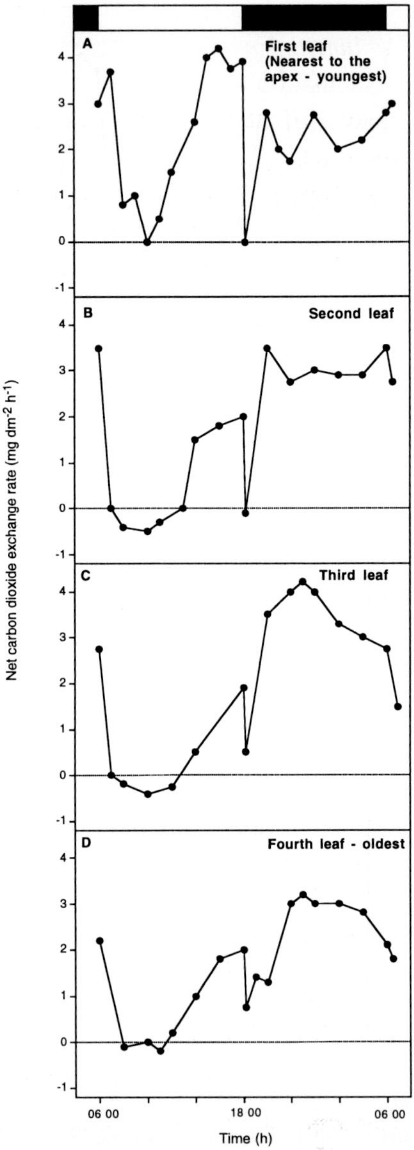

Fig. 3.20. The carbon dioxide exchange rates of *Phalaenopsis* leaves of different ages.

Redrawn from Ota, Morioka & Yamamoto (1991).

The diurnal acidity fluctuation is barely detectable in young protocorms of the CAM orchid *Dendrobium taurinum*. Young leaves of *Dendrobium* seedlings exhibit diurnal acidity fluctuation except that the magnitude is considerably lower than that of the leaves of adult plants. The data suggest that the CAM capacity increases as the seedling grows (Fig. 3.21). In contrast, protocorms and seedlings of thin-leaved orchids such as *Spathoglottis plicata* and *Arundina graminifolia* show no apparent fluctuation in acidity (Table 3.7).

The finding that in thick-leaved orchids, acidity fluctuation appears only when it reaches a certain stage of ontogeny is important. This observation

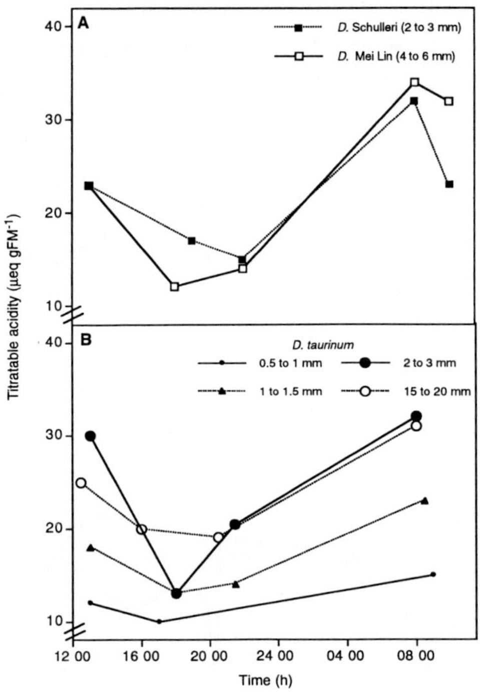

Fig. 3.21. Diurnal fluctuation of titratable acidity in young protocorms and seedlings of three *Dendrobium* orchids.
Note: (A) *Dendrobium taurinum* protocorms of different sizes (0.5–20 mm); (B) Protocorms of *Dendrobium* Schulleri (2–3 mm) and *Dendrobium* Mei Lin (4–6 mm).

Redrawn from Hew & Khoo (1980).

seems to indicate a switch in the CO_2 fixation process during development. If so, it would be of interest to examine the different CO_2 fixation pathways in these orchids. A possible change in CO_2 fixation pathway during ontogeny in orchids is not unique as it has been reported that some non-orchidaceous plants can change from C_3 to C_4 photosynthesis, C_4 to C_3 photosynthesis or C_3 to CAM during ontogeny or under certain environmental conditions. In thick-leaved orchids and other succulent plants showing CAM features, a CO_2 assimilation pathway comparable to that of C_3 photosynthesis exists in phase 4 (Fig. 3.22). The pathway of carbon fixation in young leaves and protocorms of thick-leaved orchids may be of the C_3 type.

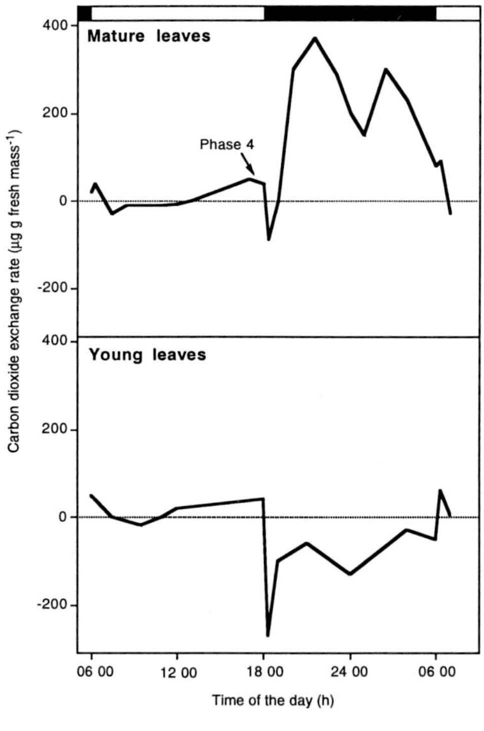

Fig. 3.22. Carbon dioxide fixation in mature and young leaves of *Arachnis* Maggie Oei. Note: Photosynthetic active radiation of 470 µmol $m^{-2}s^{-1}$ was provided during the light period.

Redrawn from Goh, Wara-Aswapati & Avadhani (1984).

Effects of water stress

CAM orchids

In CAM orchids, artificial drought can be imposed by flushing or immersing the orchids in polyethylene glycol 1000 (PEG 1000) solution with an osmotic potential of −18 bars. The relative water content of *Aranda* and *Dendrobium* leaves decreases progressively when subjected to water stress treatment. There are parallel decreases in diurnal titratable acidity fluctuation and nocturnal CO_2 uptake (Figs. 3.23, 3.24). In other CAM plants such as *Agave*, stomata close under water stress, thereby reducing CO_2 uptake. The night-time stomatal movement in CAM plants depends on the availability of stored water in the tissues, which is known to last from eight days to many months. Under severe drought for extended periods of time, *Agave* may adopt an idling mode in which organic acids fluctuate diurnally without exogenous CO_2 exchange to minimise water loss.

In *Aranda* and *Phalaenopsis*, the day-time CO_2 uptake and night-time CO_2 uptake are greatly reduced under water stress (Figs. 3.24, 3.25). After prolonged

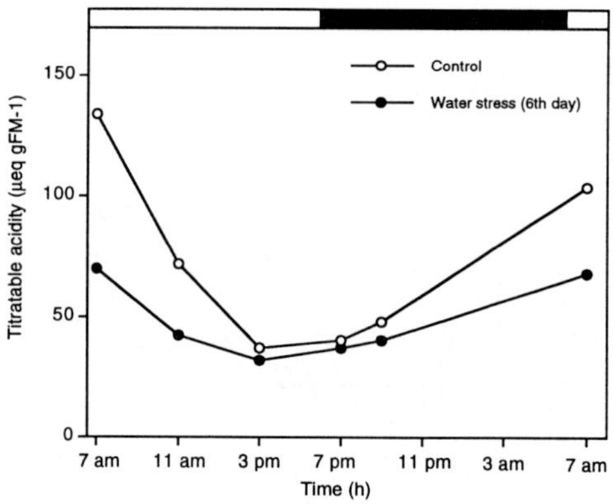

Fig. 3.23. Effect of water stress on diurnal fluctuation of titratable acidity in leaves of *Aranda* Christine 9.

Redrawn from Fu & Hew (1982).

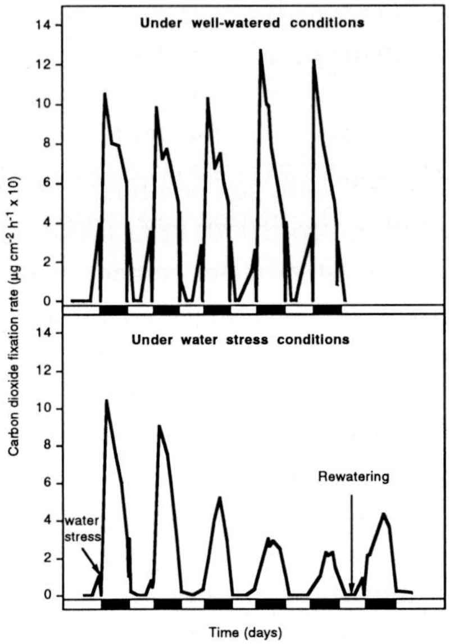

Fig. 3.24. Effect of water stress on the carbon dioxide gas exchange rate of *Aranda* Christine 9.

Redrawn from Fu & Hew (1982).

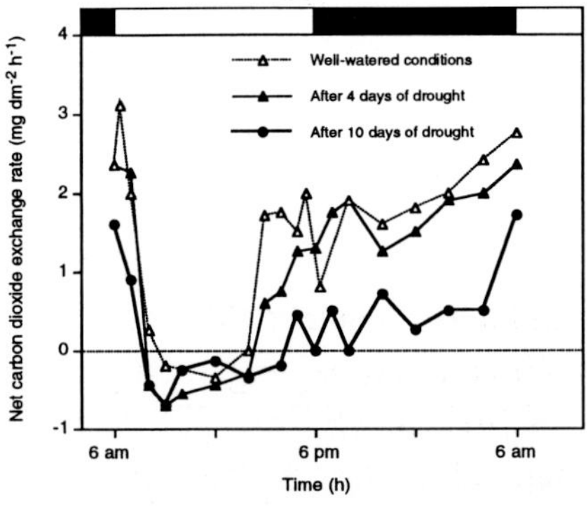

Fig. 3.25. The leaf carbon dioxide exchange rates of *Phalaenopsis* plants grown under well-watered and drought conditions.

Redrawn from Ota, Morioka & Yamamoto (1991).

water stress, CO_2 uptake is exclusively nocturnal. In *Aranda*, there is a shift in nocturnal CO_2 uptake from the peak at around 22 00 hours in the control plants to 02 00 hours in water stressed orchids (Fig. 3.24). A shift in night-time CO_2 uptake following water stress has also been reported in other CAM plants. Upon re-watering, CO_2 uptake by leaves is restored rapidly with parallel increases in titratable acidity fluctuation and leaf relative water content.

C_3 orchids

There is an immediate reduction in the leaf water potential when the C_3 orchid *Cymbidium sinense* is subjected to drought stress. In contrast, leaf transpiration remains unchanged during the first week of drought. After the first week of drought, leaf transpiration begins to decrease by an increase in stomatal resistance (Fig. 3.26). Chlorophyll content in young and mature leaves of one-year-old plants remains the same throughout the 42 days of drought. However, there is a reduction in chlorophyll content of mature leaves of two-year-old plants.

In commercial orchid nurseries, it is unlikely that orchids under cultivation are under severe water stress since watering of plants is carried out regularly on a daily basis.

Effects of temperature

CAM orchids

CAM activities in orchid leaves change with different day/night temperature. A study conducted under constant day/night temperature shows that day-time CO_2 uptake by *Phalaenopsis* leaves decreases when the temperature increase from 10°C to 30°C (Fig. 3.27). The night-time CO_2 uptake increases with an increase in temperature from 10°C to 20°C, followed by a decrease. The plasticity of CAM is well-illustrated when the CAM orchid *Phalaenopsis* is subjected to varying day/night temperature treatments (Fig. 3.28). The leaves

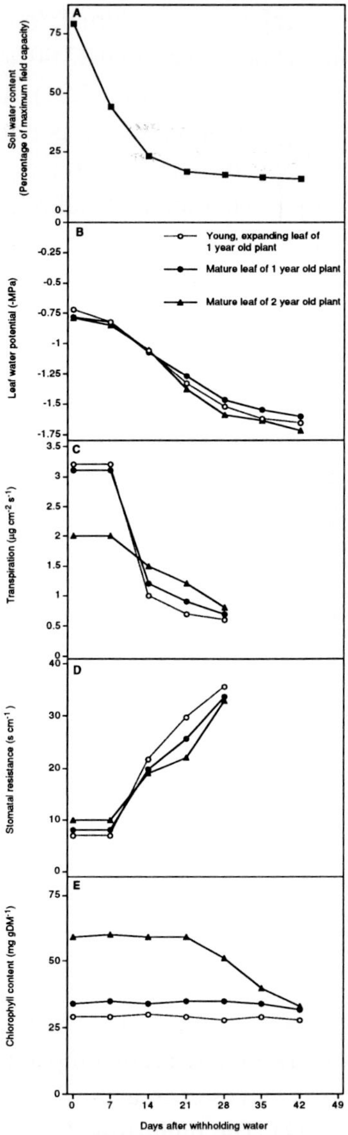

Fig. 3.26. Response of the C_3 orchid *Cymbidium sinense* to drought stress.
Note: (A) Soil water content, (B) Leaf water potential, (C) Transpiration rate, (D) Stomatal resistance and (E) Chlorophyll content.

Redrawn from Zheng, Wen, Pan & Hew (1992).

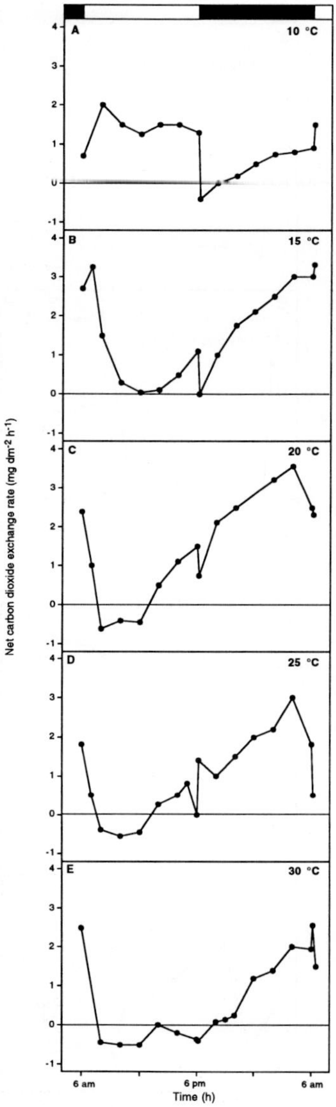

Fig. 3.27. The leaf carbon dioxide exchange rates of *Phalaenopsis* plants grown under constant day and night temperature.

Redrawn from Ota, Morioka & Yamamoto (1991).

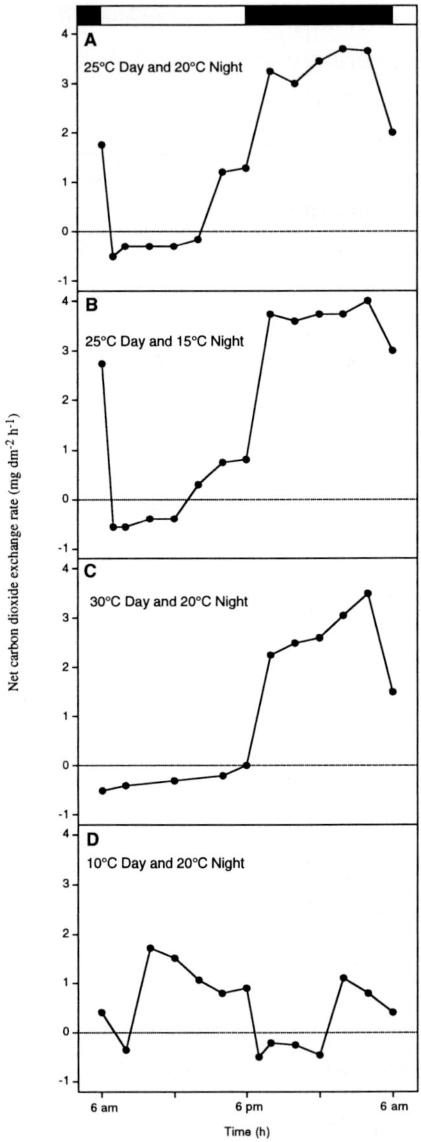

Fig. 3.28. The effects of different day and night temperatures on leaf carbon dioxide exchange rates of *Phalaenopsis* plants.

Redrawn from Ota, Morioka & Yamamoto (1991).

show normal CAM activity when the plants are grown under a day temperature of 25°C and a night temperature of 20°C. Enhanced CAM activity is observed when the plants are given a day temperature of 25°C and a night temperature of 15°C. Conversely, weak CAM activity is observed when the plants are grown in a day temperature of 10°C and a night temperature of 25°C. For most CAM plants, the optimal day and night air temperatures are about 25°C and 15°C respectively. It is noteworthy that in the tropics, day and night temperatures do not fluctuate significantly. For example, in Singapore, the day temperature is between 30°C and 33°C while the night temperature is 25°C to 27°C. Under these prevailing conditions in Singapore, it is remarkable that the thick-leaved orchids exhibit typical CAM activity.

Effects of sink demands

Numerous studies on orchids also show that leaf photosynthesis may vary in the presence of sink organs. Leaves next to flower buds have relatively higher levels of titratable acid fluctuation for the thick-leaved monopodial orchid *Vanda* Miss Joaquim (Table 3.15). Gas exchange studies on another CAM orchid *Phalaenopsis* indicate that leaves of flowering plants have significantly higher nocturnal CO_2 uptake (Fig. 3.29). New sink organs also effect leaf photo-

Table 3.15. Titratable acidity in *Vanda* Miss Joaquim leaves at different positions of the plant.

Leaf position	Titratable acidity (μequivalents gFM^{-1})
Adajacent to flower buds	190
Adjacent to opened flowers	169
Not adajacent to any flower or flower bud	156

Adapted from Avadhani, Khan & Lee (1978).

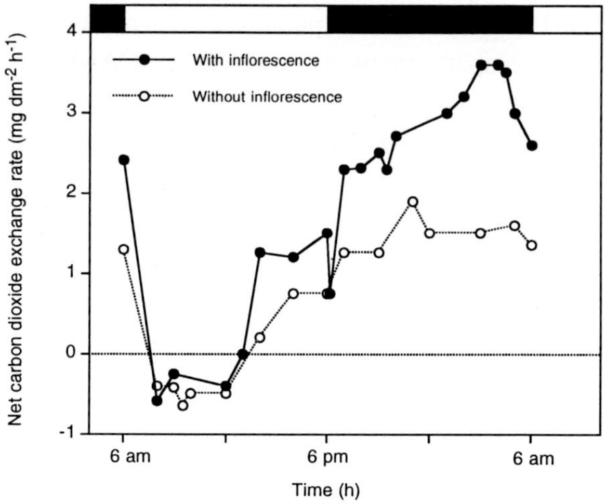

Fig. 3.29. The leaf carbon dioxide exchange rates of *Phalaenopsis* plants with or without inflorescence.

Redrawn from Ota, Morioka & Yamamoto (1991).

synthesis in a C_3 orchid *Oncidium* Goldiana (Fig. 3.30). The formation of inflorescence and axillary bud increases the photosynthetic rates of the subtending leaf (leaf L3) and main leaf (leaf L2) respectively.

Effects of pollutants

Epiphytic lichens and bryophytes are susceptible to atmospheric pollutants and have been used as a bio-indicator of atmospheric pollution. Little work has been carried out on the effect of pollutants on orchid photosynthesis. Epiphytic orchids (*Encyclia tampensis* and *Epidendrum regidum*) appear to be relatively resistant to sulphur trioxide (SO_3) and ozone (O_3) damage. In fact, CAM activity of these two orchids is enhanced with O_3 and SO_3 at 0.3 ppm and 0.45 ppm respectively. This is consistent with the observations that some CAM plants are highly resistant to air pollutants. This could be related to

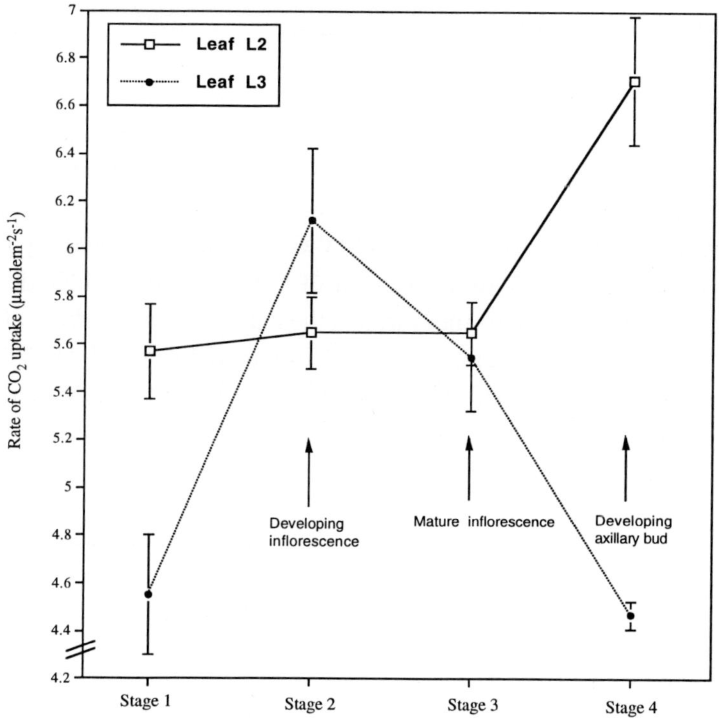

Fig. 3.30. Apparent photosynthetic rates of intact leaves at saturating light intensity for *Oncidium* Goldiana plants during the formation of inflorescence and axillary bud.
Note: Mature leaves (L2, leaf above the pseudobulb and L3, leaf subtending the inflorescence) were used for the experiments. Leaf photosynthesis was measured at photosynthetic active radiation of 200 µmol $m^{-2}s^{-1}$. Leaf temperature was maintained between 24°C and 26°C. CO_2 concentration was between 340 ppm and 360 ppm and relative humidity was kept between 80% and 95% (n = 3 to 5, ± SE).

Adapted from Hew & Yong (1994).

features associated with drought resistance and low gas exchange rates in CAM plants. The velamen, for example, may have conferred protection to orchid roots. Unlike the leaves, flowers may be vulnerable to the effects of SO_3 and O_3. For many other plants, such as tobacco and soybeans, an exposure of 0.2 ppm to 0.4 ppm of O_3 and SO_3 for half an hour causes 50% inhibition of photosynthesis.

Effects of virus infection

Reduced CAM activity as a result of virus infection has been reported for two thick-leaved orchids, *Sophrolaeliocattleya* hybrid and *Epidendrum elongatum*. Tobacco Mosaic Virus orchid strain (TMV-O) infection caused 67% and 31% reduction in leaf acidity changes in *Sophrolaeliocattleya* hybrid and *Epidendrum elongatum*, respectively (Fig. 3.31). TMV-O also changed the daily pattern of leaf nonstructural carbohydrates typical of CAM plants. A 42% decrease in nocturnal titratable acidity is measured in the leaves of *Epidendrum elongatum* infected with both Cymbidium Mosaic Virus (CybMV) and TMV-O. In the *Sophrolaeliocattleya* hybrid, CybMV infection inhibits CAM activity and induced an accumulation of glucans in the leaves. Ultrastructural evidence

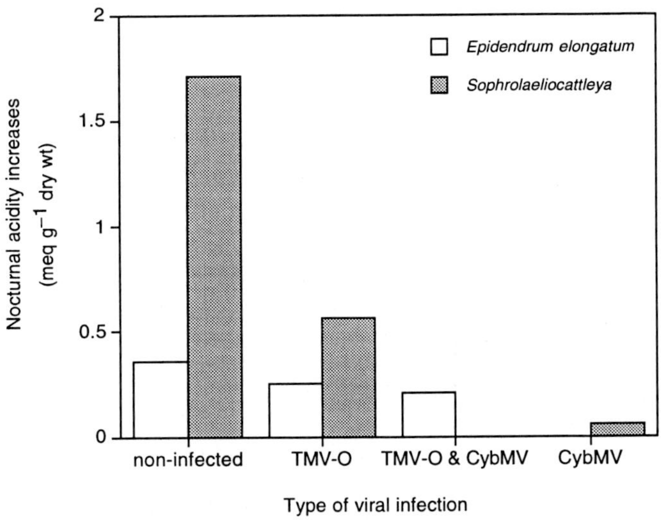

Fig. 3.31. Effect of virus infection on CAM activity in mature leaves of *Epidendrum elongatum* and *Sophrolaeliocattleya* hybrid.

Note: Values for nocturnal acidity increases are obtained by subtracting titratable acidity values at 9 am with corresponding values obtained at 6.30 pm. TMV-O = Tobacco Mosaic Virus-orchid strain; CybMV = Cymbidium Mosaic Virus.

Redrawn from Izaguirre-Mayoral, Uzcategui & Carballo (1993).

suggests that infection by TMV-O or CybMV on the *Sophrolaeliocattleya* hybrid causes an increase in chloroplast volume and the distortion of grana due to high glucan accumulation. In contrast, TMV-O infection in *Epidendrum elongatum* induces a lesser degree of damage in the cell structure of the leaves. Some tropical orchids grown for cut-flowers are infected by CybMV and ORSV (Odontoglossum Ringspot Virus) but there are no report on how these viruses affect the rate photosynthesis in both CAM and C_3 orchids.

Effects of elevated carbon dioxide

An area deserving of greater attention and study is the effects of elevated CO_2 on orchid leaf photosynthesis. CO_2 enrichment decreases photorespiration and increases the net photosynthesis in C_3 plants. Oxygen competes against CO_2 uptake by RUBPC, leading to the occurrence of photorespiration. By increasing CO_2 to higher levels, photorespiration is suppressed due to the increase in CO_2/O_2 ratio. This is the basis for increased growth rates in many horticultural plants caused by elevated CO_2 at both low and high light levels (Mortensen, 1987). Water-use-efficiency (WUE) would also be expected to increase with high CO_2.

There are relatively few reports on the effects of elevated CO_2 on the rate of photosynthesis in orchid leaves. Photosynthetic rate in mature *Arundina* leaves increases with an increase in CO_2 concentrations from 0 ppm to 350 ppm (Fig. 3.32). Similarly, CAM activity in young *Mokara* White plantlets is increased by using 0.327% to 2.8% range of CO_2 concentrations (See Chap. 9 on Advances in Orchid Tissue Culture).

Inflorescence growth and size in the C_3 orchid *Oncidium* Goldiana are promoted by supplying the plants with elevated CO_2 (1% and 10%). There is an average of 50% increase in inflorescence dry mass and 94% increase in dry matter accumulation in pseudobulbs of current shoot and first back shoot for *O.* Goldiana plants grown in elevated CO_2. However, no measurement of leaf photosynthesis is made under these favourable conditions. A search of literature reveals that there is no report of elevated CO_2 on the photosynthesis of mature thick-leaved orchids.

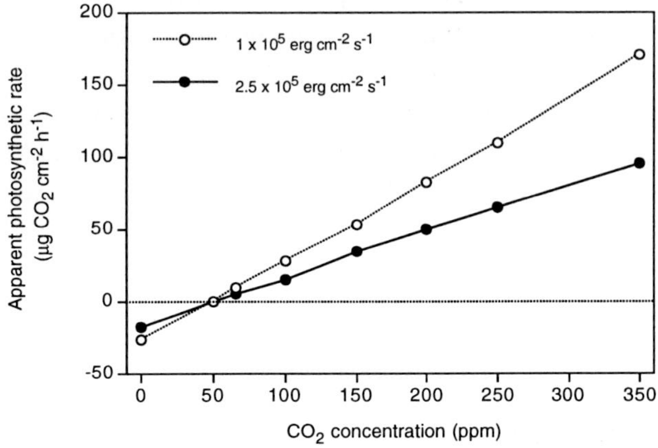

Fig. 3.32. Effect of carbon dioxide concentration on apparent photosynthesis of *Arundina graminifolia* leaves at two light intensities.

Redrawn from Wong & Hew (1973).

3.7. Concluding Remarks

The patterns of carbon fixation in orchids have been extensively studied. Carbon dioxide is fixed either through the C_3 or CAM pathways. Conclusive evidence for the presence of C_4 pathway in orchids is still lacking. Considerable advances have been made in the understanding of non-foliar green organ (roots, flowers, pseudobulbs and fruit capsules) photosynthesis and the factors affecting photosynthesis of thin- and thick-leaved orchids. However, we lack information on the photosynthesis of tropical orchids under field cultivation, particularly at a community level. This information is crucial in the optimisation of the growth and yield of orchids in commercial farms.

More information is needed to understand the response of orchid leaf photosynthesis to both short-, medium- and long-term effects of elevated carbon dioxide. The long-term effects of elevated carbon dioxide on orchid photosynthesis look at the inevitable increase in global atmospheric carbon dioxide from an ecological perspective. The concentration range of carbon dioxide is between 360 ppm and 1,000 ppm. On the other hand, higher concentrations of carbon dioxide will probably be used soon on a short-term to medium-term basis in commercial farms to increase growth and flowering of orchids through an increase in photosynthesis. The probable concentration range of carbon dioxide used is between 600 ppm (or 0.06%) and 10,000 ppm (or 1.0%).

3.8. Summary

1. Tropical orchids have either CAM or C_3 mode of photosynthesis, and these are usually associated with thick leaves and thin leaves respectively. Thick-leaved orchids fix CO_2 through the CAM pathway while the thin-leaved orchids fix CO_2 through the Calvin's cycle. Conclusive evidence for the occurrence of C_4 pathway in orchids is lacking.
2. The carbon fixation pathway of an orchid is determined by the following observations: $\delta^{13}C$ value, chlorophyll a/b ratio, identity of products after short-term $^{14}CO_2$ fixation, PEPC/RUBPC ratio and gas exchange studies.
3. Photosynthetic characteristics of various non-foliar green organs (roots, flowers, pseudobulbs, fruit capsules) of orchids have been studied. The common feature among these non-foliar green organs of leafy orchids is their inability to exhibit net photosynthesis. This unique phenomenon could be because these organs solely perform regenerative photosynthesis in the presence of well-developed leaves. Only under special conditions (e.g., in shootless orchids where roots become the sole photosynthetic organ) is net photosynthesis observed.
4. Photosynthesis of orchid leaves is affected by both physiological and environmental factors, including leaf age, light, temperature, water stress, sink demand, virus infection and carbon dioxide.

General References

Avadhani, P. N., Goh, C. J., Rao, A. N. and Arditti, J., 1982, "Carbon fixation in orchids," in *Orchid Biology: Reviews and Perspectives,* Vol. II, ed. J. Arditti (Cornell University Press, Ithaca, New York), pp. 173–193.

Basra, A. S. and Malik, C. P., 1985, "Non-photosynthetic fixation of carbon dioxide and possible biological roles in higher plants," *Biological Review* **60**: 357–401.

Bidwell, R. G. S., 1979, *Plant Physiology*, Second Ed. (MacMillan Publishing Co., New York), 726 pp.

Black, C. C., 1973, "Photosynthetic carbon fixation in relation to net CO_2 uptake," *Annual Review of Plant Physiology* **24**: 253–286.

Canvin, D. T., 1990, "Photorespiration and CO_2 concentrating mechanisms," in *Plant Physiology, Biochemistry and Molecular Biology*, eds. D. T. Dennis and D. H. Turpin (Longman Scientific and Technical, United Kingdom), pp. 263–273.

Edwards, G. and Walker, D., 1989, *C_3, C_4: Mechanisms, and Cellular and Environmental Regulation, of Photosynthesis* (University of California Press, Berkeley).

Farquhar, G. D., Ehleringer, J. R. and Hubick, K. T., 1989, "Carbon isotope discrimination and photosynthesis," *Annual Review of Plant Physiology and Plant Molecular Biology* **40**: 503–537.

Herold, A., 1980, "Regulation of photosynthesis by sink activity — the missing link," *New Phytologist* **86**: 131–144.

Hew, C. S., 1976, "Patterns of CO_2 fixation in tropical orchid species," in *Proc. of the Eighth World Orchid Conference*, Frankfurt (1975), ed. K. Senghas, pp. 426–430.

Hew, C. S., 1987, "Respiration in orchids," in *Orchid Biology: Reviews and Perspectives, Vol. IV*, ed. J. Arditti (Cornell Univ. Press, Ithaca), pp. 229–259.

Hew, C. S., 1995, "Advances in photosynthesis and partitioning of assimilates in orchids," in *Proc. of the Nagoya International Orchid Show* (1995), pp. 40–47.

Kluge, M. and Ting, I. P., 1978, "Crassulacean acid metabolism: Analysis of an ecological adaptation," *Ecological studies*, Vol. 30 (Springer-Verlag, Berlin, Heidelberg, New York).

Lerman, J. C., 1975, "How to interpret variations in the carbon isotope ratio of plants: Biologic and environmental effects," in *Environmental and Biological Control of Photosynthesis*, ed. R. Marelee (Dr. W. Junk Publishers, The Hague), pp. 323–336.

Taiz, L. and Zeiger, E., 1991, *Plant Physiology* (Benjamin/Cummings Publishing Co., Inc., USA), 559 pp.

Mortensen, L. M., 1987, "Review: CO_2 enrichment in greenhouses. Crop responses," *Scientia Horticulturae* **33**: 1–25.

Osmond, C. B., 1978, "Crassulacean acid metabolism: A curiosity in context," *Annual Review of Plant Physiology* **29**: 379–414.

Sinclair, R., 1990, "Water relations in orchids," in *Orchid Biology: Reviews and Perspectives,* Vol. V, ed. J. Arditt (Timber Press, Portland, Oregon), pp. 63–119.

References

Ando, T., 1982, "Occurrence of two different modes of photosynthesis in *Dendrobium* cultivars," *Scientia Horticulturae* **17**: 169–175.

Ando, T. and Ogawa, M., 1987, "Photosynthesis of leaf blades in *Laelia anceps* Lindl. is influenced by irradiation of pseudobulb," *Photosynthetica* **21**: 588–590.

Arditti, J. and Dueker, J., 1968, "Photosynthesis by various organs of orchid plants," *American Orchid Society Bulletin* **37**: 862–866.

Avadhani, P. N. and Goh, C. J., 1974, "CO_2 fixation in the leaves of *Bromheadia finlaysoniana* and *Arundina graminifolia* (Orchidaceae)," *Journal of the Singapore National Academy of Science* **4**: 1–4.

Avadhani, P. N., Khan, I. and Lee, Y. T., 1978, "Pathways of carbon dioxide fixation in orchid leaves," in *Proc. of the Symposium on Orchidology*, ed. E. S. Teoh (Orchid Society of South-East Asia, Singapore), pp. 1–12.

Benzing, D. H. and Ott, D. W., 1981, "Vegetative reduction in epiphytic Bromeliaceae and Orchidaceae: Its origin and significance," *Biotropica* **13**: 131–140.

Benzing, D. H. and Pockman, W. T., 1989, "Why do non foliar green organs of leafy orchids fail to exhibit net photosynthesis?" *Lindleyana* **4**: 53–60.

Cockburn, W., Goh, C. J. and Avadhani, P. N., 1985, "Photosynthetic carbon assimilation in a shootless orchid, *Chiloschista usneoides* (Don.) Ldl," *Plant Physiology* **77**: 83–86.

Dogane, Y. and Ando, T., 1990, "An estimation of carbon evolution during flowering and capsule development in a *Laeliocattleya* orchid," *Scientia Horticulturae* **42**: 339–349.

Donovan, R. D., Arditti, J. and Ting, I. P., 1984, "Carbon fixation by *Paphiopedilum insigne* and *Paphiopedilum parishii* (Orchidaceae)," *Annals of Botany* **54**: 583–586.

Dycus, A. M. and Knudson, L., 1957, "The role of velamen in the aerial roots of orchids," *Botanical Gazette* **119**: 78–87.

Endo, M. and Ikusima, I., 1992, "Changes in concentrations of sugars and organic acids in the long-lasting flower clusters of *Phalaenopsis*," *Plant and Cell Physiology* **33**: 7–12.

Erickson, L. C., 1957, "Respiration and photosynthesis in *Cattleya* roots," *American Orchid Society Bulletin* **26**: 401–402.

Fu, C. F. and Hew, C. S., 1982, "Crassulacean acid metabolism in orchids under water stress," *Botanical Gazette* **143**: 294–297.

Goh, C. J., Avadhani, P. N., Loh, C. S., Hanegraaf, C. and Arditti, J., 1977, "Diurnal stomatal and acidity rhythms in orchid leaves," *New Phytologist* **78**: 365–372.

Goh, C. J., Arditti, J. and Avadhani, P. N., 1983, "Carbon fixation in orchid aerial roots," *New Phytologist* **95**: 367–374.

Goh, C. J., 1983, "Rhythms of acidity and CO_2 production in orchid flowers," *New Phytologist* **93**: 25–32.

Goh, C. J., Wara-Aswapati, O. and Avadhani, P. N., 1984, "Crassulacean acid metabolism in young orchid leaves," *New Phytologist* **96**: 519–526.

Hew, C. S., 1978, "Crassulacean acid metabolism in young orchid seedlings," in *Proc. of the Symposium on Orchidology, Singapore* (1978), The Orchid Society of South-East Asia (Stamford College Press, Singapore), pp. 13–17.

Hew, C. S. and Khoo, S. I., 1980, "Photosynthesis of young orchid seedlings," *New Phytologist* **86**: 349–357.

Hew, C. S., Ng, Y. W., Wong, S. C., Yeoh, H. H. and Ho, K. K., 1984, "Carbon fixation in orchid aerial roots," *Physiologia Plantarum* **60**: 154–158.

Hew, C. S., Ye, Q. S. and Pan, R. C., 1989, "Pathway of carbon fixation in some thin-leaved orchids," *Lindleyana* **4**: 154–157.

Hew, C. S., Ye, Q. S. and Pan, R. C., 1991, "Relation of respiration to CO_2 fixation by *Aranda* orchid roots," *Environmental and Experimental Botany* **31**: 327–331.

Hew, C. S. and Yong, J. W. H., 1994, "Growth and photosynthesis of *Oncidium* Goldiana," *Journal of Horticultural Science* **69**: 809–819.

Hew, C. S., Hin, S. E., Yong, J. W. H., Gouk, S. S. and Tanaka, M., 1995, *"In vitro* CO_2 enrichment of CAM orchid plantlets," *Journal of Horticultural Science* **70**: 721–736.

Hew, C. S., Ng, C. K. Y., Gouk, S. S., Yong, J. W. H. and Wong, S. C., 1996, "Variation in $\delta^{13}C$ values for different plant parts of an *Oncidium* orchid," *Photosynthetica* **32**: 135–139.

Ho, K. K., Yeoh, H. H. and Hew, C. S., 1983, "The presence of photosynthetic machinery in aerial roots of leafy orchids," *Plant and Cell Physiology* **24**: 1317–1321.

Hocking, C. G. and Anderson, J. W., 1986, "Survey of pyruvate, phosphate dikinase activity of plants in relation to the C_3, C_4 and CAM mechanisms of CO_2 assimilation," *Phytochemistry* **25**: 1537–1543.

Izaguirre-Mayoral, M. L., de Uzcategui, R. C. and Carballo, O., 1993, "Crassulacean acid metabolism in two species of orchids infected by Tobacco Mosaic Virus-orchid strain and/or Cymbidium Mosaic Virus," *Journal of Phytopathology* **137**: 272–282.

McWilliams, E. L., 1970, "Comparative rates of dark CO_2 uptake and acidification in the Bromeliaceae, Orchidaceae and Euphorbiaceae," *Botanical Gazette* **131**: 285–290.

Neales, T. F. and Hew, C. S., 1975, "Two types of carbon assimilation in tropical orchids," *Planta* **123**: 303–306.

Nyman, L. P., Benzing, D. H., Temple, P. J. and Arditti, J., 1980, "Effects of ozone and sulfur dioxide on two epiphytic orchids," *Environmental and Experimental Biology* **30**: 207–213.

Ota, K., Morioka, K. and Yamamoto, Y., 1991, "Effects of leaf age, inflorescence, temperature, light intensity and moisture conditions on CAM photosynthesis in *Phalaenopsis*," *Journal of the Japanese Society for Horticultural Science* **60**: 125–132.

Sanders, D. J., 1979, "Crassulacean acid metabolism and its possible occurrence in the plant family Orchidaceae," *American Orchid Society Bulletin* **48**: 796–798.

Sinclair, R., 1984, "Water relations of tropical epiphytes. III. Evidence for Crassulacean Acid Metabolism," *Journal of Experimental Botany* **35**: 1–7.

Winter, K., Wallace, B. J., Socker, G. C. and Roksandic, Z., 1983, "Crassulacean acid metabolism in Australian vascular epiphytes and some related species," *Oecologia* **57**: 129–141.

Winter, K., Medina, E., Garcia, V., Mayoral, M. L. and Muniz, R., 1985, "Crassulacean acid metabolism in roots of a leafless orchid, *Campylocentrum tyrridion* Garay & Dunsterv," *Journal of Plant Physiology* **118**: 73–78.

Wong, S. C. and Hew, C. S., 1973, "Photosynthesis and photorespiration in some thin-leaved orchid species," *Journal of the Singapore National Academy of Science* **3**: 150–157.

Yong, J. W. H., 1995, "Photoassimilate partitioning in the sympodial thin-leaved orchid *Oncidium* Goldiana," M.Sc. dissertation. Department of Botany, The National University of Singapore, 132 pp.

Yong, J. W. H. and Hew, C. S., 1995, "The patterns of photoassimilate partitioning within connected shoots for the thin-leaved sympodial orchid *Oncidium* Goldiana during different growth stages," *Lindleyana* **10**: 92–108.

Zheng, X. N., Wen, Z. Q., Pan, R. C. and Hew, C. S., 1992, "Response of *Cymbidium sinense* to drought stress, *Journal of Horticultural Science* **67**: 295–299.

Zettler, F. W., Ko, N. J., Wister, G. C., Elliott, M. S. and Wong, S. M., 1990, "Viruses of orchids and their control," *Plant Disease* **74**: 621–626.

Chapter 4

Respiration

4.1. Introduction

When Withner reviewed the physiology of orchids in 1959, he listed only two publications on respiration. Over the last three decades, considerable advances have been made in our understanding of respiration in orchids (Hew, 1987).

This chapter aims to give a brief introduction to the processes involved in respiration and a survey on the respiratory processes in orchids. Emphasis will be placed on understanding respiration as an internal metabolic control of senescence in orchid flowers.

4.2. Respiratory Processes

Respiration generally refers to the processes of "dark respiration" that may occur in the light and darkness. The whole process of respiration involves the catabolism of sugar or some other substrates, the production of CO_2 and the consumption of O_2. Two important products are produced as a result of respiration: Reduced nucleotides (NADH and $FADH_2$) and ATP. These products are constantly regenerated during the catabolic phase of metabolism. Other intermediates produced serve as building blocks for various biosynthetic processes and are used during the anabolic phase of metabolism. Respiration, in its essence, transforms the substrates derived from photosynthesis into important intermediates and useful energy necessary for growth and maintenance of living tissues.

Two distinct processes are involved in the oxidation of hexose molecule. The first series of reactions (glycolysis) take place in the cytoplasm. This is also known as the Embden Meyerhoff Parnass (EMP) pathway. During glycolysis, a molecule of hexose is converted to two molecules of pyruvate (Fig. 4.1) which is subsequently decarboxylated. The remaining two-carbon fragment (Acetyl CoA) is oxidised in the Kreb's Cycle (Tricarboxylic acid cycle). The second series of reactions occur in the mitochondria (Fig. 4.1) where the acetyl CoA is further metabolised to carbon dioxide. The energy released is stored in ATP. In the absence of oxygen, fermentation occurs and pyruvate is converted to ethanol and CO_2. The EMP pathway and Kreb's cycle form the main respiratory pathway in plants. Another important pathway that

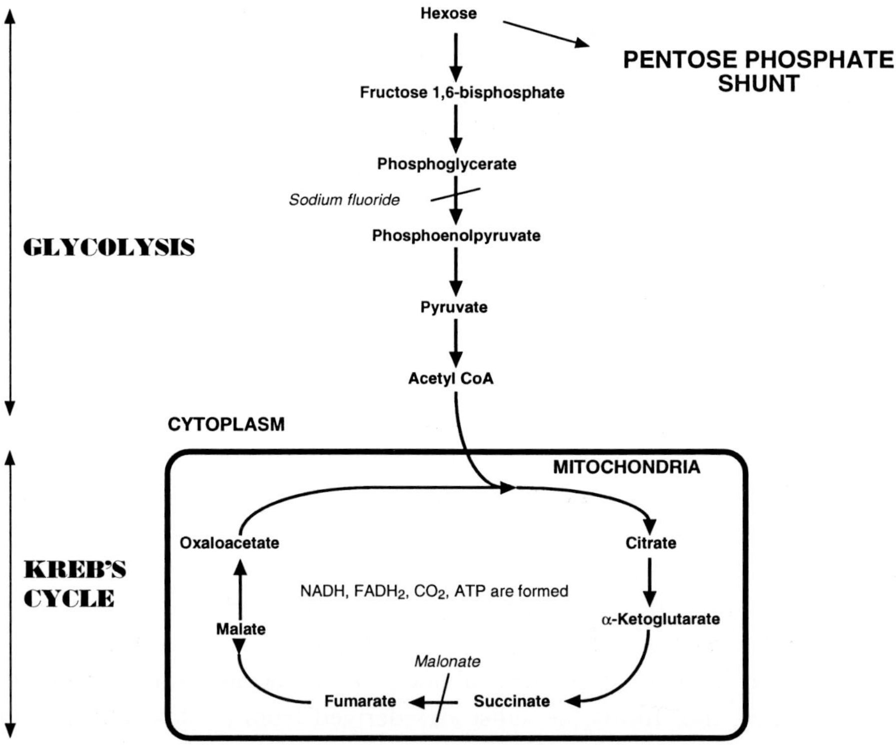

Fig. 4.1. Diagrammatic representation of the key respiratory processess.

bypasses the EMP pathway is the pentose phosphate shunt. Through the pentose monophosphate shunt, the glucose molecule is converted into triose phosphate and CO_2. The sugar-phosphate is first oxidised (dehydrogenated) by glucose-6-phosphate dehydrogenase to form 6-phosphogluconate. This is oxidatively decarboxylated to form ribulose-5-phosphate by 6-phosphogluconate dehydrogenase. In contrast to the glycolytic pathway, only one molecule of CO_2 is produced per glucose molecule metabolised. The rest of the carbon skeleton undergoes complex reorganisation. The two respiratory pathways have been well discussed by Beevers (1961) and Salisbury and Ross (1991).

Aerobic respiration in plants is strongly inhibited by certain negative ions such as cyanide and azide. In some plant tissues, the poisoning of cytochrome oxidase by such inhibitors has only minimal effects on respiration. The respiration that continues in this situation is said to be cyanide-resistant respiration. Cyanide-resistant respiration is also known as alternative respiration (Fig. 4.2). It is part of the normal electron transport chain. Ubiquinone is believed to be the site where electrons are diverted to the cyanide-resistant pathway. Electrons move faster in the cyanide-resistant pathway but at the expense of producing lesser ATPs per atom of oxygen used. The alternative oxidase has a much lower affinity with oxygen than does cytochrome oxidase and is strongly inhibited by salicylhydroxamic acid (SHAM). The significance of the cyanide-resistant pathway has been discussed (Lambers, 1985).

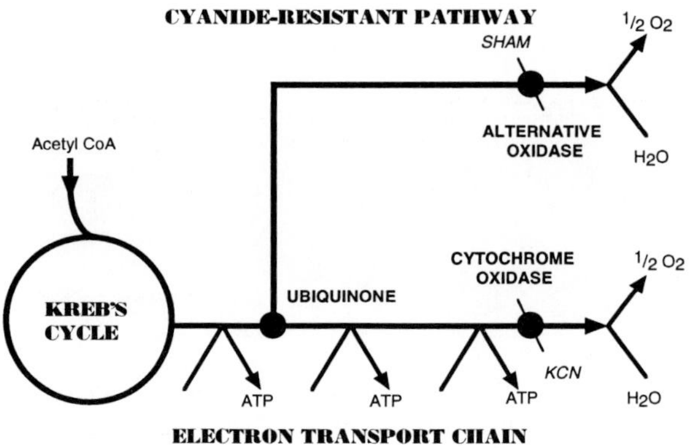

Fig. 4.2. The cyanide-sensitive and cyanide-resistant respiratory pathway.

4.3. Respiration in Plant Parts

Protocorms and seedlings

Orchid seeds are very minute and are divided into two groups with respect to their embryos. Fewer than 10 species have a rudimental cotyledon. The majority of orchid embryos are relatively undifferentiated when mature and have no endosperm or cotyledon. Apart from the small starch grain within the proplastids, there is no other carbohydrate reserves in these seeds. It has been shown that all cells in the embryos of *Cattleya aurantica* are packed with food reserves in the form of lipid bodies (Fig. 4.3). However, glyoxysomes (the organelles responsible for fat metabolism) have not been found at any time

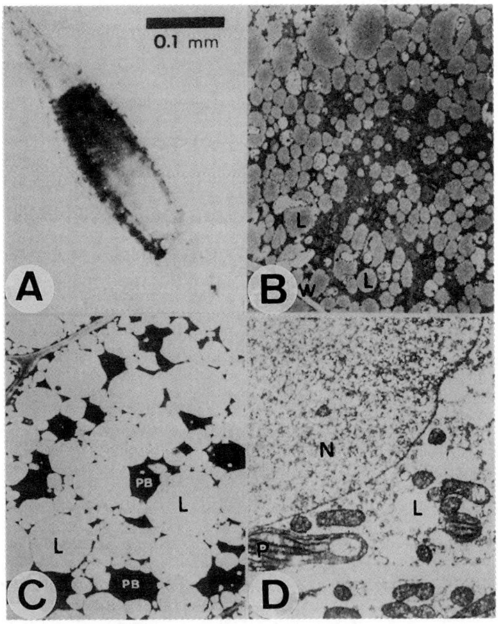

Fig. 4.3. Seed and embryo of *Cattleya aurantiaca*.

Note: (A) Whole seed showing embryo; (B) & (C) Cells of embryos in ungerminated seeds [1,950 X & 2,770 X, respectively]; (D) Basal cell in an ungerminated embryo [10,530 X]. Explanation of symbols: L, lipid bodies; N, nucleus; P, proplastid; PB, protein body; W, cell wall.

Reproduced from Harrison (1977), courtesy of Botanical Gazette.

during orchid seed germination. For example, glyoxysomes are absent in mature seeds of *Cattleya aurantica*, *Disa polygonoides* and *Disperis fanniniae*. This apparent lack of metabolic machinery severely hampers the utilisation of fat reserves and their subsequent conversion to carbohydrates. This may also account for the very low respiration rate of orchid embryos. In *Cattleya* seeds, the lipid bodies are closely associated with or enveloped by the mitochondria. The role of mitochondria in lipid breakdown is not clear. It has been shown that the embryo could convert approximately 3% of the label from acetate-2-^{14}C into sugars. It seems that the lipid reserves are used slowly in nature for the maintenance of protocorms until an appropriate endophytic fungal infection is established. Only then do the protocorms develop into leaf-bearing seedlings.

Lipolysis does occur in orchid seeds in the presence of an external source of sucrose or following mycorrihzal infection. The need for fungal infection of orchid seeds appears to be due to an impaired ability of the seeds to metabolise polysaccharides and lipids. The fungus may act by supplying the embryos with simple sugars as an energy source thus facilitating synthesis of the necessary enzyme systems and development of glyoxysomes. The involvement of glyoxysomes in the conversion of lipid to carbohydrate in germinating seeds is well-established (Beevers, 1961). Alternatively, the fungus may supply directly enzyme precursors or coenzymes, or precursors of NAD and NADP, thereby enabling hydrolysis of the orchid seed reserves to go on. It needs to be established how many genera of orchids lack glyoxysomes in their seeds. Further research on the genetic control of glyoxysomes development may provide an answer to its absence in orchid seeds.

Germinating orchid seed may respire anaerobically at some stage in its development. This is based on the depletion of oxygen in protocorms growing in an enclosed culture system. Assuming an average of 50 μl of oxygen per g fresh mass per hour, 240 μl of oxygen will be used in 24 h by a protocorm weighing 0.2 g. In a 500 ml culture flask containing 125 ml of culture medium, there is a remaining 375 ml of atmosphere and approximately 75 ml of it would be oxygen. The oxygen in an airtight flask will be depleted very quickly and the germinating seeds may be in an anaerobic condition. Very little oxygen replenishment arises from photosynthesis since protocorms have limited photosynthetic capacities. Further investigation involving time course measurement of pyruvate dehydrogenase activity and ethanol formation by germinating seeds in an enclosed vessel would be interesting.

Seeds of some orchid species are known to germinate better in airtight containers than in flasks with ample gas exchange. Seeds of an underground achlorophyllous orchid, *Galeola septentrionalis*, could only germinate in airtight vessel. Under such condition, O_2 level will be considerably reduced whereas CO_2 level will be high. It is possible that the low O_2 and high CO_2 levels simulate conditions for underground germination. There is no information about the respiration of this orchid. For other orchids, it is assumed that respiration of orchid protocorm like the other plant tissues, is dependent on the availability of O_2. The protocorms of *Aranda* Christine and *Aranthera* James Storie grown in liquid Vacin and Went medium with continuous aeration increase six- to eightfold in fresh mass over the control in 25 days. The process of aeration would increase CO_2 and O_2 levels in the culture medium. Since orchid protocorms have limited photosynthetic capacity, the increase in O_2 level accompanying aeration appears to be more important for respiration. However, one cannot rule out pH effect as there was no buffer in the medium for these experiments. Satisfactory explanation for the atmospheric requirements of germinating orchid seeds is possible only when we have a better understanding of the respiratory metabolism during seed germination.

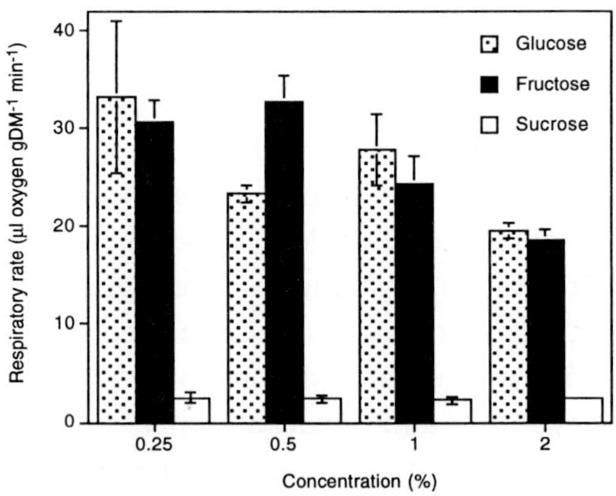

Fig. 4.4. Effects of different concentrations of glucose, fructose and sucrose on the respiratory rates of *Dendrobium* Multico White tissues after one month in culture.

Redrawn from Hew, Ting & Chia (1988).

Respiration in orchid protocorms also depends on the type of substrate supplied. For *Dendrobium* protocorms and calli, the respiration rates are higher when grown in medium with fructose or glucose than those in medium containing sucrose (Fig. 4.4). When sucrose is used as a carbon source, it must be hydrolyzed by invertase to give fructose and glucose. The difference in protocorm respiration rates grown in different carbon sources cannot be attributed to an osmotic effect because appropriate concentrations of mannitol are included in medium during the experiments.

Leaves

Orchid leaves are either thin or thick. Generally, epiphytic orchids have thick and fleshy leaves while terrestrial orchids have thin leaves. Thin-leaved orchids fix carbon dioxide through the Calvin's cycle and this renders the measurement of leaf respiration simple. In contrast, measurements of CO_2 evolution by thick-leaved orchids in the dark are hampered by the massive nocturnal CO_2 fixation. As such, measurement of O_2 uptake is preferable and easier. Oxygen uptake continues at a fairly steady rate in the dark when organic acids are formed. The uptake that continues during acidification is involved exclusively in the oxidative catabolism of carbohydrates. Carbon dioxide produced in this way is drawn into carboxylation reaction, leading to the formation of malate. The breakdown of starch or glucan in the dark provide the source of phosphoenolpyruvate (PEP) for β carboxylation. During deacidification in the day, malate is decarboxylated and starch or glucan is formed.

The rate of O_2 uptake by orchid leaves varies from 50 to 91 μl O_2 g fresh $mass^{-1}$ h^{-1}. In contrast, the respiration rates of leaves expressed in term of CO_2 are more variable (Table 4.1). A Q_{10} of two is observed for leaf respiration. Respiration of *Cattleya* leaves changes with age. Youngest leaf has the highest rate and the rate decreases with leaf age. On the other hand, no significant difference in respiration rates of leaf L2, leaf L6 and leaf L21 (counting from the apex) for *Aranda* is observed. The chlorophyll content, fresh and dry mass of *Aranda* leaves remain fairly constant with age. This may imply that leaf senescence sets in rather late for *Aranda*.

Table 4.1. Respiration of some orchid leaves.

Orchid	Temperature	Mean respiratory rate ($\mu g\ CO_2\ gFM^{-1}h^{-1}$)	($\mu g\ CO_2\ gDM^{-1}h^{-1}$)	($mg\ CO_2\ gFM^{-1}h^{-1}$)	($\mu l\ O_2\ gFM^{-1}h^{-1}$)
Aranda Christine 130					
Leaf L2 (youngest)	28°C	–	–	–	73.0
Leaf L6	28°C	–	–	–	91.0
Leaf L21 (oldest)	28°C	–	–	–	83.0
Arachnis Maggie Oei					
Young leaves	25°C	50.0–100.0			
Cattleya					
Young leaf	25°C	–	–	–	74.0
Old leaf	25°C	–	–	–	50.0
Coelogyne sp.	20°C	–	–	0.6	–
Cymbidium Oiso	20°C	–	–	1.5	–
Dendrobium Nodoka	20°C	–	–	0.5	–
Oncidium Goldiana	20°C	–	–	0.5	–
Paphilopedilum villosum	20°C	–	–	0.8	–
Paphilopedilum venustum	10°C	–	50.0	–	–
Spathoglottis plicata	25°C	–	–	1.0	–

Redrawn from Hew (1987).

To date, there is no systematic study on the biochemical and physiological details associated with respiration during leaf development. We have no information on the respiratory pathway operating in thin- and thick-leaved orchids. There is also little information on the factors affecting leaf respiration. There is one report that indicates that *Aranda* leaves respond to Physan (a quaternary ammonium compound used to control algal growth in orchid nurseries) differently from that of protocorms and flowers. Physan inhibits CAM leaf respiration but not the respiration of protocorms and flowers. The reason remains unclear. In contrast, the respiration of *Brassica* leaves, a C_3 plant, is stimulated by Physan.

Flowers

Respiration rates of orchid flowers vary with species. All young flowers generally respire at higher rates than older ones. An inverse relationship between respiration rates and flower longevity has been observed. *Arundina* flower, for example, has the highest respiration rate, and the shortest life span, of all orchids studied (Table 4.2).

Respiration of orchid flowers has been studied in relation to temperature and pollination (Table 4.3). A Q_{10} of 2 has been reported for *Cattleya mossiae*, *Cattleya skinneri*, *Oncidium* Goldiana and *Aranda* Wendy Scott (Table 4.4).

The major physiological and morphological changes in orchid flowers following pollination have been extensively studied and commonly referred to as the "Post-pollination phenomena" (see Arditti [1992] for a detailed description of the post-pollination phenomena). Marked increase of respiration following pollination is observed in *Cymbidium* flowers, especially in the gynostemium. The increase in respiration starts within an hour after pollination or auxin application. Respiration in the gynostemia reaches an initial peak 50 h after pollination and a second one 170 h later. In the perianth, respiration peaked at the 50th hour.

Respiration in the gynostemia of *Cymbidium lowianum* increases three-folds 8 h after pollination. Gynostemia of *Coelogyne mooreana* and *Cattleya bowringiana* exhibit a twofold increase in respiration. A very large proportion

Table 4.2. The relationship between respiration and longevity of orchid flowers.

Orchids	Respiratory rate ($\mu l\ CO_2\ gFM^{-1}h^{-1}$) Young flower	Mature flower	Longevity (days)
Aranda Wendy Scott	188.7	158.2	28
Aranthera James Storie	190.9	158.2	–
Arundina graminifolia	365.5	338.2	5
Dendrobium Louisae Dark	212.7	114.6	44.5
Oncidium Goldiana	255.3	163.6	–
Vanda Ruby Prince	261.8	180.0	–
Vanda Tan Chay Yan	250.9	169.1	27.5

Adapted from Hew (1980).

Table 4.3. Respiratory quotient of orchid flowers.

Orchid	Temperature (°C)		RQ
Coelogyne mooreana	25	Control, column	0.90
		Pollinated, column	0.95
Cymbidium lowianum	25	Control, perianth	0.96
		Pollinated, perianth	1.01
		Control, column	0.96
		Pollinated, column	0.93
Aranda Christine 130	28	Bud	0.5
		Newly-opened flower	0.7
		Fully-opened flower	1.0
		Mature flower	1.0

Adapted from Hsiang (1951) and Hew & Yip (1987).

Table 4.4. Q_{10} of some orchid flowers.

Orchid	Plant part	Temperature (°C)	Q_{10} Q_{10}
Cattleya mossiae	Flower segments	5–15	2.4
		15–25	1.0
Cattleya skinneri	Flower segments	5–15	7.0
Oncidium Goldiana	Lips and sepals	10–20	2.0
		20–30	2.0
Aranda Wendy Scott	Lips and sepals	10–20	2.0
		20–30	2.0

Adapted from Sheehan (1954) and Hew (1980).

of the increases in respiration following pollination is detected in the placental tissues. This information suggests a reduction of activities in senescent organs with a concomitant increase in metabolism in tissues that become the centre of new developmental events.

There is no report on the change in respiratory pathways following pollination. The RQ of pollinated mature flower remains as one (Table 4.3). This would indicate that the substrate for respiration is carbohydrate during pollination. As ethylene production is induced following pollination, there is probably an interaction between ethylene and respiration during pollination and post-pollination.

A circadian rhythm of CO_2 production by orchid flowers has been reported (Fig. 4.5). The occurrence of rhythmic CO_2 production by orchid flowers is widespread (Table 4.5). The rhythms are circadian (i.e., 24 h periodicity) and start as soon as the flowers open. They occur under constant illumination and temperature. Continuous darkness dampens the amplitude but it does not affect the rhythmicity. The dampening effects on the amplitude can be alleviated in part by an exogenous supply of sucrose. Emasculation and pollination seem to stimulate respiration particularly in the second cycle after treatment. Pollination does not change the respiratory rhythm, but it does affect the amplitude of the

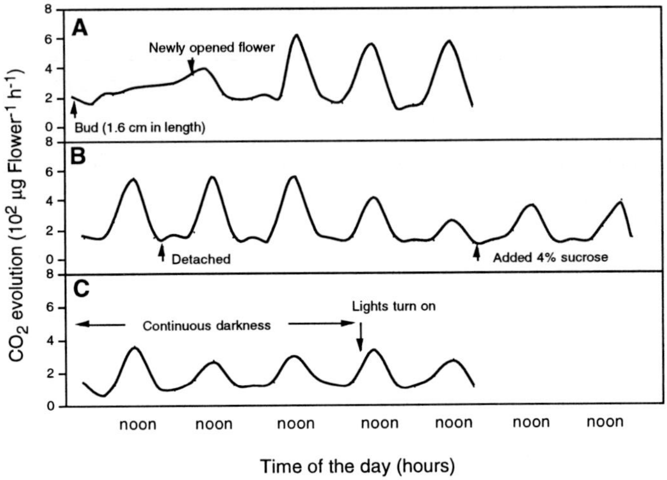

Fig. 4.5. Rhythmic production of carbon dioxide by orchid flowers.
Note: (A) Effect of flower developmental stage on carbon dioxide production by *Vanda* Tan Chay Yan. Plant was illuminated at 14 mJ $cm^{-2}s^{-1}$ from 8 am to 6 pm everyday; (B) Effect of detachment; (C) dark treatment on carbon dioxide production by *Vanda* Tan Chay Yan flowers. In detachment experiment, the plant was illuminated at 14 mJ $cm^{-2}s^{-1}$ from 8 am to 6 pm everyday. In dark treatment, the whole plant was placed in continuous darkness.

Adapted from Hew, Thio, Wong & Chin (1978).

peaks. The rhythmic CO_2 production occurs in intact flowers (i.e., on the inflorescence), detached flowers and in isolated gynostemia. These observations indicate that the circadian respiratory rhythm in orchid flowers is controlled by an endogenous oscillation system within the flowers.

Most of the orchid flowers (for example, *Aranda*, *Vanda*) which exhibit rhythmic CO_2 production have succulent leaves. It was suggested that the rhythmic CO_2 production by these flowers could be a manifestation of CAM, because the flowers' acidity fluctuation and dark $^{14}CO_2$ fixation are similar to that of the leaves. The flowers of the C_3 orchid *Oncidium* Goldiana, like its leaves, do not exhibit acidity fluctuation. However, an earlier study has shown that the flowers of *Oncidium* Goldiana have a noticeable rhythm of CO_2 production. Hence it appears that the rhythmicity of CO_2 production in orchid

Table 4.5. Rhythmic carbon dioxide production by orchid flowers.

Orchid	Rhythmic CO_2 production
Aeridachnis Bogor	+
Aranda Hilda Galistan	+
Aranda Wendy Scott	+
Aranda Deborah	+
Aranthera James Storie	+
Arachnis Maggie Oei	+
Arachnis hookeriana var. luteola	–
Brassavola nodosa	+
Cattleya intermedia	+
Dendrobium taurinum	+
Dendrobium Field King	+
Dendrobium Lam Soon	+
Dendrobium Louisae Dark x *Dendrobium* Peggy Shaw	+
Dendrobium Pompadour	+
Dendrobium Mary Mak	+
Oncidium Goldiana	+
Oncidium haematochilum	+
Phalaenopsis cornu cervi	–
Phalaenopsis Doris	+
Vanda Dearie	+
Vanda Patricia Low	+
Vanda Rothschildiana	+
Vanda Ruby Prince	+
Vanda Tan Chay Yan	+

Note: +, noticeable rhythmic CO_2 production; –, no noticeable rhythmic CO_2 production.

Redrawn from Hew, Thio, Wong & Chin (1978), Goh (1983) and Hew & Lim (1984).

flowers could not be a manifestation of CAM and is independent of CAM activity.

In the study of the rhythmic production of CO_2 by orchid flowers, it is noted that the respiratory peak of all species studied (except *Brassavola nodosa*) occurred at noon. In *Brassavola nodosa* flowers, the peak occurs at midnight and this flower is known to produce fragrance at night. A correlation is thought to exist between the rhythmic CO_2 production and fragrance production that may be relevant to pollination ecology. However, further study shows that scentless orchid flowers such as *Oncidium* Goldiana and *Aranda* Wendy Scott also exhibit rhythmic CO_2 production. Rhythmic CO_2 production by orchid flowers appears not to be correlated with fragrance production. Nevertheless, it is necessary to point out that fragrance of flowers is detected organoleptically. More careful measurements of fragrance production using gas chromatography should be carried out using both scented and scentless orchid flowers.

Roots

In *Cattleya* roots, the highest rate of respiration is detected at the root tip and the respiratory rate decreased sharply in the first 4 cm behind the tip. Beyond the fourth centimetre region, a more gradual decline in respiration is observed with increasing distance from the tip. The same process has been observed in aerial and terrestrial roots of *Aranda* Wendy Scott (see Chap. 3 on Photosynthesis; Fig. 3.10) *Arachnis* Maggie Oei, *Aeridachnis* Bogor, and *Aranthera* James Storie. This agrees with the observations made with roots of other plant species.

As discussed earlier (see Chap. 3 on Photosynthesis), the absence of net carbon dioxide fixation by aerial roots of leafy orchids, which possess chloroplasts, is not due to insufficient PEP carboxylase activity or poorly developed chloroplasts. The net CO_2 fixation in aerial roots is being masked by a relatively high rate of root respiration. This is evident from the effects of temperature on the CO_2 exchange rate of *Aranda* aerial roots at two carbon dioxide concentrations (Fig. 4.6). Respiration of *Aranda* roots increases with increasing temperature from 15 °C–35°C (Table 4.6). Net CO_2 fixation in roots is observed only at 15°C and 350 ppm of CO_2. Under these conditions, the net carbon fixation increases with light intensity and becomes saturated at 300 μmol $m^{-2}s^{-1}$ (Table 4.7). At 25°C, the roots begin to evolve CO_2. This drop in CO_2 fixation can be reversed by lowering the temperature to 15°C.

The contribution of photorespiration to CO_2 evolution in light in *Aranda* aerial root is negligible because no apparent photorespiration and glycolic acid oxidase activity are detected. This postulation is supported by the observation that the rates of CO_2 evolution in light by aerial roots are similar at 21% and 100% oxygen although photorespiration is affected by different oxygen partial pressures (Table 4.8).

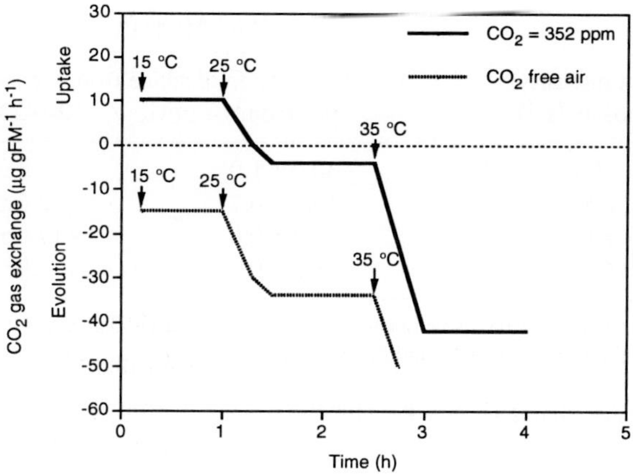

Fig. 4.6. Effect of temperature on carbon dioxide gas exchange of aerial root segments of *Aranda* Tay Swee Eng in ambient and CO_2-free air under saturating light intensity.

Redrawn from Hew, Ye & Pan (1991).

Table 4.6. Dark respiration in aerial root segments of *Aranda* Tay Swee Eng at various temperatures.

Temperature (°C)	Dark respiration (µg CO_2 gFM^{-1}h^{-1})
15	18.1 ± 1.5
20	21.5 ± 1.1
25	28.3 ± 1.2
30	40.1 ± 1.4
35	53.2 ± 0.8

Note: Mean of three replicates, ± SD.

Redrawn from Hew, Ye & Pan (1991).

Table 4.7. CO_2 gas exchange in aerial root segments of *Aranda* Tay Swee Eng at various light intensities.

Light intensity (μmol $m^{-2}s^{-1}$)	CO_2 gas exchange (μg CO_2 $gFM^{-1}h^{-1}$) Root section distance from the root tip 0–10 cm	20–30 cm
100	−30.0 ± 1.0	+ 3.3 ± 1.0
200	−10.9 ± 0.7	+12.6 ± 1.4
300	−7.9 ± 0.6	+17.6 ± 0.9
400	−7.8 ± 0.9	+17.9 ± 1.7

Note: The experiments were carried out at 15°C using 350 ppm of CO_2. (+) CO_2 fixation, (–) CO_2 evolution. Mean of three replicates, ± SD.

Redrawn from Hew, Ye & Pan (1991).

Table 4.8. CO_2 gas exchange in aerial root segments of *Aranda* Tay Swee Eng at various CO_2 and O_2 concentrations.

Root section distance from the root tip (cm)	CO_2 and O_2 concentration	CO_2 gas exchange (μg CO_2 $gFM^{-1}h^{-1}$) Light	Dark
0–10	CO_2-free air	−36.2 ± 0.8	−66.2 ± 1.0
	Ambient air	−10.5 ± 0.4	−40.2 ± 2.5
20–30	CO_2-free air	−17.6 ± 1.4	−49.4 ± 2.2
	Ambient air	+11.8 ± 1.2	−22.3 ± 0.8
	Oxygen (100%)	−14.3 ± 2.1	−52.1 ± 3.1

Note: The experiments were carried out at 15°C using 350 ppm of CO_2. (+) CO_2 fixation, (–) CO_2 evolution. Mean of three replicates, ± SD.

Redrawn from Hew, Ye & Pan (1991).

4.4. Respiratory Drift During Flower Development

Respiration of plant organ changes with development. The drift in respiration is well-studied during fruit and leaf development but less so in flower. Respiration, for example, is highest during fruit growth and the rate falls to a steady state during the maturity of the fruit. There is often a climacteric increase (or a brief rise to a new high level) during ripening of fruit, signaling the onset of irreversible processes of degeneration that marks the senescence and death of the fruit. A typical pattern of respiratory drift similar to that in ripening fruits has been observed in carnation cut-flowers.

Respiration in orchid flowers also changes with flower development. In *Aranda* Wendy Scott, the highest respiration rate is observed in tight buds, followed by a gradual decline in the other buds and flowers (Table 4.9). An increase in respiration rate is observed in the third flower, that is followed by a decline in flowers further down the inflorescence. Fresh mass, dry mass and anthocyanin content of *Aranda* flowers increase as the flowers mature, reaching a constant value in the third fully opened flower (see Chap. 8 on Flower Senescence and Postharvest Physiology).

The respiratory drift in developing orchid flowers has been studied more extensively in *Aranda* Christine. There is a shift in respiratory substrates, respiratory pathways and electron transport systems during *Aranda* flower

Table 4.9. Respiratory rates of various flowers at different positions along an inflorescence of *Aranda* Wendy Scott.

Flower position	Mean respiratory rate (μg CO_2 gFM^{-1} h^{-1})
Tight bud	278.6
First flower	132.2
Second flower	71.5
Third flower	92.6
Fourth flower	64.3
Fifth flower	53.9

Adapted from Hew (1980).

development (Table 4.10). Tight buds have respiratory quotient (RQ) of 0.5. The RQ increases to 0.7 in the first flower and reaches 1.0 in the mature flowers. A RQ of 1.0 indicates that carbohydrate is the respiratory substrate. The complete oxidation of fat molecules will yield a RQ of 0.7. However, if the fat molecules are partially converted to sugar using oxygen but without carbon dioxide evolution, the RQ will be about 0.57. The indication of high lipid content in orchid buds is interesting because it resembles the situation in cells of orchid embryos with many lipid bodies. If the fatty acid is partially oxidised and converted to sugar, a concomitant rise in RQ would follow. This may explain the increase of RQ to 0.5 in the first flower (newly opened flower) and eventually to 1.0 in the fifth flower (fully opened flower) when the carbohydrate becomes the sole respiratory substrate. However, one cannot rule out the possibility that other substrates such as amino acids, organic acids or others are being used in respiration. The possibilities of incomplete oxidation, utilisation of multi-substrates in different proportions, and the involvement of more than one chain of reactions in the breakdown of substrates prevent an accurate interpretation of the observed RQ.

The drift in RQ during flower development indicates a change in respiratory pathway. Carbohydrate metabolism in the mature *Aranda* flowers proceeds predominantly through the EMP pathway. There appears to be a non-glycolytic pathway contribution besides the EMP pathway in the tight buds and newly

Table 4.10. Respiratory metabolism in developing *Aranda* flowers.

	Developmental stage			
	Bud	Newly-opened	Fully-opened	Mature
Respiratory Quotient	0.5	0.7	1.0	1.0
Cyanide-resistant respiration	++	+	–	++
Ethylene production	++	+	+	+++

Note: – = absent; + = detectable; ++ = normal activity; +++ = very high activity.

Redrawn from Hew & Yip (1987, 1991), and Yip & Hew (1988).

opened flowers. This is clear from the studies of enzymes involved in pentose phosphate pathway and in studies involving the use of metabolic inhibitors on respiration of isolated *Aranda* petal cells. For example, higher activities of pentose phosphate shunt enzymes such as glucose-6-phosphate dehydrogenase and phosphogluconate dehydrogenase are observed in the buds than in the fully opened flowers (Fig. 4.7). Sodium fluoride (NaF) and malonate completely inhibit respiration of petal cells isolated from the newly opened flower and fully opened flower. In contrast, there is only 52% inhibition of respiration by NaF in the tight buds (Table 4.11). NaF and malonate are metabolic inhibitors that inhibit enolase and succinic dehydrogenase, respectively (see Figs. 4.1 and 4.2). Accompanying flower development, there is also a shift from cyanide-resistant respiration in the tight buds to cyanide-sensitive respiration in the fully opened flowers. A high degree of cyanide resistance is also observed in the mitochondria isolated from the tight buds. These

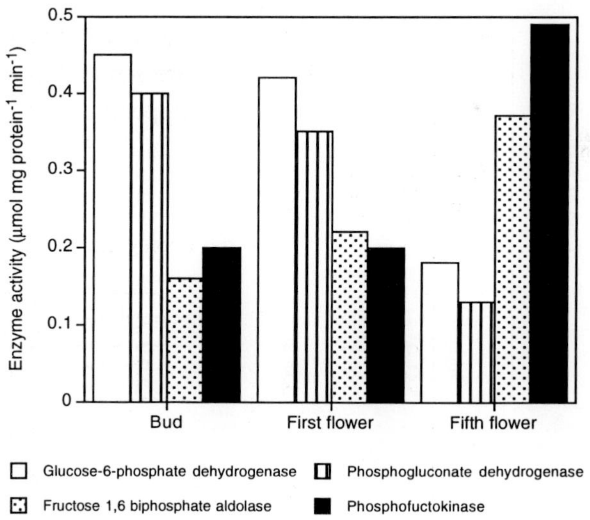

Fig. 4.7. Activity of glycolytic and pentose phosphate pathway enzymes isolated from bud, first flower and fifth fully-opened flower of *Aranda* Christine 130.

Redrawn from Yip (1990).

Table 4.11. The effects of various metabolic inhibitors on respiration of isolated *Aranda* petal cells at different developmental stages.

	Rate of respiration ($\mu l\ O_2\ mg\ protein^{-1} h^{-1}$)			Percentage of control		
Treatment	Bud	Newly-opened flower	Fully-opened flower	Bud	Newly-opened flower	Fully-opened flower
Control	541.9	693.1	284.6	–	–	–
KCN (5 mM)	386.1	389.3	0	71.3	56.2	0
SHAM (0.25 mM)	353.2	507.0	263.6	65.2	73.2	92.6
KCN (5 mM) + SHAM (0.25 mM)	11.5	13.4	0	2.1	1.93	0
NaF (100 mM)	280.9	0	0	51.8	0	0
Malonate (100 mM)	0	0	0	0	0	0

Note: Potassium cyanide = KCN; Sodium fluoride = NaF; Salicylhydroxamic acid = SHAM. NaF inhibits the conversion of phosphoglycerate to phosphoenolpyruvate while malonate competitively inhibits succinate dehydrogenase in the Kreb's cycle.

Adapted from Yip (1990).

mitochondria have low P/O ratio (the number of ATP formed per half molecule of oxygen) and respiratory control (RC, respiration rate at state 3/respiration rate at state 4) (Table 4.12). In contrast, P/O and RC ratios in mitochondria of mature flowers are high.

Ethylene production is closely associated with the triggering of cyanide-resistant respiration in storage organs of plants. The demonstration of changes in cyanide-resistant respiration and ethylene production in developing orchid flowers is interesting. A high rate of ethylene production is observed in buds of *Aranda*. Ethylene evolution increases with bud growth and reaches a peak in half-opened flowers (Fig. 4.8). This evolution rate increases again when the flowers show signs of senescence. The ethylene production profile of *Aranda* flowers is a reminiscent of the climacteric rise observed in fruits (see Chap. 8 on Flower Senescence and Postharvest Physiology).

It is generally believed that ethylene is produced at the later stages of flower development and the gas plays an important role in controlling flower senescence. By contrast, production of ethylene at the early stages of flower development has received very little attention. The high rate of ethylene production in orchid buds is related to bud opening. Aminooxyacetic acid (AOA) inhibits

Table 4.12. Respiratory Control (RC) and P/O ratios of mitochondria isolated from *Aranda* flower petal cells.

Developmental stage along the axis of an inflorescence	P/O	RC
Bud	1.19 ± 0.04	2.5 ± 1.4
First flower (newly-opened flower)	1.80 ± 0.01	3.0 ± 0.8
Fifth flower (fully-opened flower)	2.50 ± 0.01	5.5 ± 0.5

Note: P/O (equivalent to ADP/O) is the ratio of ATP formed over half an oxygen molecule consumed; RC is the rate of mitochondrial respiration at state 3 over the rate of mitochondrial respiration at state 4. The mitochondria were isolated from flower petals at different developmental stage using a Percoll gradient.

Adapted from Hew & Yip (1991).

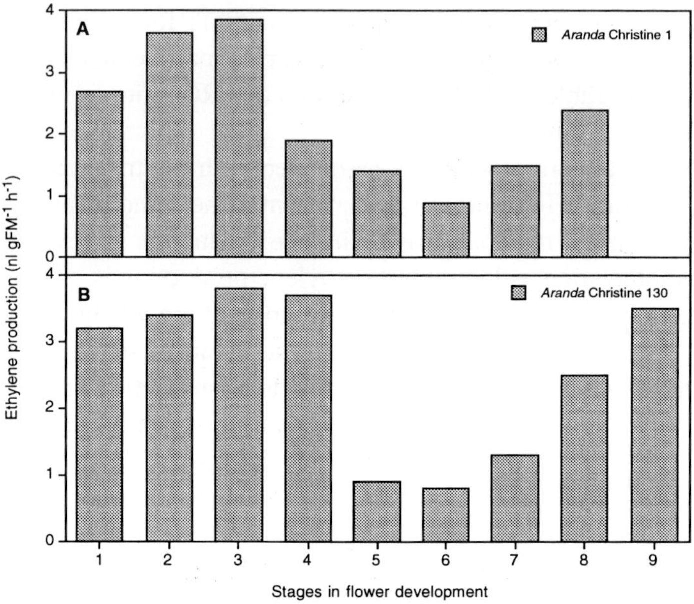

Fig. 4.8. Ethylene production by *Aranda* flowers and buds.
Note: (A) *Aranda* Christine 1; (B) *Aranda* Christine 130. Stage 1 = tight bud; Stage 2 = 'loose bud'; Stage 3 = half-opened flower; Stages 4–9 = mature flowers.

Redrawn from Yip & Hew (1988).

ethylene production as well as the expansion of *Aranda* buds. The elongation or expansion process in orchid flowers could have been mediated by a stimulation of respiration.

The pattern of ethylene production in *Aranda* buds and flowers coincides with a drift in respiration and the response to cyanide. This agrees with the view that there is a close relationship between ethylene production and the cyanide-resistant electron transport pathway in orchid flowers as reported in other plant tissues.

The development of cyanide-resistant respiration is influenced by the number of factors, including ethylene. Respiration of isolated *Aranda* orchid petal cells increases markedly after the flowers are treated with ethylene (Fig. 4.9). An increase in respiration is observed 15 to 20 h after ethylene

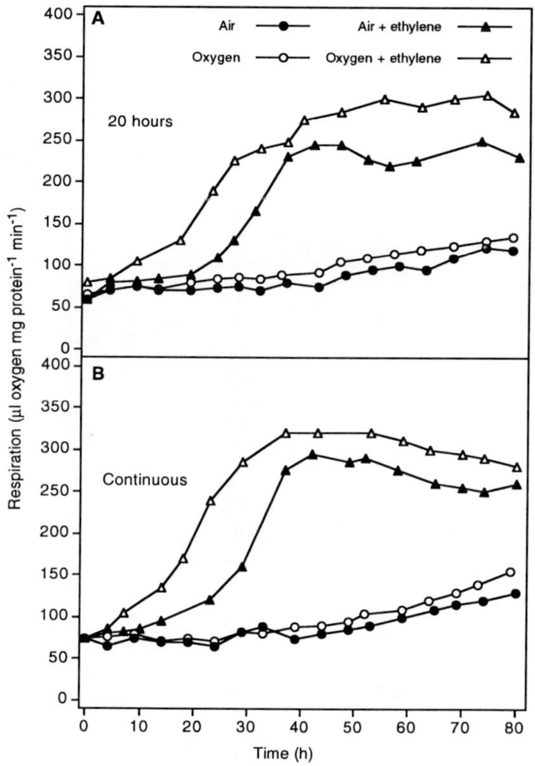

Fig. 4.9. The effects of ethylene on respiration of isolated *Aranda* petal cells in air and oxygen.

Note: (A) Short-term ethylene treatment (3 ppm); (B) Continuous ethylene treatment (3 ppm). Redrawn from Yip & Hew (1989).

treatment, which is further enhanced in the presence of high oxygen concentration. Ethylene gas induces the development of a cyanide-resistant pathway in fully opened orchid flower tissues where the capacity for cyanide-resistant respiration is negligible (Fig. 4.10). Similar results have also been observed in potato tuber slices and Iris bulbs.

The complete inhibition of respiration by cyanide in fully opened *Aranda* flowers is not clear, particularly when the fully opened flowers also produce ethylene but in considerably lower amounts. It appears that for the induction

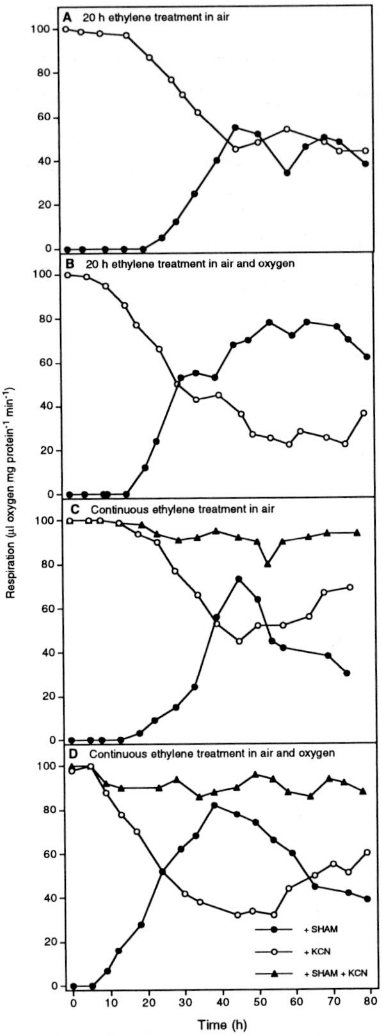

Fig. 4.10. Respiration of cells isolated from petals of *Aranda* Christine 130 flowers in the presence of ethylene and metabolic inhibitors.

Note: The effects of short-term (20 h) ethylene treatment (3 ppm) on the induction of cyanide-resistant respiration in petal cells using either (A) ethylene and air, or (B) ethylene and oxygen. Continuous ethylene (3 ppm) treatment on the induction of cyanide resistant respiration in isolated *Aranda* petal cells using either (C) ethylene and air, or (D) ethylene and oxygen.

Adapted from Hew & Yip (1987) and Yip & Hew (1989).

of cyanide-resistant respiration in orchid flower tissues, the concentration of ethylene may have to exceed a threshold value. An increase in respiration of orchid flowers is observed 20 h after exposure to ethylene. However, treatment with ethylene for a long time seems to gradually 'switch off' the cyanide-resistant respiration. Similar temporal changes in the development of cyanide resistance have also been observed in both the mitochondria and tissue slices of potato. The induction of cyanide-resistant respiration in potato takes some 6 to 9 h to begin, peaking at the 30th hour, and this is followed by a decline after prolonged ethylene exposure.

There is doubt that the augmentation of cyanide-resistant respiration by ethylene is a direct induction. The respiratory rise induced by cyanide and ethylene may be caused by a decontrol in glycolysis. The importance of supplying substrates and adenylates to mitochondria in the regulation of cyanide-sensitive and cyanide-resistant respiration has been reviewed (Lambers, 1985; Day *et al.*, 1980)

There is a striking similarity between orchid petal cells and some germinating seeds in their respiratory responses to cyanide. The importance and contribution of the cyanide-resistant pathway to the carbon and energy requirements of seed germination have been discussed (Day *et al.*, 1980; Berrie, 1984). It was suggested that during early germination, the need for ATP is sufficient but the need for carbon skeletons required for early protein synthesis may not be enough. The Kreb's cycle will produce the carbon skeletons needed and this is possible by an 'overflow' mechanism involving the alternative path (Berrie, 1984). It remains to be established whether the same phenomenon is observed for bud opening of orchid flowers.

The demonstration of a close relationship between ethylene production and bud opening in developing orchid flowers has practical implications. One may consider the use of ethylene to force orchid buds to open. Also, the high rates of ethylene production observed in buds and very young orchid flowers serve to remind orchid growers and exporters that a consideration must also be given to buds, along with flowers, during the development of post-harvest storage and handling technology for orchid cut-flowers. The attractiveness and advantages for orchid growers in ASEAN countries to harvest orchid flowers at the bud stage have been discussed (Hew, 1994).

4.5. Photorespiration

The respiratory process that takes place concurrently with photosynthesis in green leaves has generated considerable interest. This process, commonly known as photorespiration, is distinct from dark respiration (mitochondrial respiration) and light respiration. Light respiration refers to the continuation of normal dark respiratory processes in the light (i.e., respiration in the light). For photorespiration, carbon compounds formed during photosynthesis is metabolised through a C_2 (photorespiratory) cycle.

The mere fact that photorespiration takes place concurrently with photosynthesis makes accurate measurements very difficult, if not impossible, despite the use of several methods. The C_2 cycle starts with the production of phosphoglycolate in chloroplasts and continues with the oxidation of glycolate and formation of glycine in peroxisomes (Fig. 4.11). Two molecules of glycine are converted in mitochondria to serine and photorespired CO_2 molecule, and serine is converted to glycerate in the peroxisomes. Glycerate may re-enter chloroplasts and be reassimilated into the C_3 cycle as phosphoglyceric acid (PGA). The integration of the C_2 and C_3 cycles is shown in Fig. 4.12. The significance of photorespiration has been discussed (Bidwell, 1983).

Thin-leaved orchids studied so far possess the characteristics of C_3 plants. These orchids have high CO_2 compensation points (50–60 ppm), prominent post-illumination CO_2 outbursts, and active glycolic acid oxidase activity. The rate of photorespiration in orchid leaves is at least twice that of dark respiration. This agrees with the values reported for other C_3 plants.

As discussed earlier, there is no concrete evidence to indicate that C_4 photosynthesis may exist in orchids. It is now believed that the majority of C_4 plants also have photorespiration but to a much lesser extent than C_3 plants. The absence of photorespiration previously noted in C_4 plants could either be real or apparent. The absence of photorespiration in C_4 plants is due primarily to a suppression of photorespiration as a result of elevated levels of CO_2 level within the bundle sheath cells.

No study has specifically examined the photorespiration of CAM orchids. Several lines of evidence indicate that CAM plants may also photorespire: (1) the occurrence of post-illumination CO_2 outbursts, (2) oxygenase activity in RUBISCO, (3) presence of peroxisomes, (4) O_2 sensitivity of CO_2

assimilation in the light. As in C_4 plants, photorespiration in CAM plants is also suppressed by the high internal CO_2 concentration. Maintenance of high internal CO_2 concentration in C_4 and CAM plants is achieved through a unique CO_2 concentrating mechanism involving PEPC.

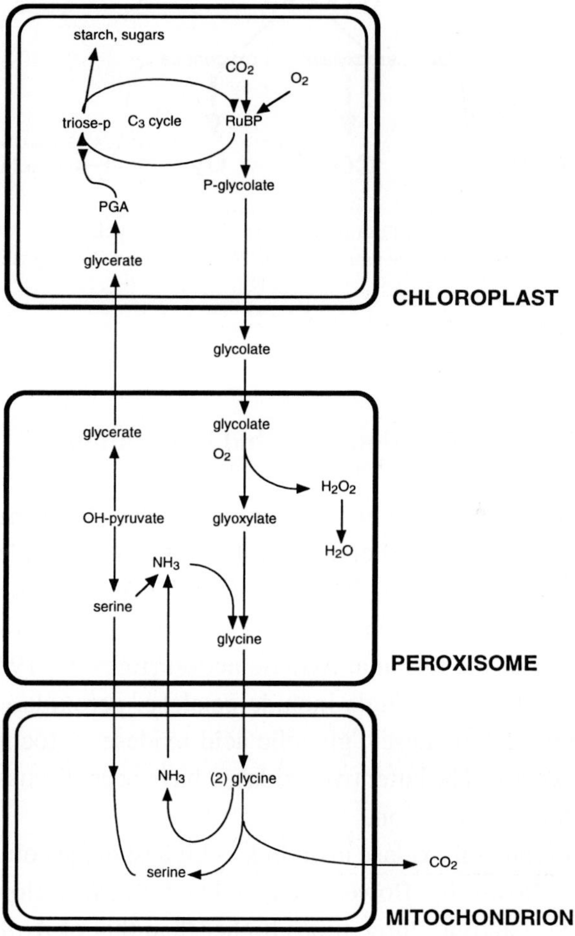

Fig. 4.11. The C_2 Photorespiratory cycle.

Redrawn from Bidwell (1979).

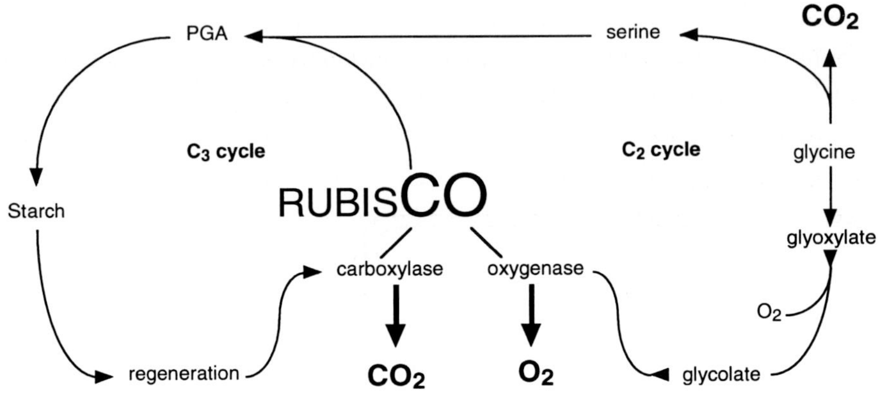

Fig. 4.12. Integration of the C_2 and C_3 cycles.
Note: The C_2 cycle is so-called because the product of RuBP oxygenase is a C_2 compound, as are glyoxlate and glycine.

Redrawn from Bidwell (1979).

4.6. Other Oxidases in Relation to Orchid Respiration

During mitochondrial respiration, cytochrome oxidase is normally the final electron acceptor. In addition, there are oxidases that are capable of oxidizing substrates using atmospheric O_2. The possible relevance of these oxidases to respiratory O_2 uptake depends on the ability of hydrogen-donating systems to reduce the product of the terminal oxidase action (Beevers, 1961). Oxidases that have been studied in orchids include catalase, peroxidase, polyphenol oxidases, ascorbic acid oxidases, glycolic acid oxidase, cytochrome oxidase and alternative oxidase. The latter two oxidases have been discussed in relation to cyanide resistance respiration.

Activity of polyphenol oxidase in orchids is highest in the column, followed by the aerial root, flower lip, flower petal, and leaf. *Cattleya* flowers exhibit a polyphenol oxidase activity three times higher than that of *Arachnis* Maggie Oei flowers. The difference could in part be due to the difference in flower longevity. *Arachnis* flowers have a relatively longer vaselife. After pollination and emasculation (depollination), a rise in polyphenol oxidase activity is evident particularly in the column. The O_2 uptake by *Arachnis* columns reaches a peak 7 h after pollination and 21 h after emasculation. The first increase in O_2 uptake is evident 1 h after pollination. Interestingly, a pollination-induced ethylene production in *Vanda* flowers is also evident after 1 h. The similarity in response between ethylene production and polyphenol oxidase activity after pollination suggests a close relationship. In fact, ethylene has been shown to stimulate polyphenol oxidase activity in tobacco flowers.

The infection of protocorms of *Dactylorhiza purpurella* and a *Cymbidium* hybrid by an endophytic fungus (*Rhizoctonia* sp.) is accompanied by a fourfold increase in the rate of respiration of the host, measured in terms of O_2 uptake. Marked increases in the activity of polyphenol oxidase, ascorbic acid oxidase and catalase are observed after infection. The peak of O_2 uptake by *Dactylorhiza* protocorms coincides with the formation and digestion of pelotons and the peak activities of the three oxidases. The general enhancement of metabolism after infection of orchid tissue by an endophytic fungus is associated with defense reactions rather than with the death of cells in the host and/or autolysis of the fungus, as has been observed in *Rhizoctonia*-infected bean hypocotyls.

An increase in ascorbic acid oxidase, peroxidase and polyphenol oxidase after infection has also been reported in other plant tissues, but in all cases oxidase activities increased with degenerative changes in the leaf. Clearly, one must distinguish between the initial responses in early stages of infection and the oxidative metabolism associated with degenerative changes. Equally important is the localisation of enzymes following infection. Activation of enzyme(s) can occur in the host or within the symbiont. Cytochemical localisation studies of polyphenol oxidases in *Rhizoctonia*-infected *Ophyrs* roots show that the fungus is able to synthesise or activate polyphenol oxidases. The enzymes synthesised in the fungal cytoplasm are "translocated across the plasma membrane and the cell wall of the fungus and accumulated in the interface close to the host plasmalemma where they are likely to promote the oxidation of phenols from the host." The production of phenolic phytoalexins with antifungal activity in orchids after fungal infection has been reviewed (Hadley, 1982; Arditti, 1992).

In *Vanda* seedlings, the highest peroxidase activity is observed during early stages of development while the lowest activity occurs during differentiation.

In *Encyclia* fruit development, peroxidase activity increases linearly with fruit diameter and size and is highest in the portion of the fruit containing the developing ovule.

Electrophoretic studies show that the amount of isoperoxidase varies with developmental stages in seedlings. Changes in isoenzyme patterns of peroxidases have also been reported in *Arundina graminifolia*, *Cymbidium sinense* and *Phalaenopsis amabilis* flowers at different stages of development. In these orchid flowers, peroxidase activity rises markedly with the onset of senescence. Similar pattern of changes in peroxidase activity is reported in tobacco corollas. There is evidence to indicate that the sharp rise in peroxidase activity in aging orchid flowers is caused by an increase in ethylene production during senescence (Avadhani *et al.*, 1994).

A dramatic increase in catalase activity is observed in the columns and petals of *Cymbidium lowianum* and *Dendrobium nobile* after pollination. This increase precedes the stimulation of respiration and the increase in catalase activity and respiration is affected by NAA (an auxin) treatment. These results have led to the conclusion that catalase activity may play a significant role in the chain of reactions that takes place in the columns of pollinated flowers.

4.7. Concluding Remarks

In the past two decades, considerable progress has been made in our understanding of orchid respiration but many questions remain unanswered. Respiratory metabolism in germinating seeds and the rhythmic nature of CO_2 production by orchid flowers are some examples. We have little information about the respiratory processes associated with growth and maintenance of orchids. The study of the respiratory drift in orchid flowers has provided valuable insight into the relationship between respiration and senescence in orchid flowers. Understanding respiration as an internal metabolic control of floral senescence is important to the development of a proper postharvest technology for cut-flowers. The study of carbohydrate level in flowers harvested at various times of day would provide useful information to the preharvest quality of flowers. We have yet to study the carbohydrate metabolism of orchid cut-flowers during and after storage under various conditions.

4.8. Summary

1. Respiration of different plant parts of orchids has been studied. By comparison, the respiration of seeds, roots and flowers have received more attention.
2. Majorities of the undifferentiated orchid embryos have no endosperm and orchid embryos are heavily packed with lipids as food reserves. The biochemistry of lipid metabolism in germinating orchid seeds remains unclear. No glyoxysome is detected in orchid seeds.
3. Flower respiration varies with orchid species and hybrids and seems to correlate well with flower longevity. Many flowers exhibit a circadian rhythm of carbon dioxide evolution. Root respiration is highest at the root tip and decreases markedly with increasing distance from the root tip. Net photosynthesis in roots of leafy orchids is masked by high respiration.
4. A respiratory drift involving changes in substrates, carbohydrate metabolic pathways and electron transport chain have been observed in developing *Aranda* flowers. Carbohydrate metabolism in mature flowers proceeds predominantly through the EMP pathway and Kreb's cycle. There appears to be a non-glycolytic pathway contribution along with the EMP pathway in the tight buds and newly opened flowers.
5. There is a shift from cyanide-resistant respiration in the tight buds to cyanide-sensitive respiration in the fully opened flowers of *Aranda*. Cyanide-resistant respiration in mature flowers is induced by ethylene. There is a close relationship between ethylene production and respiration in developing *Aranda* flowers.
6. Photorespiration is present in the leaves of C_3 orchids. No study has specifically examined photorespiratory processes in CAM orchids.
7. Polyphenol oxidase, ascorbic acid oxidase, peroxidase and catalase in orchids have been studied in relation to aging, pollination and fungal infection.

General References

Arditti, J., 1979, "Aspects of the physiology of orchids," *Advances in Botanical Research* **7**: 421–655.

Arditti, J., 1992, *Fundamentals of Orchid Biology* (John Wiley and Sons, New York), 691 pp.

Arditti, J. and Ernst, R., 1984, "Physiology of germinating seeds," in *Orchid Biology: Reviews and Perspectives, Vol. III*, ed. J. Arditti (Cornell Univ. Press, Ithaca), pp. 179–222.

Beevers, H., 1961, *Respiratory Metabolism in Plants* (Harper and Row, New York), 232 pp.

Berrie, A. M. M., 1984, "Germination and dormancy," in *Advanced Plant Physiology*, ed. M. B. Wilkins (Pitman, London), pp. 111–126.

Bidwell, R. G. S., 1979, *Plant Physiology*, Second ed. (MacMillan Publishing Co., New York), 726 pp.

Bidwell, R. G. S., 1983, "Carbon nutrition of plants: Photosynthesis and respiration," in *Plant Physiology: A Treatise, Vol. 7. Energy and Carbon Metabolism*, eds. F. C. Steward and R. G. S. Bidwell (Academic Press, New York), pp. 287–457.

Day, D. A., Arron, G. P. and Laties, G. G., 1980, "Nature and control of respiratory pathways in plants: The interaction of cyanide-resistant with cyanide-sensitive pathway," in *The Biochemistry of Plants — A Comprehensive Treatise, Vol. 2. Metabolism and Respiration*, eds. P. K. Stumpf and E. E. Conn (Academic Press, New York), pp. 198–243.

Hadley, G., 1982, "Orchid mycorrhiza," in *Orchid Biology: Reviews and Perspectives, Vol. II*, ed. J. Arditti (Cornell Univ. Press, Ithaca), pp. 299–307.

Halevy, A. H. and Mayak, S., 1979, "Senescence and postharvest physiology of cut flowers, Part 1," in *Horticultural Reviews 1*, ed. J. Janick (AVI Publishing, West Point, Conn.), pp. 204–236.

Hew, C. S., 1987, "Respiration in orchids," in *Orchid Biology: Reviews and Perspectives, Vol. IV*, ed. J. Arditti (Cornell Univ. Press, Ithaca), pp. 229–259.

Salisbury, F. B. and Ross, C. W., 1991, *Plant Physiology*, Fourth ed. (Wadsworth Publishing, Belmont, California), 682 pp.

Lambers, H., 1985, "Respiration in intact plants and tissues: Its regulation and dependence on environmental factors, metabolism and invaded organisms," in *Encyclopedia of Plant Physiology, New Series*, eds. R. Douce and D. A. Day (Springer-Verlag, Berlin), pp. 418–473.

Lambers, H., 1990, "Oxidation of mitochondrial NADH and the synthesis of ATP," in *Plant Physiology, Biochemistry and Molecular Biology*, eds. D. T. Dennis and D. H. Turpin (Longman, Scientific & Technical, London), pp. 124–143.

Zelitch, I., 1971, *Photosynthesis, Photorespiration and Plant Productivity* (Academic Press, New York), 347 pp.

References

Arditti, J. and Ernst, R., 1981, "Metabolism of germinating seeds of epiphytic orchids: An explanation for the need for fungal symbiosis," in *Proc. 10th World Orchid Conference*, eds. J. Stewart and C. N. van der Merwe (L. Backhouse Pte. Ltd., Pietermaritzburg), pp. 263–267.

Burg, S. P. and Dijkman, M. J., 1967, "Ethylene and auxin participation in pollen induced fading of *Vanda* orchid blossoms," *Plant Physiology* **42**: 1648–1650.

Bredemeizer, G. M. M., 1973, "Peroxidase activities and peroxidase-isoenzyme patterns during growth and senescence of the unpollinated style and corolla of tobacco plants," *Acta Botanica Neerlandica* **22**: 40–48.

Cheng, Y. W. and Chua, S. E., 1982, "The use of air-flow system in plant tissue and organ culture," in *Proc. COSTED Symp. on Tissue Culture of Economically Important Plants* (Singapore, 1981), pp. 210–212.

Eng, P. S., Yeoh, H. H., Khoo, S. I. and Hew, C. S., 1983, "Effect of Physan 20 on respiration, photosynthesis and growth of orchid plants," *Singapore Journal of Primary Industries* **11**: 76–83.

Erickson, L. C., 1957, "Respiration and photosynthesis in *Cattleya* roots," *American Orchid Society Bulletin* **26**: 401–402.

Goh, C. J., 1983, "Rhythms of acidity and CO_2 production in orchid flowers," *New Phytologist* **93**: 25–32.

Harrison, C. R., 1977, "Ultrastructural and histochemical changes during the germination of *Cattleya aurantiaca* (Orchidaceae)," *Botanical Gazette* **138**: 41–45.

Harrison, C. R. and Arditti, J., 1978, "Physiological changes during the germination of *Cattleya aurantiaca* (Orchidaceae)," *Botanical Gazette* **139**: 180–189.

Hew, C. S., 1980, "Respiration of tropical orchid flowers," in *Proc. 9th World Orchid Conference*, ed. M. R. Sukshom Kashemsanta (Bangkok, 1978), pp. 191–195.

Hew, C. S., Thio, Y. C., Wong, S. C. and Chin, T. Y., 1978, "Rhythmic production of CO_2 by tropical orchid flowers," *Physiologia Plantarum* **42**: 226–230.

Hew, C. S. and Lim, B. S., 1984, "Biological clocks in orchid flowers," *Malayan Orchid Review* **18**: 18–19.

Hew, C. S. and Yip, K. C., 1987, "Respiration metabolism in isolated orchid petal cells," *New Phytologist* **105**: 605–612.

Hew, C. S., Ting, S. K. and Chia, T. F., 1988, "Substrate utilisation by *Dendrobium* tissues," *Botanical Gazette* **149**: 153–157.

Hew, C. S. and Mah, T. C., 1989, "Sugar uptake and invertase activity in *Dendrobium* tissues," *New Phytologist* **111**: 167–171.

Hew, C. S., Tan, S. C., Chin, T. Y. and Ong, T. K., 1989, "Influence of ethylene on enzyme activities and mobilisation of materials in pollinated *Arachnis* orchid flowers," *Journal of Plant Growth Regulation* **8**: 121–130.

Hew, C. S. and Yip, K. C., 1991, "Respiration of orchid flower mitochondria," *Botanical Gazette* **152**: 289–295.

Hew, C. S., Ye, Q. S. and Pan, R. C., 1991, "Relation of respiration to CO_2 fixation by *Aranda* orchid roots," *Environmental and Experimental Botany* **31**: 327–331.

Hew, C. S. and Yip, K. C., 1991, "Ethylene and respiration in orchid flowers," *Proc. of the Nagoya International Orchid Show* (1991), pp. 117–121.

Hsiang, T. H. T., 1951, "Physiological and biochemical changes accompanying pollination in orchid flowers. I. General observations and water relations," *Plant Physiology* **26**: 441–455.

Hsiang, T. H. T., 1951, "Physiological and biochemical changes accompanying pollination in orchid flowers. II. Respiration, catalase activity, and chemical constituents," *Plant Physiology* **26**: 708–721.

McWilliams, E. L., 1970, "Comparative rates of dark CO_2 uptake and acidification in the Bromeliaceae, Orchidaceae and Euphorbiaceae," *Botanical Gazette* **131**: 285–290.

Manning, J. C. and Van Staden, J., 1987, "The development and mobilisation of seed reserves in South African orchids," *Australian Journal of Botany* **35**: 343–353.

Maxwell, D. P. and Bateman, D. F., 1967, "Changes in the activities of some oxidases in extracts of *Rhizoctonia*-infected bean hypocotyl in relation to lesion maturation," *Phytopathology* **57**: 132.

Roebuck, K. I. and Steinhart, W. L., 1978, "Pollination ecology and the nocturnal scent response in the genus *Brassavola*," *American Orchid Society Bulletin* **47**: 507–511.

Sheehan, T. J., 1954, "Respiration of cut-flowers of *Cattleya mossiae*," *American Orchid Society Bulletin* **23**: 241–246.

Stahmann, M. A., Clare, B. G. and Woodbury, W., 1966, "Increased disease resistance and enzyme activity induced by ethylene and ethylene production by black rot infected sweet potato tissue," *Plant Physiology* **41**: 1505–1515.

Tan, S. C. and Hew, C. S., 1973, "Polyphenol oxidase activity in orchid flowers," *Journal of the Singapore National Academy of Science* **3**: 282–296.

Wong, S. C. and Hew, C. S., 1975, "Do orchid leaves photorespire?" *American Orchid Society Bulletin* **44**: 902–906.

Yip, K. C., 1990, "Respiratory metabolism of orchid flower," M.Sc. dissertation. Department of Botany, The National University of Singapore, 278 pp.

Yip, K. C. and Hew, C. S., 1988, "Ethylene production by young *Aranda* orchid flowers and buds," *Plant Growth Regulation* **7**: 217–222.

Yip, K. C. and Hew, C. S., 1989, "Ethylene induced cyanide resistant respiration in orchid petal cells," *Plant Growth Regulation* **8**: 365–373.

Chapter 5

Mineral Nutrition

5.1. Introduction

Orchid hybrids grown for their cut-flowers have similar characteristics as their parents that are epiphytic in origin. The epiphytic orchids grow on the canopies of trees in the tropical rain forest and this presents a unique problem regarding water and nutrient supply. A general account for mineral nutrition of orchids has been reviewed by Poole and Sheehan (1982) and Benzing (1990).

This chapter aims to provide a basic understanding of mineral requirements and nutrition of tropical orchids. Hopefully, this will lead to the development of a proper fertiliser programme for tropical orchid cultivation. The discussion in Chap. 5 will focus mainly on the applied aspects of mineral nutrition in tropical orchids.

5.2. Mineral Requirements and Tissue Analysis

As in other plants, the orchid plant requires various essential elements for normal growth. Some essential elements are needed in larger quantities (macro-elements) while others (micro-elements) are needed in trace amounts. Nutrient deficiency of an element may develop when the concentration of the element drops below a level necessary for optimal plant growth. The concentrations of macro- and micro-nutrient elements in most plant tissues have been extensively studied and the concentrations at levels considered to be adequate are well-documented (Table 5.1).

Table 5.1. Concentrations of nutrient elements in plants considered to be at the adequate levels.

	Concentration in dry matter	
Element	μmole g^{-1}	ppm or %
Micro-elements		ppm
Molybdenum	0.001	0.1
Copper	0.10	6
Zinc	0.30	20
Manganese	1.0	50
Iron	2.0	100
Boron	2.0	20
Chlorine	3.0	100
Macro-elements		%
Sulphur	30	0.1
Phosphorous	60	0.2
Magnesium	80	0.2
Calcium	125	0.5
Potassium	250	1.0
Nitrogen	1,000	1.5
Oxygen	30,000	45
Carbon	40,000	45
Hydrogen	60,000	6

Adapted from Epstein (1972).

The composition of minerals in plant tissues is determined by tissue analysis. The values given in Table 5.1 serve only as a guide because the level of elemental content varies with the different plant parts and stages of plant development. There exists a relationship between plant growth (or yield) and the mineral content of plant tissue (Fig. 5.1). When the nutrient content in a tissue sample is low, growth is reduced. In the deficiency zone of the curve, an increase in the concentration of mineral in tissue increases growth or yield. The nutrient level in tissue samples will reach a point where further increase in mineral content will no longer bring about an increase in growth. This region is often referred to as the adequate zone. The narrow transition between the deficiency and adequate zones gives the critical concentration of the minerals that may be defined as the minimal tissue concentration of mineral that is correlated with maximal growth. The critical concentration of nutrient in tissue is at the point where there is 10% reduction in maximum growth. When the tissue nutrient content increases beyond the adequate zone, plant growth or yield begins to decline due to toxicity.

Orchids are similar to the other plants in their requirements except that they may take a longer time to show mineral deficiency. For *Vanilla* growing in gravel culture, nitrogen (N) deficiency occurs within three weeks while phosphorus (P) and potassium (K) deficiencies appear only after more than three months. *Cattleya* seedlings grown in purified quartz and nutrient solution without iron (Fe) fail to demonstrate deficiency symptoms after seven months of growth. Under similar conditions, many rapidly growing horticultural plants would have exhibited symptoms of Fe deficiency in a few days. It has been reported that *Dendrobium phalaenopsis* is severely affected by the omission of N, P, K, calcium (Ca) or magnesium (Mg) in nutrient solution and the leaves drop before deficiency symptoms appear. Descriptions of N, P and Ca deficiency symptoms do exist in the orchid literature, but not in full details (Poole and Sheehan, 1982). Reports of micro-element deficiency symptoms

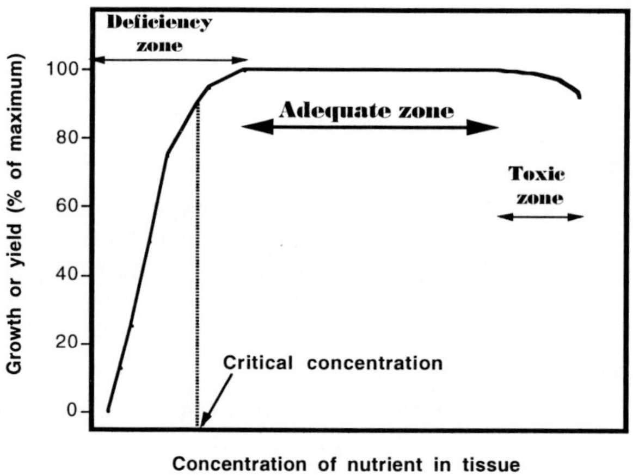

Fig. 5.1. The relationship between plant growth and mineral content of plant tissue.

in orchids are rare though well-defined toxicity symptoms of iron, zinc and boron are encountered. The slow development of deficiency symptoms in orchids is related to their remarkable ability to remobilise minerals from older leaves and other storage organs such as pseudobulbs, to meet new growth requirement. This 'efficient-recycling' phenomenon observed in most tropical orchids may be attributed to its epiphytic origin where the supply of minerals is scanty and unpredictable.

Considerable work has been done on tissue analysis of plant parts of *Phalaenopsis* and *Aranda* (Table 5.2). There is variation between the different plant parts. Mineral composition of *Laeliocattleya* grown in different medium is also different (Table 5.3). Clearly, the tissue mineral content of an orchid would depend on the growing media, genera, age and fertiliser programme.

Generally, the values for macro- (N, P, K, Ca, Na, Mg) and micro-elements (Fe, Mn, Zn, B, Cu, Mo) in *Cattleya, Cymbidium, Phalaenopsis* and *Aranda* are above the adequate range (Table 5.4). By comparison, orchid leaves generally have higher levels of calcium. Among the trace elements, iron in orchid roots is four times higher than the average value. For *Aranda*, the values for N, K and Mg are generally lower than the levels found in *Cattleya*,

Table 5.2. Elemental composition of orchid plant parts.

	Element (mg plant^{-1})				
	N	P	K	Ca	Mg
Phalaenopsis Dos Pueblos					
Leaf	2.3–2.6	0.12–0.20	3.8–4.5	1.8–2.1	0.70–0.72
Stem	1.9	0.25	2.0	1.58	0.59
Roots	2.9–3.3	0.31–0.54	1.73–2.93	0.40–0.53	0.57–0.71
Flowers	2.02	0.21	5.12	0.53	0.41
Aranda Noorah Alsagoff					
Leaf & Stem	0.74–0.88	0.17–0.20	0.64–0.95	2.38–2.86	0.21–0.28
Roots	0.88–0.99	0.26–0.39	0.34–0.44	0.77–0.91	0.13–0.15
Flowers	1.52	0.26	1.92	0.47	0.13

Adapted from Khaw & Chew (1980) and Poole & Sheehan (1982).

Table 5.3. The effects of media on elemental composition of leaves and roots in *Laeliocattleya* Aconcagua.

Plant part	Medium	% dry mass					ppm, dry mass			
		N	P	K	Ca	Mg	Fe	Mn	Zn	Cu
Leaves	Tree fern	1.85[b]	0.07[a]	1.94[a]	1.05[a]	1.11[a]	311[a]	842[a]	88[a]	12[a]
	Tree fern & redwood	1.78[ab]	0.08[a]	2.10[a]	1.63[b]	1.11[a]	295[a]	760[a]	90[a]	13[a]
	Fir bark	1.68[a]	0.06[a]	2.72[b]	1.60[b]	0.99[a]	405[a]	1047[b]	87[a]	13[a]
	Peat & perlite	1.80[ab]	0.06[a]	2.77[b]	1.18[a]	1.43[b]	352[a]	724[a]	145[b]	15[b]
Roots	Tree fern	1.24[a]	0.06[b]	0.77[a]	1.12[a]	0.81[b]	270[a]	351[b]	117[a]	19[a]
	Tree fern & redwood	1.20[a]	0.07[b]	0.91[b]	1.08[a]	0.77[ab]	283[a]	332[b]	98[a]	18[a]
	Fir bark	1.11[a]	0.03[a]	0.94[b]	1.38[b]	0.70[a]	293[a]	458[c]	107[a]	16[a]
	Peat & perlite	1.29[a]	0.06[b]	0.79[a]	1.31[b]	1.11[c]	212[a]	250[a]	203[b]	28[b]

Note: Means within a vertical column for each tissue followed by the same letter are not statistically different at the 5% level.

Adapted from Poole & Sheehan (1977).

Table 5.4. Elemental composition of some orchids.

Plant part	Orchid	% dry mass					ppm, dry mass				
		N	P	K	Ca	Mg	Fe	Mn	Zn	Cu	B
Leaves	*Cattleya*	1.8	0.2	4.2	1.3	0.5	66	79	28	10	41
	Cymbidium	2.3	0.3	2.9	1.0	0.3	133	54	46	12	48
	Phalaenopsis	2.0	0.3	7.1	3.0	0.5	97	210	23	5	47
	Aranda	0.9	0.2	1.0	2.4	0.3	110	102	350	63	34
Roots	*Cattleya*	2.0	0.3	2.2	0.8	0.8	440	28	118	25	19
	Cymbidium	2.3	0.7	3.8	0.9	0.8	546	–	116	16	17
	Phalaenopsis	3.9	0.3	3.5	1.2	0.7	502	30	86	6	8
	Aranda	0.8	0.3	0.4	0.8	0.2	430	14	285	161	12

Note: *Cattleya, Cymbidium* and *Phalaenopsis* are grown in solution culture, and *Aranda* is grown in pot with charcoal.

Adapted from Khaw & Chew (1980) and Poole & Sheehan (1982).

Cymbidium and *Phalaenopsis*. The difference may be because *Cattleya, Cymbidium* and *Phalaenopsis* are grown in solution culture whereas *Aranda* is potted in charcoal. For *Phalaenopsis* Dos Pueblos and *Aranda* Noorah Alsagoff, potassium is generally higher in the inflorescence whereas phosphorus is present in larger quantities in the roots (Table 5.2). These findings may have practical implications in fertiliser formulation and application.

There are many reports of orchids growing in various media. For example, vegetative growth and flowering of *Aranda* Kooi Choo are influenced by the growing medium (Table 5.5). However, it is difficult to compare the effects of potting media on growth and mineral composition in the tissues because the growing conditions, media composition, fertiliser program and orchid genera are either not mentioned or different. In addition, in many of these studies, no measurement of growth is carried out. Table 5.6 shows the effects of media on the growth of *Laeliocattleya*. Mericloned plants of *Laeliocattleya* Aconcagua are potted in tree fern fiber, a commercial mix of 60% tree fern fiber plus 40% redwood bark, fir bark or a mix of 50% peat moss plus 50% perlite by volume. Media affect all growth responses (number of leaves, new leads, dry mass,

Table 5.5. The effects of planting media on inflorescence and vegetative yield of *Aranda* Kooi Choo.

Treatment	Number of inflorescence (per 4 plants per month)	Average length of inflorescence (cm)	Number of leaves (per 4 plants per month)	Average plant height (cm)
Dried lallang leaves	3.9[a]	38.1[a]	6.0[a]	54.4[a]
Broken bricks	3.9[a]	38.0[a]	5.5[a]	52.2[a]
Oil palm kernal waste	2.7[b]	38.1[a]	5.9[a]	52.8[a]
Empty pot	2.0[c]	36.7[a]	5.3[a]	49.5[a]
Coconut husks	2.0[c]	37.5[a]	5.2[a]	51.6[a]

Note: Figures with a different letter differ significantly based on Duncan's Multiple Range Test at $P = 0.05$.

Adapted from Khelikuzzaman (1992).

Table 5.6. The effects of media on growth responses of *Laeliocattleya* Aconcagua.

Plant part	Number of leaves	Number of leads	Dry mass (g)			Leaf/root ratio
			Leaves	Pseudobulbs	Roots	
Tree fern	8.6[a]	1.1[a]	2.9[a]	0.7[a]	2.7[a]	1.1[a]
Tree fern & redwood	7.2[a]	0.9[a]	3.0[a]	0.9[a]	2.8[a]	1.1[a]
Fir bark	6.7[a]	0.8[a]	2.4[a]	0.9[a]	2.2[a]	1.1[a]
Peat & perlite	10.7[b]	2.0[b]	5.6[b]	2.2[b]	4.0[b]	1.4[b]

Note: Means within a vertical column for each tissue followed by the same letter are not statistically different at the 5% level.

Adapted from Poole & Sheehan (1977).

leaf/shoot ratio). Peat and perlite mixture give the best growth response. This is attributed to improved water relationships and nutrient retention. Plants grown in fir bark are of the lowest quality. Chemical composition of orchid plants is influenced mainly by the media rather than by supplementary micro-element levels.

It should be pointed out that in many of these studies, there is no mention of the possible adsorption of minerals by the supporting media. Pine bark, for example, adsorbs 1.5 mg Ng^{-1} of bark when ammonium ions are leached through the bark. Adsorption of ammonium ions and the other cations is increased at pH 3.8 to pH 5.8 (Fig. 5.2).

5.3. Fertiliser Application Practices

Extensive and careful studies on nutrient requirements of a plant are required before the formulation of a fertiliser programme. Tissue analysis, the N : P : K

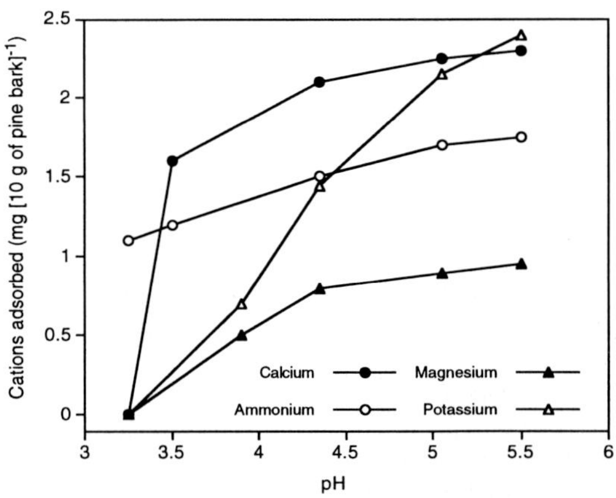

Fig. 5.2. The influence of pH on the adsorption of calcium, magnesium, ammonium and potassium ions by pine bark.

Redrawn from Foster, Wright, Alley & Yeager (1983).

ratio and nutrient requirements of various orchids at different stages of growth are needed. By comparison, the nutrient requirements of temperate orchid such as *Cattleya*, *Phalaenopsis* and *Cymbidium* are well-studied (Poole and Sheehan, 1982). Various N : P : K ratios for a number of orchid genera have been suggested (Table 5.7). In these recommended ratios, magnesium (Mg) requirement has often be ignored. The importance of Mg for orchid growth has been noted. The N : P : K : Mg ratio of 12 : 1 : 15 : 3 and 1.25 : 0.4 : 0.75 : 0.1 has been suggested for *Phalaenopsis* and *Cattleya*, respectively. Magnesium deficiency is frequently detected in tropical orchids grown under full sun in field. The N : P : K : Mg ratio of 13 : 3 : 11 : 1 has been recommended for *Aranda* Noorah Alsagoff. This ratio is rather similar to the ratios for *Phalaenopsis* and *Cattleya*.

The mineral requirement of two tropical orchids, *Aranda* Noorah Alsagoff and *Aranda* Wendy Scott has been studied. For *Aranda* Noorah Alsagoff, it is estimated that for a mature flowering plant, it requires 20.9 mg of N, 5.0 mg of P, 21.8 mg of K and 3.4 mg of Mg per week. For *Aranda* Wendy Scott, the feeding of plants on a 10 day basis is 72 mg of N, 72 mg of P and 36 mg of K. The results indicate that the two hybrids of *Aranda* have similar nutrient requirements.

Although there are inconsistencies in the fertiliser ratios recommended for tropical orchids, the results obtained so far have provided valuable information

Table 5.7. Optimal nutrient ratio of some orchids.

Orchid	N : P : K ratio
Cattleya	1.0 : 0.4 : 0.8
Cymbidium	1.0 : 0.4 : 0.8
Paphiopedium	1.0 : 0.8 : 1.0
Phalaenopsis	1.0 : 0.8 : 1.5
Aranda Wendy Scott	1.0 : 6.3 : 8.1
Aranda Noorah Alsagoff	4.3 : 1.0 : 3.7
Dendrobium Pompadour	1.5 : 1.5 : 1.0

Adapted from Penningsfeld & Fast (1970, 1973), Wong & Chua (1974), Khaw & Chew (1980), Vacharotayan & Kreetapirom (1975) and Penningsfeld & Forchthammer (1980).

on the nutritional status and requirements for *Aranda* and *Dendrobium*. The same approach should also be done for *Oncidium* and *Mokara*, which are of equal importance for cut-flower production in ASEAN countries. There are many commercial fertilisers with different ratios available in our market today (Table 5.8). However, these ratios have been formulated for general use and are not specific for tropical orchids. Also, most of the fertilisers exclude magnesium.

Effects of organic fertilisers on orchid growth

In the seventies, practically all the nurseries in Singapore and Malaysia used organic manure for growing orchids. Holttum (1964) stated that manuring, if judiciously applied, would be beneficial to orchids grown on charcoal and brick. He did not, however, give the exact rate or frequency of application for potted orchids. Application of animal waste can be done in diluted solution or in solid form. Chicken manure significantly increases the total flower yield of *Dendrobium* Lousiae Dark when compared to the chemical treated control. However in *Oncidium* Goldiana, the use of chicken manure only increases the spike length when compared with those treated with chemical fertilisers. Flower yields in *Oncidium* Goldiana decrease when high dosages of chicken manure

Table 5.8. Chemical composition of some commonly used fertilizers in orchid cultivation.

Trade names	Nitrogen	Phosphorous	Potassium	N : P : K ratio
Gaviota 63	21	21	21	1 : 1 : 1
Gaviota 67	14	27	27	1 : 2 : 2
Welgrow	15	30	15	1 : 2 : 1
Grofas	18	33	18	1 : 2 : 1
Peters	30	10	10	3 : 1 : 1
Hyponex	20	20	20	1 : 1 : 1
	7	6	19	1 : 1 : 3
	30	10	10	3 : 1 : 1

Adapted from Lee (1979).

are applied. (200 g per plant once in three months). Hence, the recommended level of chicken manure for *Dendrobium* and *Oncidium* is 100 and 50 g per plant once in 3 months, respectively (Table 5.9).

The use of pig-dung as organic manure in the cultivation of *Oncidium* Goldiana has been studied. There is no significant difference in the vegetative growth and the quality of inflorescences produced by plants grown with different levels of pig-dung. The levels of pig manure used range from 100–300 ml per month.

Chicken manure has also been used in the cultivation of monopodial orchids (Table 5.10). A comparison is made between chicken manure (64, 129 and 257 g per plant per month) and an inorganic fertiliser. *Arachnis* Maggie Oei, *Aranda* Deborah, *Aranda* Nancy and *Aranthera* James Storie, when given chicken manure, give significantly higher inflorescence yields and faster growth than those grown using inorganic fertiliser. When comparing between the chicken manure treatments, only the fast growing *Arachnis* Maggie Oei responds significantly to manuring; the other hybrids showed no significant yield differences.

In all the studies involving organic manuring, the main objective is not to determine the requirements for each element by the orchids but rather, to investigate how to use available animal wastes effectively. Nutrient content is generally low in animal manure (Table 5.11). The organic manure retains moisture and nutrients and it releases nutrient slowly. However, it has often been observed that potted orchids fertilised with manure grow well initially, but with time, the roots begin to rot, probably as a result of the disintegration of manure causing air and water blockage as well as bacterial growth. This may partly explain why manure applied at a higher level consistently decreases orchid growth and yields. In addition, there is a possibility of harbouring pest and weeds.

Effects of mulching on orchid growth

Hitherto, no study on nutrient uptake by the 'ground' orchids has been undertaken. 'Ground' orchids are different from terrestrial orchids as they are

Table 5.9. Average flower production per plant of *Dendrobium* Louisae Dark and *Oncidium* Goldiana for two years using different levels of chicken manure.

Orchid	Treatment	Quantities of major elements in the dried chicken manure (%)				Flower yield (per plant per two years)
		N	P	K	Mg	
Dendrobium Louisae Dark	Control: Gaviota solution	21	21	21	0.02	17.7^{b}
	Chicken manure (50 g per pot per 3 months)	1.17	1.08	1.19	0.32	28.3^{a}
	Chicken manure (100 g per pot per 3 months)	2.34	2.16	1.38	0.63	30.5^{a}
	Chicken manure (200 g per pot per 3 months)	4.68	4.32	2.76	1.26	26.6^{a}
Oncidium Goldiana	Control: Gaviota solution	21	21	21	0.02	10.2^{ab}
	Chicken manure (50 g per pot per 3 months)	1.17	1.08	1.19	0.32	11.6^{a}
	Chicken manure (100 g per pot per 3 months)	2.34	2.16	1.38	0.63	11.4^{a}
	Chicken manure (200 g per pot per 3 months)	4.68	4.32	2.76	1.26	7.6^{b}

Note: Figures with a different letter differ significantly at 1% level. Foliar application of Gaviota solution was applied at 25 g in 4.5 litres of water once in 10 days. The dried chicken manure was analysed for nitrogen (N), phosphorous (P), potassium (K) and magnesium (Mg) content before each application.

Adapted from Chua (1976).

Table 5.10. Inflorescence and vegetative yield of monopodial orchids grown with different levels of chicken manure.

Orchid	Treatment	Equivalent amounts applied per hectare per annum (tonnes)	Equivalent amounts of the major elements applied per hectare per annum			Average height increment over a 2-year period	Average leaf production over a 2-year period	Inflorescence yield (per plant per two years)
			N (Tonnes)	P_2O_5 (Tonnes)	K_2O (Tonnes)			
Arachnis Maggei Oei	Chicken manure (64 g per plant per month)	44.6	0.63	0.55	0.46	231^{b}	89^{b}	23^{b}
	Chicken manure (129 g per plant per month)	93.2	1.26	1.11	0.91	243^{ab}	93^{ab}	28^{c}
	Chicken manure (257 g per plant per month)	186.5	2.52	2.21	1.82	263^{a}	98^{a}	26^{ab}
	Inorganic fertilizer (3 g per plant per month)	2.3	0.36	0.18	0.13	96^{c}	37^{c}	11^{c}
Aranda Nancy	Chicken manure (64 g per plant per month)	44.6	0.63	0.55	0.46	108^{a}	54^{a}	21^{a}
	Chicken manure (129 g per plant per month)	93.2	1.26	1.11	0.91	120^{a}	58^{a}	23^{a}
	Chicken manure (257 g per plant per month)	186.5	2.52	2.21	1.82	116^{a}	56^{a}	20^{a}
	Inorganic fertilizer (3 g per plant per month)	2.3	0.36	0.18	0.13	48^{b}	25^{b}	9^{b}
Aranthera James Storie	Chicken manure (64 g per plant per month)	44.6	0.63	0.55	0.46	131^{a}	39^{b}	7^{a}
	Chicken manure (129 g per plant per month)	93.2	1.26	1.11	0.91	140^{a}	44^{ab}	7^{a}
	Chicken manure (257 g per plant per month)	186.5	2.52	2.21	1.82	151^{a}	46^{a}	6^{a}
	Inorganic fertilizer (3 g per plant per month)	2.3	0.36	0.18	0.13	76^{b}	26^{c}	3^{b}

Note: Figures with a different letter differ significantly based on Duncan's Multiple Range Test at $P = 0.05$.

Adapted from Wong & Chua (1974).

Table 5.11. Chemical composition of some organic manures.

Source	Nitrogen	Phosphorous	Potassium	Magnesium
	(% of dry matter)			
Fish emulsion	5	5	1	not determined
Blood & bone	5–6	12	not determined	not determined
Oil palm sludge (dried)	4.3	1.19	1.51	1.21
Cattle manure (dried)	2.0	1.5	2.0	1.0
Poultry manure (dried)	5.0	3.0	1.5	1.0
Goat manure (dried)	1.5	1.5	3.0	not determined

Adapted from Khaw (1982).

epiphytic orchids that have been planted on the ground. The aerial roots of 'ground' orchids (e.g., *Aranda, Arachnis, Mokara)* lose their chlorophyll when they penetrate into the mulch. The mulch, consisting mainly of woodshavings and sawdust (derived from a great variety of timber species coming mainly from the family Dipterocarpaceae), is used to retain water and nutrients. The aerial roots of 'ground' orchids branch extensively within the mulch. One area of orchid nutrition that has been neglected is the effect of mulching on the nutrition of ground orchids. Woodshavings and sawdust have a high C/N ratio and they undergo microbial breakdown in the field. Therefore, the amount of nitrogen needed to compensate for that required by microorganisms, above that needed for orchid growth, has to be worked out. However, the change in C/N ratio as a result of disintegration and the regular replenishment of new woodshavings or sawdust makes the task difficult, if not insurmountable. Woodshavings and sawdust have also been used in potted orchids. For potted orchids, the disintegrated woodshavings following decay could block water and air movement through the potting media. There is only one report on the effect of mulching on orchid growth. Mulching apparently decrease the yield and growth of potted *Aranda* Wendy Scott under various N : P : K treatments. More work is still needed in this area of orchid cultivation to clarify and evaluate the effect of mulching on growth and flowering of the other orchids.

Effects of inorganic fertilisers on orchid growth

In fertiliser formation and application, nitrogen, potassium and phosphorus are three major elements that have received greater attention. Past research effort has focused on the effect of nitrogen on growth due to its greater abundance. However, the effectiveness of nitrogen application depends on the presence of other minerals.

For *Cattleya* Trimos G grown on tree bark, the number of flowers per plant increases with increasing levels of nitrogen. In contrast, the response to varying levels of nitrogen differs in orchids grown on different barks (White fir, Red fir). In this experiment, ammonium nitrate is used as the source of nitrogen at a rate of 0, 8.6, 17.2 g of N per 2 litre of water. The types of bark and levels of phosphorus and potassium have no effect on the flowering of *Cattleya* Trimos G.

Growth of roots and leaves of *Cymbidium sinense* is considered to be fastest when the plants are grown with ammonium nitrate as the nitrogen source. Chlorophyll content of leaves is highest in plants grown with ammonium as a source of nitrogen. Moreover, the highest photosynthetic rate is observed in plants supplied with ammonium nitrate as a nitrogen source (Table 5.12).

Table 5.12. Effect of various nitrogen sources on chlorophyll content and photosynthetic rates of *Cymbidium sinense* leaves.

	Days after treatment			
	30	60	90	100
Chlorophyll content (mg gFM^{-1})				
Nitrate only (10 mM)	0.98	1.05	1.13	1.25
Ammonium only (10 mM)	1.10	1.23	1.29	1.42
Nitrate (5 mM) and ammonium (5 mM)	0.86	0.93	1.08	1.07
Photosynthetic rates (μmol m^{-2} s^{-1})				
Nitrate only (10 mM)	0.85	1.25	1.60	2.15
Ammonium only (10 mM)	0.82	1.15	1.45	2.05
Nitrate (5 mM) and ammonium (5 mM)	0.87	1.28	1.45	2.05

Redrawn from Wen & Hew (1993).

Increasing the level of nitrogen generally increases the flower size of *Vanda* Miss Joaquim (Fig. 5.3). The addition of phosphorus and potassium further increases *Vanda* flower size. The stem diameter is also influenced by the level of nitrogen supplied (Fig. 5.4). For *Aranda* Wendy Scott, both vegetative (stem length and leaf production) and reproductive (inflorescence length) growth are affected by increasing nitrogen levels (36, 72 mg per plant per application). The magnitude of increase in growth is also dependent on potassium and phosphorus levels (Fig. 5.5).

In a white-flowered *Phalaenopsis* hybrid, fertiliser level has no effect on bloom date or flower size in the first flowering season while the growing media affect on bloom date, flower number and root quality (Table 5.13). Following flowering, increasing the fertility from 0.25 g litre^{-1} to 1.0 g litre^{-1} increases flower count, stalk diameter and length, and leaf production, regardless of medium. During the second flowering season, the planting media have limited effect on growth. Increased fertility promoted earlier inflorescence emergence and blooming (Table 5.14). Higher fertiliser rates also caused a linear increase

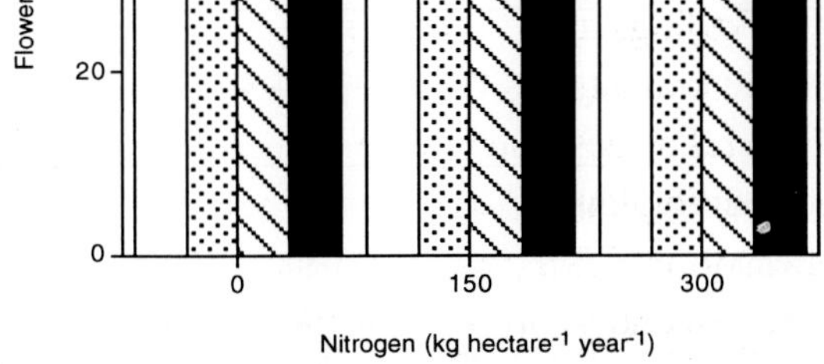

Fig. 5.3. Effects of interaction of nitrogen, phosphorous and potassium on flower size in *Vanda* Miss Joaquim.
Note: P = phosphorous; K = potassium. The rate of N, P and K application is expressed in terms of kg hectare^{-1} year^{-1}.

Redrawn from Higaki & Imamura (1987).

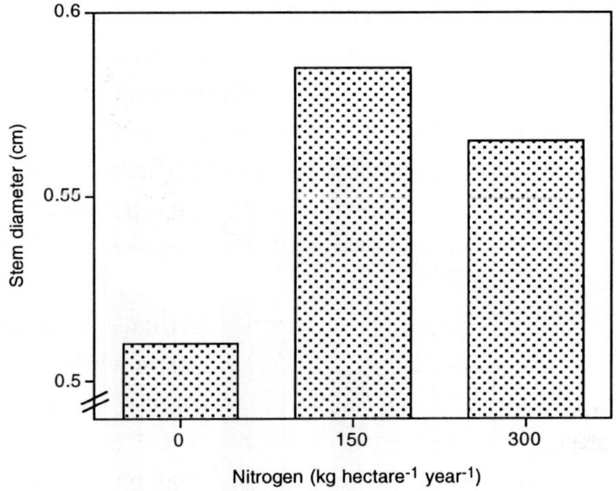

Fig. 5.4. Influence of nitrogen application on stem diameter in *Vanda* Miss Joaquim.

Redrawn from Higaki & Imamura (1987).

in the number of flowers and inflorescences per plant, and in stalk diameter, leaf number and size.

The observation that the response of *Cattleya* and *Aranda* to nitrogen is affected by the presence of another mineral is noteworthy. In *Cattleya*, the levels of foliar potassium and calcium decrease following an application of phosphorus. These findings agree with the reports of other agricultural crops. The complexity of mineral interaction in soil is well-documented. The excessive addition of one mineral affects the uptake of another mineral, often leading to deficiency of the latter in plant. The phenomenon of ion antagonism exists and orchids are no exception. This information is important in any attempt to optimise orchid growth and yield by appropriate fertiliser application.

Reports on using inorganic and organic fertilisers on orchid growth are few, even though it is widely practiced in local nurseries. There is evidence to indicate that a combination of organic and inorganic fertilisers is recommended as it generally gives better growth and flowering. The long-term effect, however, has not been carefully assessed.

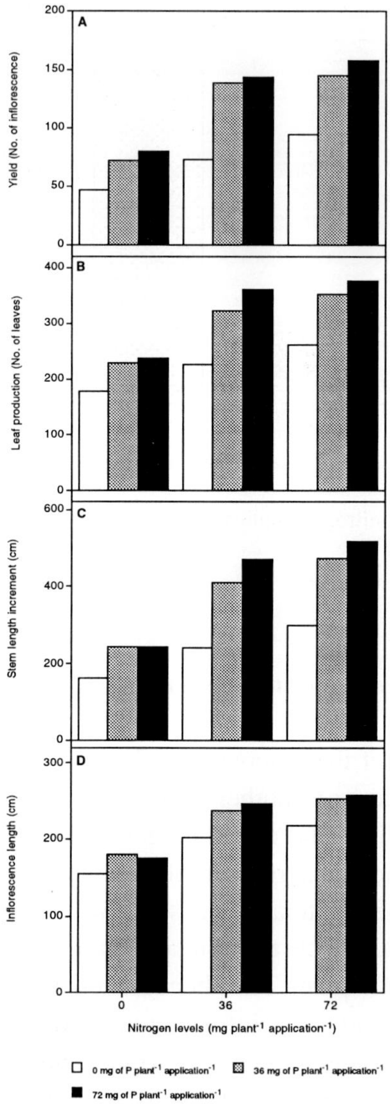

Fig. 5.5. The effects of nitrogen and phosphorous interaction on the yield and growth of *Aranda* Wendy Scott over a two-year period.

Redrawn from Wong & Chua (1975).

Table 5.13. The effect of medium and fertility on flowering and growth of *Phalaenopsis* in the first flowering season.

Treatment	First flower			Stalk[x]		Medium leachate			
	Bloom (days)	Width (cm)	Flower number	Diameter (mm)	Length (cm)	pH	EC (dS m^{-1})	Root grade[y]	New leaves
Medium									
1 (1 perlite: 1 Metro Mix 250: 1 charcoal)	123^{bc}	10.2^{a}	6.8^{b}	4.70^{a}	66.3^{a}	7.46^{a}	1.91^{b}	4.1^{a}	2.8^{a}
2 (2 perlite: 2 pine bark: 1 vermiculite)	124^{ab}	10.4^{a}	7.4^{ab}	4.75^{a}	68.6^{a}	6.42^{c}	1.64^{c}	4.3^{a}	2.9^{a}
3 (100% pine bark)	122^{bc}	10.4^{a}	7.1^{ab}	4.72^{a}	68.5^{a}	6.17^{d}	1.72^{c}	4.3^{a}	2.9^{a}
4 (3 perlite: 3 Metro Mix 250: 1 charcoal)	127^{a}	10.3^{a}	7.2^{ab}	4.78^{a}	67.8^{a}	7.43^{a}	2.16^{a}	3.4^{b}	2.2^{b}
5 (1 perlite: 1 rockwool)	119^{c}	9.8^{b}	7.5^{a}	4.49^{b}	66.7^{a}	7.00^{b}	1.87^{b}	2.9^{b}	2.8^{a}
Fertilizer (g $litre^{-1}$)[z]									
0.25	124	10.3	6.7	4.62	65.2	7.29	1.66	3.9	2.2
0.50	123	10.1	7.0	4.61	66.5	7.00	1.78	3.7	2.6
1.00	123	10.2	8.0	4.82	71.0	6.41	2.15	3.8	3.4
Significance	NS	NS	L***	L**	L***	L***	L***	NS	L**

Note: Bare-root seedling plants of a *Phalaenopsis* hybrid (*P. amabilis* × *P.* Mount Kaala 'Elegance') were grown in five potting media under three fertility levels from a water soluble fertilizer applied at every irrigation. Figures with a different letter within a column differ significantly based on Duncan's Multiple Range Test at $\alpha = 0.05$. NS, **, ***Nonsignificant or linear (L) and significant at $\alpha = 0.01$ or 0.001, respectively. There was no significant media × fertilizer rate interaction, therefore, only the main effect means are presented. [x]Diameter was measured at the middle of the fourth basal internode and the length was the distance between the base and the oldest flower; [y]Root grade: 1 = all roots dead; 2 = poor roots; 3 = some dead roots, good roots overall; 4 = few dead roots, some new roots; 5 = very few dead roots, abundant new roots; [z]Type of fertilizer used: Peters (20% N : 8.6% P : 16.6% K) water soluble fertilizer.

Adapted from Wang & Gregg (1994).

Table 5.14. The effect of medium and fertility on flowering and growth of *Phalaenopsis* in the second flowering season.

Treatment	Inflorescence emergence (Oct 1992)	Bloom date (Jan 1993)	Inflorescence emergence to bloom (days)	Number of flowers on the inflorescence			Number of lateral inflorescences	Number of inflorescence ($plant^{-1}$)
				Main	Lateral	Total		
Medium								
1 (1 perlite: 1 Metro Mix 250: 1 charcoal)	10[a]	14[a]	97[a]	11.5[a]	3.4[a]	14.9[a]	0.76[b]	1.72[a]
2 (2 perlite: 2 pine bark: 1 vermiculite)	12[a]	17[a]	98[a]	11.0[a]	3.4[a]	14.4[a]	0.75[b]	1.89[a]
3 (100% pine bark)	16[a]	20[a]	96[a]	10.4[a]	4.9[a]	15.3[a]	1.26[a]	1.69[a]
4 (3 perlite: 3 Metro Mix 250: 1 charcoal)	14[a]	18[a]	97[a]	10.6[a]	2.8[a]	13.4[a]	0.57[b]	1.85[a]
5 (1 perlite: 1 rockwool)	7[a]	9[a]	95[a]	11.7[a]	2.9[a]	14.6[a]	0.69[b]	1.85[a]
Fertilizer (g $litre^{-1}$)[x]								
0.25	21	25	96	10.2	0.5	10.7	0.13	1.53
0.50	13	15	95	10.8	2.3	13.1	0.60	1.73
1.00	2	8	99	12.0	7.8	19.8	1.72	2.12
Significance	L***	L***	L*	L***	L***	L***	L***	L***

Note: Bare-root seedling plants of a *Phalaenopsis* hybrid (*P. amabilis* × *P.* Mount Kaala 'Elegance') were grown in five potting media under three fertility levels from a water soluble fertilizer applied at every irrigation. Figures with a different letter within a column differ significantly based on Duncan's Multiple Range Test at $\alpha = 0.05$. NS, *, ***Nonsignificant or linear (L) and significant at $\alpha = 0.05$ or 0.001, respectively. [x]Type of fertilizer used: Peters (20% N : 8.6% P : 16.6% K) water soluble fertilizer.

Adapted from Wang & Gregg (1994).

5.4. Foliar Application and Root Absorption

In ASEAN orchid farms, application of fertiliser is carried out frequently by foliar feeding. Penetration and uptake of minerals through foliar feeding in orchids have been confirmed with the use of radioisotope.

Absorption and transport of phosphorus through the leaves and roots of *Phalaenopsis* have been studied. When ^{32}P is fed to the roots, 13% of ^{32}P absorbed is transported to the leaf. Conversely, when ^{32}P is fed to the leaf, 19% of ^{32}P absorbed by the leaves moved down to the roots (Table 5.15). Hence, the absorption of ^{32}P through leaves is comparable to that through the roots. When ^{32}P is applied to the second mature *Cattleya* leaves, about 34% of the amount supplied is found in the pseudobulb 24 hours later (Table 5.16). In these two studies, it is not clear whether the media has any effect on the availability of ^{32}P to the roots and the efficiency of ^{32}P uptake by the roots or leaves.

It has been suggested that foliar feeding is more effective for CAM orchids if it is done at night when the stomata are open. However, the pathway through the stomata is only one of the suggested routes whereby nutrients move into the plant system during foliar feeding. The practicality of having farm workers applying foliar feeding in the late evening in large orchid farms should be considered. We have no information on the efficiency of nutrient uptake through leaves. It has yet to be proven whether it is economically justifiable to fertilise the orchids at night. Most of the economically important tropical orchids for cut-flowers are thick-leaved orchids and the stomata are present only in the lower leaf surface in these orchids. For practical purposes, farmers spray all the foliage, particularly the under surface, besides a directed spray on the root zone.

The controversy of adopting either foliar application or root feeding as one of the method for fertiliser application has been discussed (Poole and Sheehan, 1982). The issue here is that we have little information on the efficiency of mineral uptake by these two application methods. Uptake efficiency of nutrients by *Aranda* (0.2% for P; 0.9% for Mg; 1.7% for N and 2.0% for K) has been reported. The low efficiency of nutrient uptake by *Aranda* Noorah Alsagoff potted in broken charcoal could be real or apparent. Leaching of nutrients

Table 5.15. The distribution of ^{32}P in *Phalaenopsis* after selective application to the leaf or root.

	Roots	Leaves
Root application	87%	13%
Leaf application	19%	81%

Adapted from Rahayu (1980).

Table 5.16. Percentages of ^{32}P absorbed by *Cattleya* Trimos as affected by time and the method of application.

	Hours after application				
Stage and treatment	1/2	2	12	24	120
First mature leaf					
Foliar application	0.025	0.036	0.038	0.034	0.077
Medium drench[b]	0.035	0.061	0.237	0.123	0.960
Second pseudobulb					
Foliar application	6.023	3.892	10.565	34.450	37.730
Medium drench	0.036	0.023	0.150	0.143	2.440
Third leaf					
Foliar application	0.018	0.021	0.028	0.052	0.042
Medium drench	0.024	0.025	0.103	0.138	0.480
Third pseudobulb					
Foliar application	0.028	0.069	0.265	0.163	0.337
Medium drench	0.032	0.072	0.065	0.128	0.128

Note: [a]Foliar application. Foliar spray was applied to the second mature leaf; [b]Medium drench = pot drench.

Adapted from Sheehan, Joiner & Cowart (1967).

following watering could have been partly responsible for the low efficiency of nutrient uptake observed. Increased watering is known to reduce the orchid responses to fertiliser application. Uptake of nutrient by orchid grown in solid culture media has been studied. In such a system where leaching is prevented, uptake efficiency of N and P by *Dendrobium* roots is found to be 13% and 4%, respectively (Table 5.17). The same has also been obtained for *Aranda* roots.

To be able to fully exploit the potential of either foliar application or root feeding for optimal orchid fertilisation, extensive research on the basic physiology of mineral nutrition in orchids is needed. Aspects of orchid mineral nutrition that awaits careful research are uptake, transport, distribution, storage and re-utilisation of minerals. The relative effectiveness of foliar application and root feeding also needs to be ascertained.

The rates at which foliar applied nutrients are absorbed by the leaves and translocated within the leaf is an important criterion in determining the effectiveness of foliar fertilisation. Foliar fertilisation is affected by a number of interacting factors: plant morphology and physiology, environmental conditions and the spraying solution (Table 5.18). Information on the mechanism of nutrient uptake by orchid leaves is lacking. This is probably the main reason why this method of fertilisation remains controversial. Compared with the uptake of nutrients through roots, foliar application is a much speedier

Table 5.17. Mineral uptake efficiency of some orchid roots.

	Uptake efficiency (%)			
Growing conditions	Nitrogen	Phosphorous	Potassium	Magnesium
Mature *Aranda* Noorah Alsagoff plants potted in charcoal	1.7	0.2	2.0	0.9
Young *Dendrobium* White plantlets in agar medium	12.5	4.2	not determined	not determined
Young *Aranda* Tay Swee Eng plantlets in agar medium	11.3	3.0	1.3	2.1

Redrawn from Hew (1990).

Table 5.18. Factors that determines the efficacy of foliar feeding.

Plant	Environment	Spray solution
Cuticular wax	Temperature	Concentration
Epicuticular wax	Light	Application rate and technique
Leaf age	Photoperiod	Wetting agent
Stomata	Wind	pH
Guard cells	Humidity	Polarity
Trichomes, leaf hairs	Drought	Hygroscopicity
Leaf turgor	Time of the day	Compounds used
Surface moisture	Osmotic potential of the root medium	Sticking property
Cultivar	Nutrient stress	Nutrient ratio
Growth stage		Carriers, penetrants

Adapted from Alexander (1986).

way of supplying minerals, micro-elements in particular, to the plants. It can also be used to satisfy acute plant needs quickly. Root feeding has its problems too. Orchid roots are capable of absorbing nutrients but the efficiency of uptake is comparatively low.

5.5. Ion Uptake

As discussed earlier, there is considerable information on the mineral composition of orchids and their growth in different potting media supplemented by various fertiliser programmes. However, research work on ion uptake processes is rather scanty. In this section, we will focus on the uptake of nutrients by orchids and discuss the possible mechanism involved.

Ion uptake by orchid tissues

Dendrobium tissues grown in liquid culture media show a preferential uptake of ammonium ions over nitrate. Nitrate uptake begins only when ammonium ions are depleted. The pattern of uptake of ammonium and nitrate varies according to the carbon source supplied and not on its concentration.

The preferential uptake of ammonium ions by *Dendrobium* Multico White tissues may be attributed to the favourable initial pH of the nutrient medium. As has been reported for other plant tissues, ammonium utilisation generally inhibits nitrate utilisation. Alternatively, nitrate reductase activity may be low in orchid tissues. Rates of uptake of ammonium and nitrate by callus tissue of *Aranda* Noorah Alsagoff kept at constant culture pH value have been studied. The results showed that at pH 5.0–5.5, the uptake of ammonium is preferred over nitrate.

There exists a strong correlation between the uptake of nitrate and ammonium by orchid tissues and pH changes in the media. The pH decreases sharply with ammonium uptake and increases later, when the ammonium ions in the media has been depleted and uptake of nitrate begins. These agree with the findings in other plant tissues. The decrease in pH following ammonium uptake is attributed to an efflux of protons. Similarly, the pH increase following nitrate uptake is a result of the efflux of hydroxyl ions.

Ion uptake by orchid roots

The uptake of ammonium, nitrate and phosphate by the roots of young *Dendrobium* Multico White plantlets is linear with time. Uptake of ammonium is faster than the uptake of nitrate. A good correlation in mineral uptake and growth of plantlets is observed. Both mineral uptake (Table 5.19) and growth (Fig. 5.6) are enhanced when the culture media are supplemented with sucrose. When *Dendrobium* plantlets are grown under high light intensity, the percentage of nitrate uptake increases while the pattern of uptake remains the same. The uptake patterns of various minerals (N, P, K, Ca, Mg) by roots of *Aranda* Tay Swee Eng, *Dendrobium* Multico White and *Oncidium* Gower Ramsey plantlets grown on Vacin and Went media are rather similar.

Orchid plantlets are generally grown in a flask where the amount of nutrients is finite. The rate of nutrient depletion depends on the size as well as the number of plantlets per culture flask. It is a common practice to have 25 plantlets in

Table 5.19. Percentage uptake of ammonium, nitrate and phosphate ions by *Dendrobium* Multico White plantlets growing on Vacin and Went solid medium.

	Total uptake (mg)	Percentage of uptake	Uptake per day (mg)
Ammonium ions	5.25 (sugar-free)	64.6 (sugar-free)	0.087(sugar-free)
	5.64 (+ sugar)	80.3 (+ sugar)	0.094 (+ sugar)
Nitrate ions	2.50 (sugar-free)	15.7 (sugar-free)	0.042 (sugar-free)
	3.76 (+ sugar)	22.1 (+ sugar)	0.063 (+ sugar)
Phosphate ions	0.92 (sugar-free)	21.6 (sugar-free)	0.015 (sugar-free)
	1.90 (+ sugar)	33.3 (+ sugar)	0.032 (+ sugar)

Redrawn from Hew & Lim (1989).

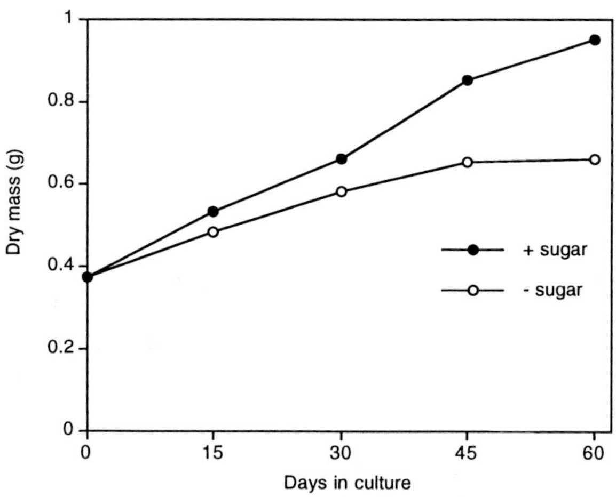

Fig. 5.6. Growth of *Dendrobium* Multico White plantlets in Vacin and Went solid medium with or without sugar added.

Redrawn from Hew & Lim (1989).

50 ml of solid medium in most commercial laboratories. Under such condition, it would be advisable not to keep plantlets longer than 60 days as there would be little mineral nutrients left for growth.

The absorption of nitrate and ammonium ions by roots varies linearly with time for adult orchid plants, for example *Bromheadia finlaysoniana* (Fig. 5.7). Unlike the orchid tissues and young orchid plantlets, there is no preferential uptake of ammonium over nitrate. Generally, the rates of nitrate and ammonium

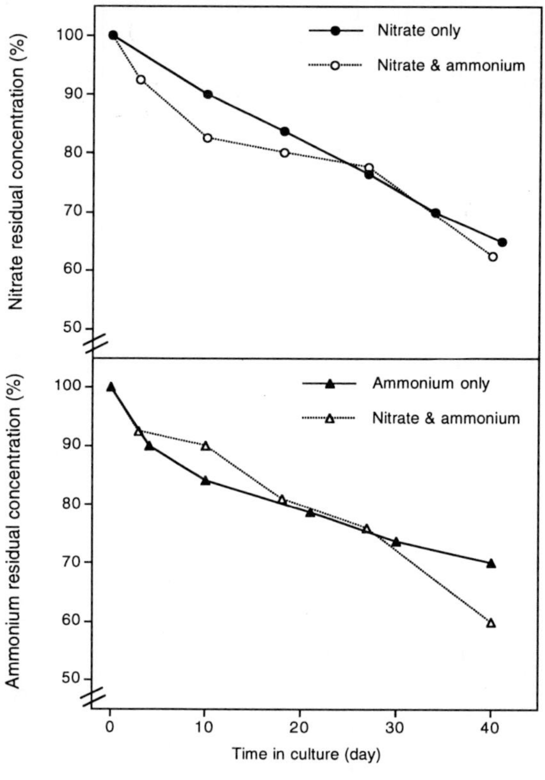

Fig. 5.7. The rate of nitrogen uptake by *Bromheadia finlaysoniana* grown in different nitrogen sources.

Note: Mature plants were grown hydroponically in culture media containing different sources of nitrogen: nitrate only; ammonium only; nitrate and ammonium.

Redrawn from Hew, Lim & Low (1993).

Table 5.20. Comparative rates of nitrate uptake and activity of nitrate reductase, glutamine synthetase and glutamate dehydrogenase in orchids.

	Rate of nitrate uptake (μmol $gFM^{-1}h^{-1}$)		Enzyme activity (μmol $gFM^{-1}h^{-1}$)		
			NR	GS	GDH
Bromheadia finlaysoniana	0.41 ± 0.05	Leaf	0.73 ± 0.03	12.8 ± 1	1.86 ± 0.84
		Root	1.06 ± 0.23	6.3 ± 2	3.18 ± 0.5
Dendrobium White Fairy	0.95 ± 0.07	Leaf	1.23 ± 0.15	267.4 ± 27	0.24 ± 0.01
		Root	0.55 ± 0.25	68.9 ± 5.8	0.70 ± 0.03
Hordeum vulgare (barley)	2.67–5.00	Leaf	14	144	3.3
		Root	5	13.7	36.7

Note: Mean ± SE.

Adapted from Rao & Rains (1976) and Hew, Lim & Low (1993).

uptake in tropical orchids are comparatively lower than those of other nonorchidaceous plants. For example, the rate of nitrogen uptake in tropical orchids is only one-third to one-seventh of the rate in barley (Table 5.20).

Key enzymes involved in nitrogen assimilation include nitrate reductase (NR), glutamine synthetase (GS) and glutamate dehydrogenase (GDH). Nitrate reductase activity is present in orchid leaves and roots but the activity is considerably lower than that of barley. GS activity is higher in orchid leaves than in roots, and is comparable to the levels found in barley leaves. GDH is also detected in orchid leaves and roots. The enzyme GS is believed to play a more important role in nitrogen assimilation as it has a very much lower K_m for ammonium ions. It appears that the same is also true for the role of orchid GS activity. Accumulation of ammonium ions is observed following the inhibition of GS activity by methionine sulfoximine (MSX, an inhibitor of GS) (Fig. 5.8).

It is interesting that nitrate is present in the roots and leaves of orchids that have been grown in medium with ammonium ions as the sole nitrogen source.

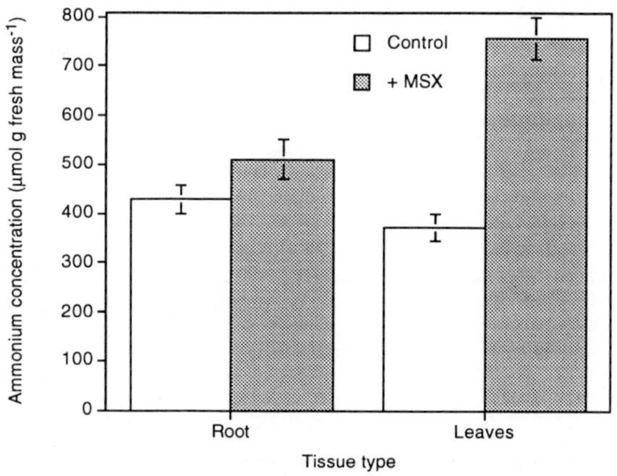

Fig. 5.8. Effect of methionine sulfoximine (MSX) on the accumulation of ammonium in leaves and roots of *Dendrobium* White Fairy grown in nutrient solutions with nitrate as the sole nitrogen source.

Redrawn from Hew, Lim & Low (1993).

This would indicate that nitrate present in the storage pool is not readily available for transport or reduction.

As observed in the other plants, the nitrate absorbed by orchid roots is utilised in the following ways (Fig. 5.9):

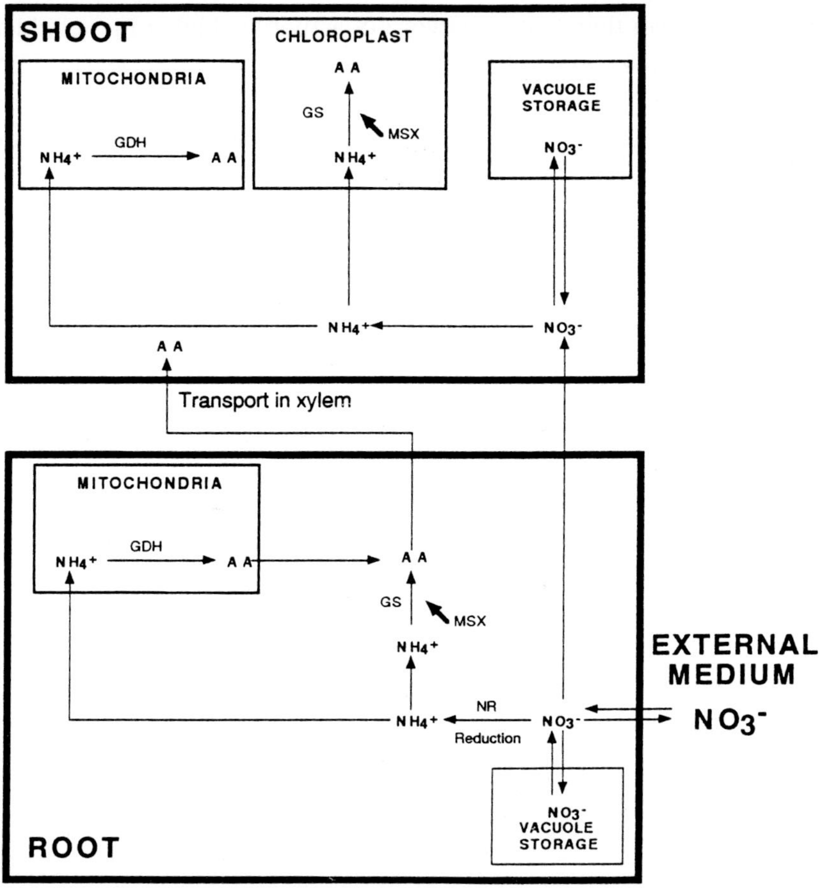

Keys : AA = amino acid; GDH = Glutamate dehydrogenase; GS = Glutamine synthetase; MSX = methionine sulfoximine, an inhibitor of GS; NH_4^+ = ammonium ions; NO_3^- = nitrate ions; NR = Nitrate reductase

Fig. 5.9. Hypothetical scheme for the assimilation and transport of nitrogen in orchids.

Adapted from Lim (1992) and Hew, Lim & Low (1993).

(1) Reduced to ammonium and incorporated into amino acids (e.g. glutamate) and subsequently transported to the leaf;
(2) Stored in the storage pool in leaves and roots;
(3) Transported to the leaf in the form of nitrate and reduced to ammonium ions in the leaves.

This is supported by the demonstration of active NR and GS in the roots and leaves of *Dendrobium*, *Cymbidium*, *Bromheadia*, and the presence of a nitrate storage pool in leaves and roots. As mentioned earlier, the rate of uptake of nitrate by orchid root is comparably low. The relatively slow uptake rate could be attributed to:

(1) Physical barrier arising from the unique structure of orchid roots;
(2) Low activity of enzymes (e.g., NR, GS) involved in nitrogen assimilation;
(3) Shortage of carbohydrate in roots.

Although the activity of NR in orchid roots is low, it is sufficient to account for the corresponding rate of nitrate uptake. GS activity is high and comparable to the levels found in other plants. It is unlikely that velamen hinders the uptake of nitrate because the rates of nitrate uptake by the roots of *Dendrobium* and *Bromheadia* are comparable despite the two orchids having different number of cell-layers in the velamen. The velamen of *Dendrobium* roots consists of 6 layers whereas the velamen of *Bromheadia* is of a single layer. It is possible that restriction to ion uptake lies in the barrier at the interface between exodermis and cortex. The exodermis consists of thin passage cells and suberised cells. A possible regulatory role in mineral movement has been assigned to the passage cells. In *Dendrobium*, the passage cells account for only 7% of the total exodermal surface. As such, the passage cells in the roots may offer considerable barrier to the influx of ions into the underlying cortex tissue. In addition, movement of ions across the endodermis into the stele may also be restricted. A fair number of epiphytic and a few terrestrial orchids have teliosomes within its roots. Tilosomes may be seen directly above the passage cells (Fig. 5.10). The intermeshing branch is believed to act as a one-way valve that allows fluids to enter while minimizing water vapour loss. The precise mechanism of the teliosome in controlling fluid and ion movement remains unclear (Pridgeon, 1987).

Another factor that may restrict ion influx into the root stele is the supply of carbohydrate as an energy source in the roots. As mentioned earlier, the uptake of nitrate is dependent on light and the supply of exogenous sugar. Therefore, the manipulation of sink activity in roots to induce more carbon supply to the roots is an area that deserves more investigation.

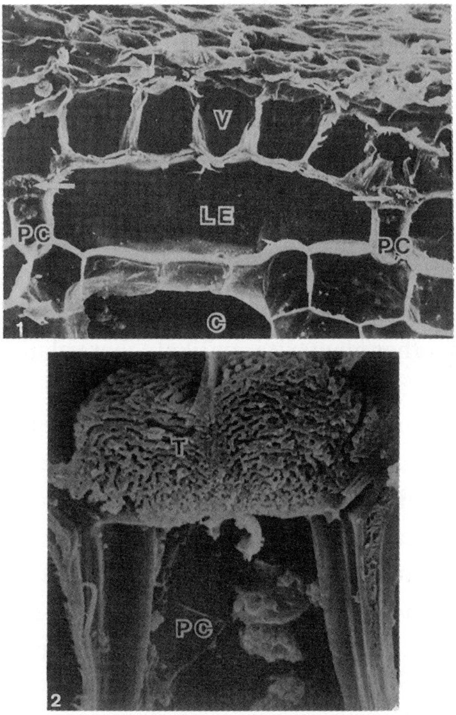

Fig. 5.10. Scanning electron micrographs of *Sobralia decora* roots illustrating exodermis and spongy tilosomes.
Note: (1) Root radial section showing velamen [V], long exodermal cell [LE], passage cells [PC] and cortical parenchyma [C]. Arrows above passage cells indicate tilosomes [450 X]; (2) Root transection showing spongy-type tilosome [T] and passage cell [PC] [2,000 X].

Reproduced from Pridgeon, Stern & Benzing (1983), courtesy of *American Journal of Botany*

5.6. Concluding Remarks

To date, studies on the mineral nutrition of tropical orchid such as *Dendrobium*, *Aranda* and *Oncidium* are limited to mature flowering plants. Poole and Sheehan (1982) have identified the critical factors affecting orchid mineral nutrition: the medium, degree of decomposition of organic materials and age of the plants. More information pertaining to mineral nutrition of orchid plants at different stages of development is needed. Nonetheless, the results obtained so far have provided a basis for formulating practical fertiliser program for tropical orchid cultivation.

To be able to fully exploit the potential of either foliar application or root feeding for optimal orchid fertilisation, extensive research on the basic physiology of mineral nutrition in orchids is needed. Recent works have shown that the uptake of minerals by orchid roots is rather slow, but the nature and factors affecting the slow uptake remains unclear. Aspects of orchid mineral nutrition that awaits careful research are uptake, transport, distribution, storage and reutilisation of minerals. We have yet to know how different the nature of mineral uptake by terrestrial and aerial roots of tropical orchids is. Equally important is the effect of mulching on mineral uptake by tropical orchids grown in the ground.

The relative effectiveness of foliar application and root feeding needs to be ascertained. One of the important criteria for foliar fertilisation is the rate at which the nutrients are absorbed by the leaves and translocated to the other plant parts. At present, we still lack information on the mechanism and factors governing mineral nutrient uptake by the orchid leaves.

5.7. Summary

1. Orchids are similar to the other non-orchidaceous plants in their requirements for minerals except that they generally take a longer time to show mineral deficiency. Analysis of tissues shows that the mineral content of orchids is in the same range as those reported for non-orchidaceous plants.

2. Organic manure (chicken and pig manure) has been used for the cultivation of tropical orchids. However, the disadvantages associated with organic manure seem to outweigh the advantages. Nitrogen, potassium and phosphorus are the three major elements that have received greater attention in the studies involving the use of inorganic fertiliser application. The effectiveness of application of one element depends on the presence of other elements. There is evidence to indicate that a combination of organic and inorganic fertilisers gives better orchid growth.
3. The controversy of adopting either foliar or root feeding of fertilisers remains unresolved. This is attributed to the fact that we have little information on the efficiency of mineral uptake using the two methods of application.
4. The rate of mineral uptake by orchid roots is relatively low when compared to the other crop plants. The slow uptake may either be attributed to the barrier encountered at the interface between exodermis and cortex, or the supply of respiratory substrates to the roots. Enzymes (nitrate reductase and glutamine synthase) responsible for nitrate assimilation are present in leaves and roots of orchids and the activities of these two enzymes are sufficient to account for the observed rates of nitrate uptake.

General References

Alexander, A., 1986, *Foliar Fertilisation: Proceedings of the First International Symposium on Foliar Fertilisation* (Martinus Nijhoff Publishers, Dordrecht), 488 pp.

Benzing, D. H., 1990, "Vascular epiphytes: General biology and related biota," *Cambridge Tropical Biology Series* (Cambridge University Press, England), 354 pp.

Epstein, E., 1972, *Mineral Nutrition of Plants: Principles and Perspectives* (John Wiley, New York), 412 pp.

Pilbeam, D. J. and Kirkby, E. A., 1990, "The physiology of nitrate uptake," in *Nitrogen in Higher Plants*, ed. Y. P. Abrol (Research Studies Press, Wiley, New York), pp. 39–64.

Poole, H. A. and Sheehan, T. J., 1982, "Mineral nutrition of orchid roots," in *Orchid Biology: Reviews and Perspectives, Vol. II*, ed. J. Arditti (Cornell University Press, Ithaca, New York), pp. 195–212.

Pridgeon, A. M., 1987, "The velamen and exodermis of orchid roots," in *Orchid Biology: Reviews and Perspectives, Vol. IV*, ed. J. Arditti (Cornell University Press, Ithaca, New York), pp. 139–192.

References

Awasthi, O. P., Sharma, E. and Palni, L. M. S., 1995, "Stemflow: A source of nutrients in some naturally growing epiphytic orchids of the Sikkim Himalaya," *Annals of Botany* **75**: 5–11.

Beaumont, J. H. and Bowers, F. A. I., 1954, "Interrelationships of fertilisation, potting media and shading on growth of seedling *Vanda* orchids," *Hawaii University Agriculture Station Technical Paper* **334**: 88–93.

Brundell, D. J. and Powell, C. L., 1985, "Environmental and nutritional factors affecting growth and development of *Cymbidium* orchids," *Proc. of the 2nd New Zealand International Orchid Conference* (Wellington, 1985), pp. 7–11.

Chin, T. T., 1966, "Effect of major nutrient deficiencies in *Dendrobium phalaenopsis* hybrids," *American Orchid Society Bulletin* **35**: 549–554.

Chua, S. E., 1976, "The effects of different levels of dried chicken manure on the growth and flowering of *Oncidium* Golden Shower (var. Caldwell) and *Dendrobium* Louisae Dark," *Singapore Journal of Primary Industry* **4**: 16–23.

Davidson, O. W., 1960, "Principles of orchid nutrition," in *Proc. of the Third World Orchid Conference* (London, 1960), pp. 224–233.

Esnault, A., Masuhara, G. and McGee, P. A., 1994, "Involvement of exodermal passage cells in mycorrihzal infection of some orchids," *Mycological Research* **98**: 672–676.

Foster, W. J., Wright, R. D., Alley, M. M. and Yeager, T. H., 1983, "Ammonium adsorption on a pine-bark growing medium," *Journal of the American Society for Horticultural Science* **108**: 548–551.

Gething, P. A., 1977, "The effect of fertilisers on the growth of orchid (*Odontoglossum*) seedlings," *Expl. Hort.* **29**: 94–101.

Hew, C. S., 1990, "Mineral nutrition of tropical orchids," *Malayan Orchid Review* **25**: 70–76.

Hew, C. S., Ting, S. K. and Chia, T. F., 1988, "Substrate utilisation by *Dendrobium* tissues," *Botanical Gazette* **149**: 153–157.

Hew, C. S. and Lim, L. Y., 1989, "Mineral uptake by orchid plantlets grown on agar culture medium," *Soilless Culture* **5**: 23–34.

Hew, C. S. and Mah, T. C., 1989, "Sugar uptake and invertase activity in *Dendrobium* tissues," *New Phytologist* **111**: 167–171.

Hew, C. S., Lim, L. Y. and Low, C. M., 1993, "Nitrogen uptake by tropical orchids," *Environmental and Experimental Botany* **33**: 273–281.

Higaki, T and Imamura, J. S., 1987, "NPK requirement of *Vanda* Miss Joaquim orchid plants," College of Tropical Agriculture and Human Resources, University of Hawaii, Research Extension Series 087, 5 pp.

Holttum, R. E., 1964, *A Revised Flora of Malaya I: Orchids of Malaya* (Government Printing Office, Singapore).

Khaw, C. H., 1982, "Mineral nutrition of orchids," *Malayan Orchid Review* **16**: 34–39.

Khaw, P. S. and Chew, P. S., 1980, "Preliminary studies on the growth and nutrient requirements of orchids (*Aranda* Noorah Alsagoff)," *Proc. 3rd ASEAN Orchid Congress* (Malaysia, 1980), pp. 49–64.

Khelikuzzaman, M. H., 1992, "Observations on the effect of planting media on flower production of orchid variety *Aranda* Kooi Choo," *Malaysian Orchid Bulletin* **6**: 73–77.

Koay, S. H. and Chua, S. E., 1979, "The appropriate utilisation of an organic manure for optimum inflorescence production of *Oncidium* Golden Shower potted in an economical and suitable granite medium," *Singapore Journal of Primary Industry* **7**: 1–8.

Franke, W., 1986, "The basis of foliar application of fertilisers with special regard to the mechanism," in: *Foliar Application*, ed. A. Alexander (Martinus Nijhoff Publishers, Dordrecht), pp. 17–25.

Lee, C. K., 1979, *Orchids: Their Cultivation and Hybridisation* (Eastern Universities Press, Singapore), 94 pp.

Lee, Y. K., Hew, C. S. and Loh, C. S., 1987, "Uptake of ammonium and nitrate in callus tissue culture of orchid *Aranda* Noorah Alsagoff," *Singapore Journal of Primary Industries* **15**: 37–41.

Lim, L. Y., 1992., "Mineral nutrition of tropical orchids," M.Sc. Dissertation. Department of Botany, The National University of Singapore. 239 pp.

Penningsfeld, F., 1985, "Soilless propagation and cultivation of orchids. Possibilities, advantages and disadvantages," *Soilless Culture* **1**: 55–66.

Penningsfeld, F. and Fast, G., 1970, "Ernahrungstragen bei Paphiopedilum callosum," *Gartenwelt* **9**: 205–208.

Penningsfeld, F. and Fast, G., 1973, "Ernahrungstragen bei Disa uniflora," *Die Orchidee* **24**: 10–13.

Penningsfeld, F. and Forchthammer, L., 1980, "Ergebnisse neunjahriger Cymbidien-Ernährungsversuche," *Die Orchidee* **31**: 11–19.

Poole, H. A. and Seely, J. G., 1978, "Nitrogen, potassium and magnesium nutrition of three orchid genera," *Journal of the American Society for Horticultural Science* **103**: 485–488.

Poole, H. A. and Sheehan, T. J., 1977, "Effects of media and supplementary microelements fertilisation on growth and chemical composition of *Cattleya*," *American Orchid Society Bulletin* **45**: 155–160.

Powell, C. L., Caldwell, K. I., Littler, R. A. and Warrington, I., 1988, "Effect of temperature regime and nitrogen fertiliser level on vegetative and reproductive bud development in *Cymbidium* orchids," *Journal of the American Society for Horticultural Science* **113**: 552–556.

Pridgeon, A. M., Stern, W. L. and Benzing, D. H., 1983, "Tilosomes in roots of Orchidaceae. I. Morphology and systematic occurrence," *American Journal of Botany* **70**: 1365–1377.

Rahayu, S., 1980, "Absorption and transport of phosphorous through *Phalaenopsis* leaf and root," *Proc. 3rd ASEAN Orchid Congress* (Malaysia, 1980), pp. 37–48.

Rao, K. P. and Rains, D. W., 1976, "Nitrate absorption by barley. I. Kinetics and energetics," *Plant Physiology* **57**: 55–58.

Sheehan, T. J., Joiner, J. N. and Cowart, J. K., 1967, "Absorption of ^{32}P by *Cattleya* 'Trimos' from foliar and root applications," *Proc. of the Florida State Horticultural Society* **80**: 400–404.

Tanaka, T., Matsuno, T., Masuda, M. and Gomi, K., 1988, "Effects of concentration of nutrient solution and potting media on growth and chemical composition of a *Cattleya* hybrid," *Journal of the Japanese Society for Horticultural Science* **57**: 85–90.

Tanaka, T., Kanto, Y., Masuda, M. and Gomi, K., 1989, "Growth and nutrient uptake of a *Cattleya* hybrid grown with different composts and fertilisers," *Journal of the Japanese Society for Horticultural Science* **57**: 674–684.

Vacharotayan, S. and Kreetapirom, S., 1975, "Effects of fertilisers upon growth and flowering of *Dendrobium* Pompadour," *Proc. 1st ASEAN Orchid Congress* (Bangkok, Thailand 1975), pp. 138–156.

Wang, Y. T. and Gregg, L. L., 1994, "Medium and fertiliser affect the performance of *Phalaenopsis* orchids during two flowering cycles," *HortScience* **29**: 269–271.

Wen, Q. S and Hew, C. S., 1993, "Effects of ammonium and nitrate on photosynthesis, nitrogen assimilation and growth of *Cymbidium sinense*," *Journal of Singapore National Academy of Science* **20/21**: 21–23.

Withner, C. L. and Van Camp, J., 1948, "Orchid leaf analyses," *American Orchid Society Bulletin* **17**: 662–663.

Wong, Y. K. and Chan, W. F., 1973, "Agronomic practices in the cultivation of ground orchids," *Singapore Journal of Primary Industry* **1**: 12–19.

Wong, Y. K. and Chua, S. E., 1974, "The use of chicken manure in the cultivation of ground orchids," *Singapore Journal of Primary Industry* **2**: 6–15.

Wong, Y. K. and Chua, S. E., 1975, "Yield and growth responses of *Aranda* Wendy Scott to manurial treatments with NPK and sawdust mulch," *Singapore Journal of Primary Industry* **3**: 75–106.

Yamaguchi, S., 1979, "Determination of several elements in orchid plant parts by neutron activation analysis," *Journal of the American Society for Horticultural Science* **104**: 739–742.

Yoneda, K., 1989, "Effects of fertiliser application on growth and flowering of orchid (*Epidendrum radicans* Pavon)," *Bull. Coll. Agr. and Vet. Med., Nihon University (Japan)* **46**: 69–74.

Chapter 6

Control of Flowering

6.1. Introduction

Flowering is an important stage of plant development. Through flowering, sexual reproduction can be effected, resulting in the production of fruits and seeds. Flowering is genetically controlled, but it can be induced by environmental stress such as low temperature and water stress. Orchids are generally grown for their flower except for the Oriental cymbidiums, where the beauty of the leaves, flowers as well as the fragrance of the plant is appreciated. It is therefore not surprising that there have been considerable interests in studying flower induction of orchids. Literature pertaining to flowering in orchids is plentiful (Rotor, 1952; Goh & Arditti, 1985), but we have no information on the nature of flowering in orchids.

This chapter aims to provide a general account of flowering in orchids and discuss the possible ways to control flowering to meet market demands.

6.2. Differentiation of Flower Bud

Flower evocation is considered to be a multifactorial process (Bernier, 1988). The process of flowering in tropical orchids can be separated into two processes: Flower induction (or flower initiation) and floral development (Fig. 6.1). Induction of flower is influenced by genetic, environmental and physiological factors. Following induction, the flower bud will grow and its subsequent growth will depend on the supply of photoassimilates from various sources

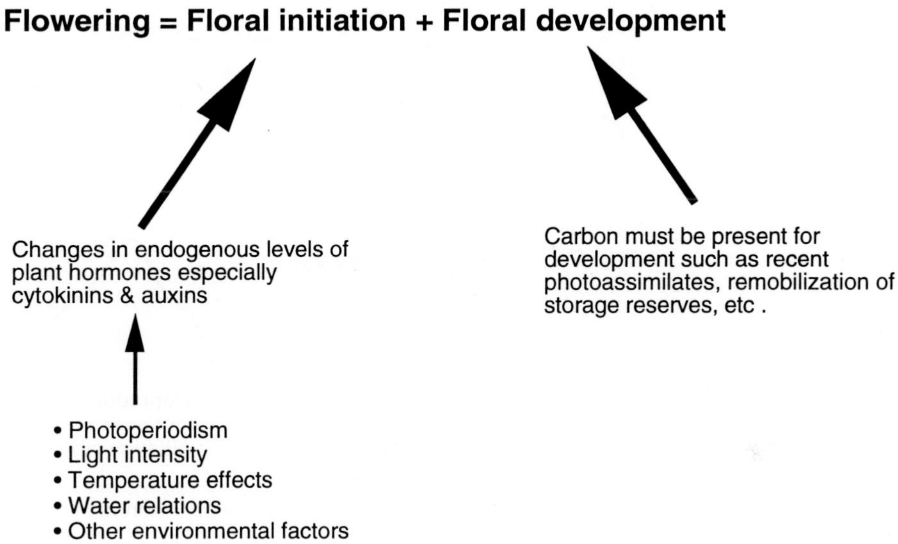

Fig. 6.1. The process of flowering in tropical orchids.

and from its own photosynthesis. The supply of assimilates from the leaves to the flowers depends on source–sink relationship (see Chap. 7 on Partitioning of Assimilates).

Orchid inflorescence is either terminal or axillary in nature. Upon induction, the bud/apical meristem is changed from vegetative to reproductive phase. Rotor (1952) has divided orchids into two groups according to their position and numbers of differentiated bud primordia that are capable of developing into flower shoots. Orchid hybrids with terminal bud primordium such as *Cattleya* and *Paphiopedilum* belong to the first group. In these orchids, bud differentiation is associated with new growth (pseudobulb). Only the apical bud primordium of a new pseudobulb is capable of developing into an inflorescence. The bud primordium before induction is slightly convex in form and it elongates after induction (Fig. 6.2). The second group consists of orchid species/hybrids with several axillary bud primordia and these include cymbidiums and dendrobiums. For *Cymbidium,* dormant axillary buds at the base of the pseudobulb will develop into an inflorescence. For *Dendrobium*,

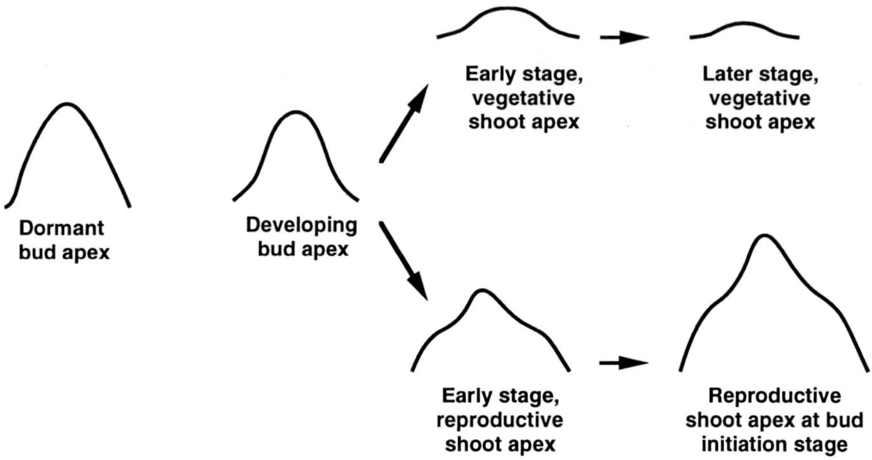

Fig. 6.2. The development of the vegetative and reproductive meristems from the dormant bud stage in orchids.

Redrawn from Rotor (1952).

(e.g., *Dendrobium nobile* and *Dendrobium phalaenopsis*), bud primordia are formed at the axils of the leaves. For *Dendrobium nobile*, upon flower initiation, practically all the bud primordia arranged alternatively on opposite side of the pseudobulb develop into inflorescences almost simultaneously. *Dendrobium phalaenopsis* produces flowers on both new and old pseudobulbs. In a new developing pseudobulb, the axillary bud primordia are located on opposite side of the axis, with the youngest nearest the apex. The youngest is the first bud primordia to develop into an inflorescence (Fig. 6.3). The types of flowering characteristics of some orchids are schematically represented in Fig. 6.4.

6.3. Factors Affecting Flower Induction

Juvenility, vernalisation and photoperiodism are three important factors that determine when the plants will flower with respect to ontogeny and season.

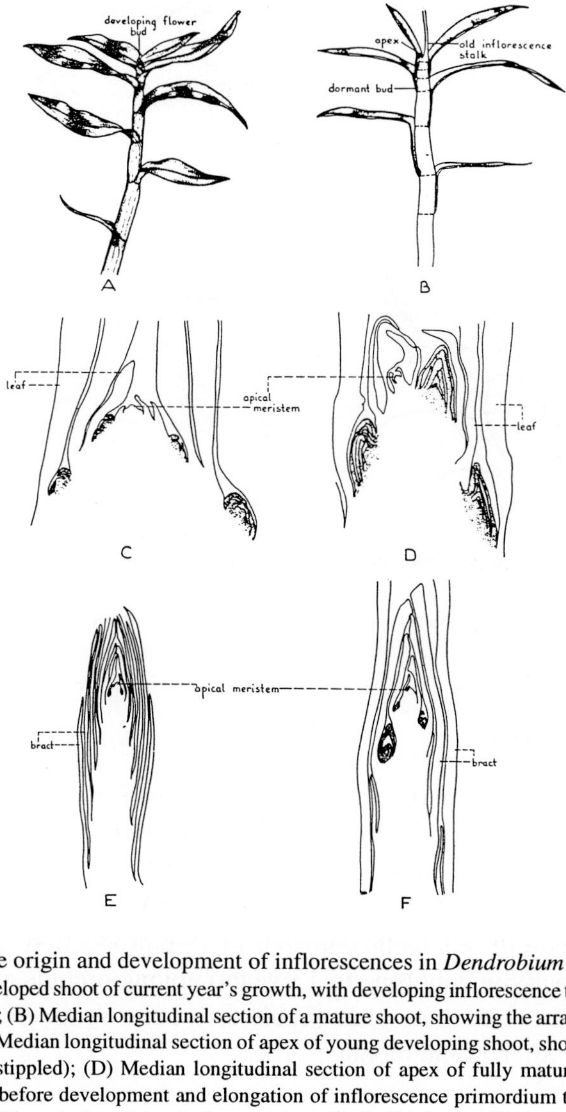

Fig. 6.3. The origin and development of inflorescences in *Dendrobium phalaenopsis*.
Note: (A) Fully developed shoot of current year's growth, with developing inflorescence that appears terminal but is really axillary; (B) Median longitudinal section of a mature shoot, showing the arrangement of dormant bud primordia; (C) Median longitudinal section of apex of young developing shoot, showing differentiation of bud primordia (stippled); (D) Median longitudinal section of apex of fully matured shoot of current year's growth, just before development and elongation of inflorescence primordium that precedes flower bud differentiation. The apical meristem is degenerating; (E) Median longitudinal section of inflorescence shoot at time of flower bud differentiation. The axis elongates before the first floral primordia are differentiated. (F) Median longitudinal section of the upper part of inflorescence shoot, showing further differentiation and development of flower buds.

Reproduced from Rotor (1952), courtesy of New York State College of Agriculture, Cornell University.

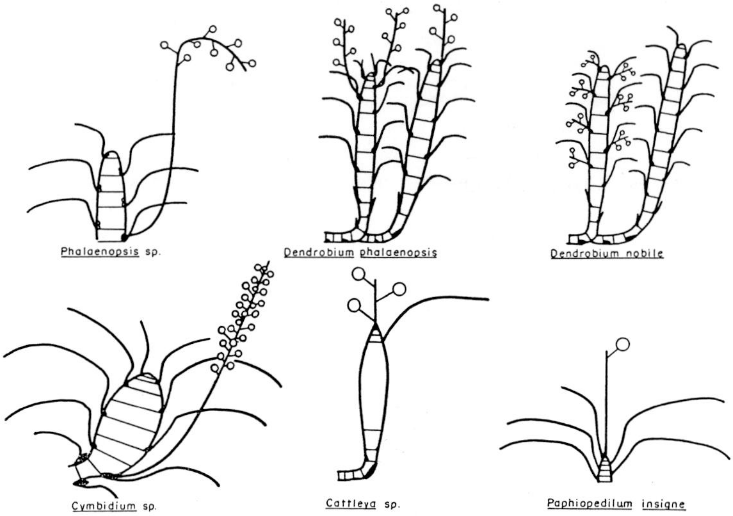

Fig. 6.4. Flowering characteristics of some orchids.

Reproduced from Rotor (1952), courtesy of New York State College of Agriculture, Cornell University.

Juvenility in orchids

Juvenility refers to the early phase of plant growth during which flowering cannot be induced by any treatment. It is an important phase that controls the changes from vegetative to reproductive growth. Physiologists believe that it is a device to ensure that flowering does not occur until the plant is large enough to support the energetic demands of seed production. The duration of juvenility can vary widely (one to 13 years) among orchids and the average time is between two to three years (Table 6.1). Most commercially important hybrids flower after 12–36 months.

Response to low temperature

Flower induction under low temperature in non-orchidaceous plants is well-documented. Tropical orchids are no exception, though the number of cases

Table 6.1. Time from seed sowing to flowering in some orchid hybrids.

Orchid hybrid	Juvenile period (years–months–days)
Arachnopsis Eric Holttum	7–5–2
Aranda Hilda Galistan	4–8–8
Aranda Lucy Laycock	13–3–0
Aranda Wendy Scott	7–10–6
Aranthera Anne Block	5–10–5
Aranthera Beatrice Ng	6–1–3
Burkillara Henry	5–9–22
Cymbidium Faridah Hashim	5–0–20
Dendrobium Sarie Marijs	3–4–10
Dendrobium Lin Yoke Ching	8–2–12
Holttumara Cochineal	8–0–24
Laeliocattleya Cheah Chuan Keat	6–7–14
Paphiopedilum Shireen	8–5–0
Renantandra Storiata	9–3–1
Spathoglottis Penang Beauty	2–11–13
Vanda Miss Joaquim	3–1–5
Vanda Ruby Prince	3–4–16
Vanda Tan Chin Tuan	8–4–2

Adapted from Wee (1971).

studied is less than that of other plants. *Cymbidium* hybrids, *Phalaenopsis schilleriana* and *Polystachya culiviformis* are some of the best examples that require low temperature for flower bud initiation. In Malaysia, orchids such as

Cymbidium roseum and *Paphiopedilum barbatum* grown naturally in highlands rarely flower when grown in the lowlands. In Singapore Botanic Gardens, *Paphiopedilum barbatum* plants collected from various parts of Malaysia highlands (986 m to 1903 m above sea level) flower profusely in a cold house kept at 23.8°C to 26.8°C and relative humidity of 75% to 91%. They flower throughout the year; there are at least two flower stalks on each plant and, in some cases, there are four to six flower stalks on a plant.

Dendrobium crumenatum (dove or pigeon orchid) is one of the best known example of tropical orchids which flowers under low temperature induction. The terminal inflorescence of *Dendrobium crumenatum* produces flower buds that develop until the anther is almost fully grown and all other segments are formed. This flower buds then undergo dormancy. Development resumes in the dormant flower buds after a sudden drop in temperature of 5°C. In South East Asia, this is often provided by a rain storm. Flowering takes place nine days after the rainstorm. The reason for this induction remains unclear. Hydration of flower buds and low temperature induction are two possible reasons. Unlike some other non-orchidaceous plants, the cold treatment requirement for orchids cannot be replaced by gibberellin (GA). There are quite a number of orchids that respond in a similar manner as *Dendrobium crumenatum* to low temperature. The difference between these orchid species only lies in the number of days to flowering following a rain storm.

The duration of low temperature treatment and the difference in day/night temperature are important consideration for flower initiation. Thermoperiodic flower induction in orchids is best illustrated by *Cymbidium* and *Phalaenopsis*. Hybrids of *Cymbidium* require a period of cool night and warm days for flower induction. *Phalaenopsis amabilis*, in contrast, requires more pronounced day/night temperature fluctuation.

Response to low temperature in orchids is further complicated by an interaction between temperature and light. Rotor (1952) conducted pioneering research on flower induction in orchids. He found that cymbidiums would flower only at 21°C and the degree of response to low temperature is determined by light intensity. Response to low temperature induction in *Dendrobium nobile*, on the other hand, is not affected by light.

So far most of the reports dealt with the phenomena of flower induction in response to low temperature treatment. Detailed study of the nature of induction

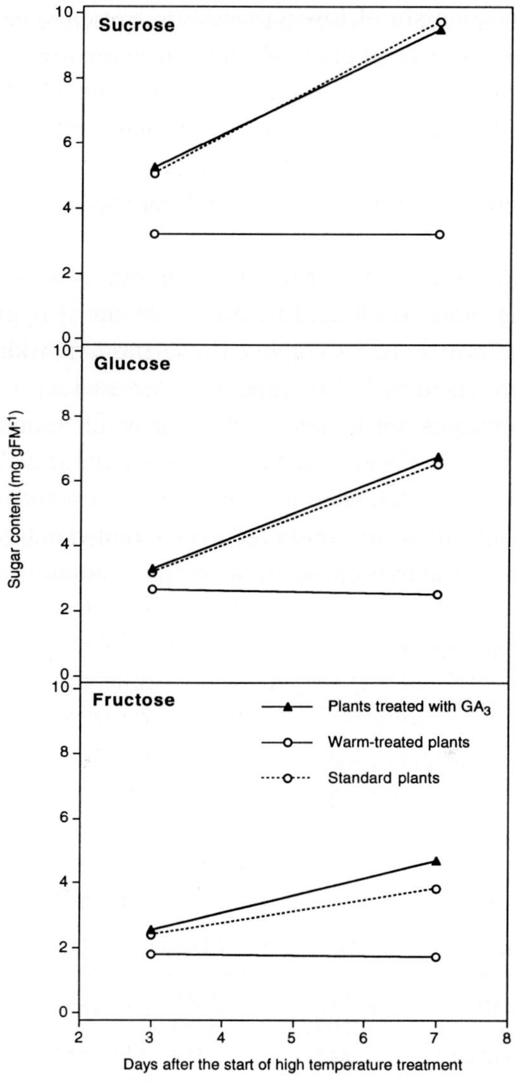

Fig. 6.5. Sucrose, glucose and fructose content in the buds of *Phalaenopsis amabilis* infloresecences.

Note: In warm treated plants (30°C day and 25°C night), flowering is fully inhibited unless given GA_3 treatment. Standard plants refer to the flowering plants grown under optimal conditions (25°C day and 20°C night).

Redrawn from Chen, Liu, Liu, Yang & Chen (1994).

is lacking. The mechanism of low temperature response remains unclear. A study on the endogenous level of cytokinins following low temperature would be interesting. In non-orchidaceous plants, for example, in *Boronia*, it has been reported recently that a transient increase in cytokinin (zeatin riboside and dehydrozeatin riboside) concentration occurs in roots and tissues of *Boronia* within days of transferring the plants to cool inductive conditions (17°C day and 9°C night).

Sugar may play a role in flowering of *Phalaenopsis*. Flowering of *Phalaenopsis amabilis* is induced by cold treatment. When it is grown under high temperature (30°C day and 23°C night), flowering is blocked. However, this blockage can be reversed by GA treatment. Associated with the GA treatment under high temperatures are increases in sugar levels (sucrose, glucose and fructose) (Fig. 6.5). It has been concluded that sucrose translocation from the source leaves to the inflorescence increases the sink activity. An increase in sucrose synthase activity is also observed in GA treated plant (Table 6.2). This observation supports the hypothesis of hormonal mediated nutrient diversion as a mode of action by which exogenously applied GA could bring about flowering in *Phalaenopsis*.

Table 6.2. Activities of sucrose synthase and acid invertase in the apical buds of *Phalaenopsis amabilis* inflorescences.

	Sucrose synthase (nmol min^{-1} gFM^{-1})		Acid invertase (nmol min^{-1} gFM^{-1})	
Treatment	3 Days	7 Days	3 Days	7 Days
Treated with GA_3	3513	5688	434	466
Standard plant	3609	5783	412	440
Warm-treated plant	1442	1453	421	421

Note: In warm treated plants (30°C day and 25°C night), flowering is fully inhibited unless given GA_3 treatment. Standard plants refer to the flowering plants grown under optimal conditions (25°C day and 20°C night).

Redrawn from Chen, Liu, Liu, Yang & Chen (1994).

Photoperiodic response

Garner and Allard (1923) first established that response to daylength is one of the major controlling factors in flowering. Since then, photoperiodism has been a subject of intensive research. Plant response to daylength is divided into three groups: Short day plants (SDP), long day plants (LDP) and day neutral plants (DNP). Like other plants, orchids can also be classified as SDP, LDP and DNP.

According to Sanford (1974), tropical plants of equatorial origin are believed to be more sensitive to small differences in daylength than those from temperate regions. Such sensitivity would confer an evolutionary advantage since the daylength differences is less pronounced in the tropics. However, tropical orchid hybrids like *Arachnis* Maggie Oei, *Aranda* Deborah, *Aranda* Wendy Scott, *Vanda* Miss Joaquim, as well as several *Dendrobium* hybrids, are all day neutral plants and are indifferent to daylength.

Hormonal control

Many *Aranda* hybrids exhibit a flowering gradient. Decapitation of shoot apex in these orchids produces axillary shoots. The nature of these axillary shoots, whether vegetative or reproductive, is correlated with their position along the stem axis. Buds near the apex usually develop into inflorescences, whereas those situated further away from the apex develop into vegetative shoots. Therefore, there exists a gradient in which the flowering capacity is greatest near the apex and the capacity diminishes basipetally along the monopodial stem axis. Among the *Aranda* hybrids, the only difference lies in the extent of flowering capacity down the gradient. For *Aranda* Lucy Laycock, and *Aranda* Meiling, decapitation of the 15th or 16th node will produce an inflorescence; for *Aranda* Hilda Galstan and *Aranda* Nancy, decapitation of the 18th–20th node will produce an inflorescence; for *Aranda* Deborah, decapitation of the 25th node will also produce an inflorescence. It is not known whether this difference among *Aranda* hybrids is due to the genetic or nutritional differences. Flowering gradient has also been observed in other monopodial orchids such as *Holttumara* Maggie Mason and *Aranthera* James Storie. The occurrence of flowering gradient in monopodial orchids appears to be widespread. However, flowering gradient in monopodial orchids is not unique as the same has been reported in other plants such as strawberry.

There is evidence to indicate that flowering gradient is hormonal in nature. Auxin has been implicated to play an important role. Application of an anti-auxin, e.g., Triodobenzoic acid (TIBA), growth retardant (B-nine) or a cytokinin (6-benzylaminopurine, BAP), releases the buds from apical dominance. The effect of cytokinin on flowering is enhanced by gibberellin in some orchids.

Substantial evidence for endogenous hormones in flowering of thick leaved monopodial orchids is shown recently when Zhang and coworkers (1995) identified that root tips of aerial roots constitute an important source of cytokinins (mainly isopentenyladenosine), auxin and abscisic acid. Higher levels of endogenous cytokinin are detected in the root tips of flowering plants than those in non-flowering plants of *Aranda* Noorah Alsagoff (Fig. 6.6).

Further evidence of hormonal control on flower induction in orchids comes from three sources. Application of cytokinin (10^{-4} M of BAP) also induces flowering of sympodial orchids such as *Dendrobium*. Success with gibberellins on flower induction of orchids such as *Cattleya* and *Cymbidium* is more variable. The fact that orchids such as *Dendrobium* can be induced to flower by either cytokinin treatment or following an exposure to low temperature suggests strong correlation between the two as mentioned earlier. Similarly, non-orchidaceous plants, which require low temperature as a stimulus for flowering, also show increases in endogenous cytokinin levels following low temperature treatment.

The third source of evidence for hormonal control of flowering comes from the effect of light intensity on flowering of *Vanda* Miss Joaquim. A correlation between light intensity and flowering is observed (see Chap. 7 on Partitioning of Assimilates and Table 7.13). Under high irradiance, there are more flowers produced. The analysis of endogenous levels of auxin by bioassay shows that the levels of endogenous auxin are lower in *Vanda* plants growing in high irradiance, while plants growing in lower irradiance have higher levels of auxin. However, simultaneous measurement of changes in endogenous cytokinins under varying sunlight has not been carried out in this experiment. It is the ratio of cytokinin/auxin rather than the absolute amount of auxin that is an important determinant of flowering.

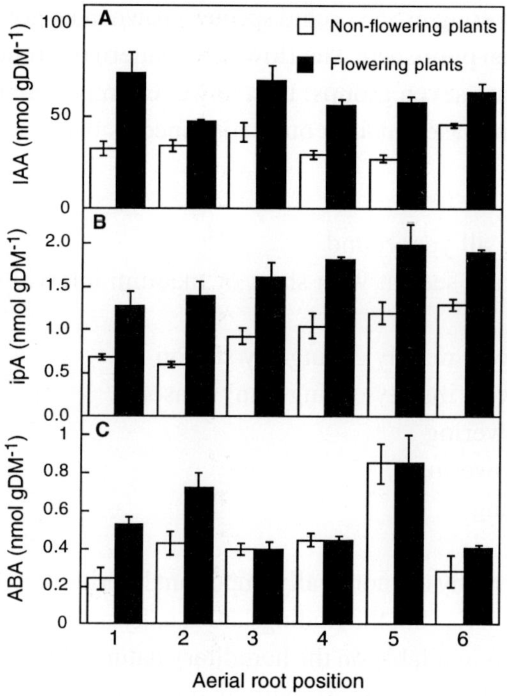

Fig. 6.6. The level of endogenous plant hormones in aerial root tips at different positions along the stem of *Aranda* Noorah Alsagoff at the flowering and non-flowering stage.
Note: (A) IAA; (B) iPA; (C) ABA. Mean of five replicates ±SE. Roots at six positions along the stem were selected and harvested at 1700 h.

Redrawn from Zhang, Yong, Hew & Zhou (1995).

6.4. Seasonality in Flowering

An important aspect of commercial orchid cut-flower production is to have flower production coincide with market demand. Induction of flowering and seasonality of flowering are important factors that determine the pricing and marketability of a popular cut-orchid flower. A detailed record of the flowering months of orchids in the northern hemisphere for over 150 years has been compiled by Hamilton (1990). More recently, a comprehensive survey of seasonality of flowering in 553 orchid species grown in Bogor Botanic Gardens, Indonesia, has been published. The flowering pattern of orchids studied can be broadly divided into seven groups. It is, however, important to note that many of these studies are based mainly on orchid species that are not grown for the cut-flower market.

1. Free flowering all year round
2. A long flowering season with short or medium interval of non-flowering period
3. Seasonal, flowers mainly during dry season
4. Seasonal, flowers mainly during rainy season
5. Regular in flowering
6. Sporadic in flowering
7. Rarely blooming

Seasonality of many economically important hybrids may have been studied in commercial nurseries but the findings are usually not published. Surprisingly, little information is available on the hereditary nature of flowering seasonality. Often, the period of study of seasonality in orchid flowering is one year. There is evidence to indicate that flowering peaks change with years of cultivation.

The growth and flowering production of *Aranda* Christine cultivar 130 grown under tropical conditions has been studied. The large population of 25,703 *Aranda* plants grown over a period of three years (1980–1983) gives a fairly reliable record of the seasonality of flowering in this orchid hybrid. Top cuttings that measured 60 cm with 2–3 roots are planted in a single row. The planting density is about 22 plants per 4.2 m^2, a common practice adopted by most nurseries. On the average, *Aranda* plants produce 2.3 leaves and gain 3.4 cm in height per month for the first 19 months after transplanting. Flowering begins five months after planting in October 1980. Sizable production of flowers is only evident in June 1981. There are three major peaks of flowering in 1981: April, July and October. For 1982, there is a shift in the peaks: April, June and December. For 1983, a peak at March–April is observed (Table 6.3). However, the large-scale field study on *Aranda* is concluded prematurely in May 1983 because of land redevelopment.

Table 6.3. Large scale flower production of *Aranda* Christine 130 under field conditions.

	Flower production (inflorescence $month^{-1}$)		
Month	1981	1982	1983
January	3,000	12,901	7,003
February	2,000	5,232	5,181
March	556	12,765	24,123
April	23,259	29,360	24,335
May	5,393	6,370	
June	5,278	40,200	
July	15,067	4,842	
August	5,992	3,830	
September	9,380	3,937	
October	19,137	5,602	
November	9,222	11,647	
December	7,163	25,696	
Total	105,447	162,382	60,645

Note: Total number of plants were estimated to be 25703 and these were planted in June 1980. The experiment was terminated in May 1983.

Redrawn from Hew & Lee (1989).

The flowering season of the sympodial orchid hybrid *Oncidium* Goldiana over a period of two years has also been reported. In this study, 500 pots of plants are used. There is a shift in the monthly peak flower production for *Oncidium* Goldiana during 1974 and 1975 (Fig. 6.7). Monthly flower production in *Oncidium* Goldiana fluctuates and there is a tendency for high flower production to be followed by a period of low flower production. This is consistent with the suggestion that the low flower production following high production in *Oncidium* Goldiana is attributed to depletion of storage reserves. Moreover, it has been shown that partitioning of assimilates to the inflorescence in *Oncidium* Goldiana is source-limited (see Chap. 7 on Partitioning of Assimilates).

More recently, various factors (daylength, temperature, solar radiation) controlling flowering in *Dendrobium* Jacquelyn Thomas flowering are evaluated

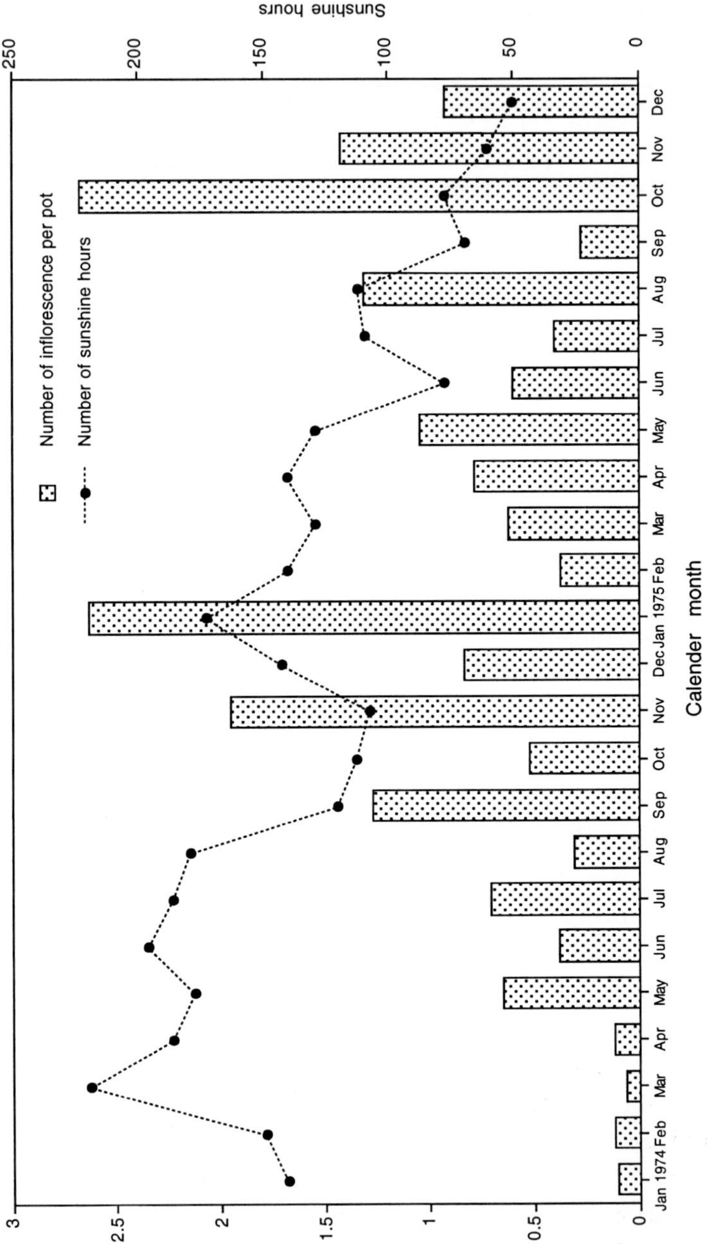

Fig. 6.7. Monthly flower production of *Oncidium* Goldiana and the total number of sunshine hours in Serdang, Selangor, Malaysia during 1974 and 1975.

Note: 500 pots of plants were used in this experiment.

Redrawn from Ding, Ong & Yong (1980).

over a period of five years in Hawaii. The number of flower spikes per plant increases as the plants age, reaching a maximum at 3–4 years and then declines. Within a year, flowering in *Dendrobium* peaks in the summer and late summer periods. The seasonality in flowering is not very consistent over the five years. Double flowering peaks are not uncommon. The peak of flowering in late summer becomes less pronounced as the plants age

6.5. Application of Flower Induction at the Commercial Level

From a commercial point of view, inducing an orchid plant to flower is not the only aspect for consideration. In order for flower induction to be commercially viable, the following conditions must be satisfied.

(1) The method must be simple, economical and give reproducible results.
(2) The quantity and quality of flowers must not be affected.
(3) There should not be any adverse effects on the plant or on subsequent flowering.

Various attempts have been made to put into practice the known scientific methods of flower induction in orchids. Some of the methods suggested for control of flowering in selected tropical orchids are listed below.

(a) Decapitation/incision
(b) Application of chemicals
(c) Bud removal
(d) Environmental control

Decapitation is used to control flowering of monopodial orchids (Goh and Arditti, 1985). For example, *Aranda* hybrids could be decapitated 10 weeks before the desired blooming date. Alternatively, an incision can be made halfway through the stem and the two parts held together with tape. Decapitation is only suitable for plants due for replanting as this method caters to a one-harvest crop. This method is impractical for normal flower induction on orchid farms.

By comparison, the regulation of flower production by the incision method is deemed to be more appropriate by some researchers as there is a possibility of repeated incisions at scheduled intervals for production to meet market demands. The incision method works well for some monopodial orchids (*Aranda* Peter Ewart) but not with others such as *Aranda* Christine 1. For the latter, there is no significant difference in the number of inflorescences and vegetative shoots produced between treatments. The reason for the discrepancy between the two *Aranda* hybrids is not clear. It could be related to the physiological state of the plants. *Aranda* plants produce inflorescences after treatment with BAP only during or just before the flowering season. At other times, exogenous application of BAP causes the plants to produce vegetative shoots. Therefore, due consideration must be given to the seasonality of flowering. The seasonality of flowering in orchids may be attributed to the level of assimilates/carbon available for flower development. The overall carbon status of an orchid may be low after a flowering period and time is needed for replenishment of carbon through photosynthesis. This postulation is supported by the observation of periodic flowering in *Aranda*, *Dendrobium* and *Oncidium*.

The failure of buds initiated by decapitation or incision to develop into flowers cannot be overlooked. About 20% to 30% of buds initiated following decapitation do not develop to maturity. Vase-life and marketability of flowers produced by decapitation or the incision methods are also not known. The long-term effects of incision on flowering, such as flower quality, quantity and seasonality of flowering, are not clear.

Chemical regulation of flowering in orchids has been explored. Application of hormones is done by injection, lanolin paste or water spray. The spraying method is found to be most practical for large-scale operations. Injection of hormones using hypodermic syringe is considered too laborious and impractical for commercial growers. Lanolin paste is also not suitable due to the sticky nature of lanolin. Among the various plant growth regulators, BAP appears to give a consistent effect on flower induction in orchids. BAP stimulates flowering of *Aranda* Deborah, *Dendrobium* Lousiae Dark and *Aranthera* James Storie. More recently, the control of flowering by BAP has also been extended to other monopodial orchids, such as the *Aranda* Kooi Choo, *Holttumara* Loke Tack Tip, *Mokara* Chark Kuan, *Aranthera* Beatrice Eng and sympodial orchids such as *Dendrobium* Mary Mak, *Dendrobium* Madam Uraiwan, *Dendrobium* Jacquelin Concert × Jester and *Oncidium* Grower Ramsey. In the latter study, spraying is carried out at three intervals. For the monopodial orchids, the highest response is obtained in the first spray (Table 6.4). Poor response is obtained in spray 2 (day 163) and spray 3 (day 330). This is attributed to the lack of new bud available for flower initiation as most of the bud has already been initiated. In contrast, the response of sympodial orchids to hormone application increases with increasing number of spray application, the highest response being during the last spray (Table 6.5). In the sympodial orchids, there are dormant buds at certain nodes that are only initiated if the main bud is damaged. For example, the bud subtended by leaf L4 of *Oncidium* Goldiana may develop into an inflorescence. It therefore appears that the nature of response to BAP spray depends on the growth habit of orchids.

To improve bud initiation and the number of developed bud in both the monopodials and sympodials, a vigorous vegetative growth is highly desirable. The number of new shoots and maturity of the pseudobulb are equally important. Special attention must be given to proper fertiliser application. The timing and rotational application of fertiliser with high nitrogen or high potassium and/or phosphate are critical. Some *Aranda* (e.g., *Aranda* Christine 130) have been observed to change from a very young flower shoot to vegetative shoot after high nitrogen application. Evidently, the C/N ratio needs to be considered during flowering.

It is noteworthy that application of BAP with concentration higher than 200 mg $litre^{-1}$ tends to increase the frequency of deformed flowers in the sympodial orchids. Also, the percentage of buds that developed to flower sprays is generally low, i.e., less than 50% of all the BAP treated plants. It is interesting that buds are induced to flower by BAP only when the control also produces flowers. This is true for both the monopodial and sympodial orchids. For *Dendrobium*, BAP does not induce floral bud initiation on the newly developing pseudobulb where vegetative growth is taking place. These studies highlight some of the problems met in chemical induction of flowering in orchids and emphasise the importance of the physiological state of the plants during experiments. It is well documented that the levels of endogenous plant hormones in plants are affected by physiological and environmental factors.

Table 6.4. Effect of Cytokinin (BAP) on flowering in monopodial orchids.

Spraying date	Parameter	BAP concentration			
		0 ppm	200 ppm	400 ppm	800 ppm
Spray 1 (day 1)	No. of initiated buds per plant	3.5[b]	7.5[a]	7.6[a]	8.6[a]
	No. of developed sprays per plant	1.5[a]	1.8[a]	2.0[a]	2.5[a]
	Percentage of buds which developed to sprays	51[a]	30[a]	33[ab]	33[ab]
	Average spray length (cm)	46[a]	48[a]	46[a]	46[a]
	No. of flowers per spray	14.2[a]	14.3[a]	14.2[a]	13.7[a]
Spray II (day 163)	No. of initiated buds per plant	1.9[a]	1.9[a]	1.8[a]	1.5[a]
	No. of developed sprays per plant	0.9[a]	0.2[a]	0.3[a]	0.4[a]
	Percentage of buds which developed to sprays	55[a]	27[a]	25[a]	27[a]
	Average spray length (cm)	46[a]	38[b]	48[a]	42[ab]
	No. of flowers per spray	15.3[a]	11.2[a]	13.1[a]	13.1[a]
Spray III (day 330)	No. of initiated buds per plant	1.3[a]	1.1[a]	1.2[a]	1.7[a]
	No. of developed sprays per plant	0.8[a]	0.5[a]	0.5[a]	0.6[a]
	Percentage of buds which developed to sprays	66[a]	40[a]	49[a]	38[a]
	Average spray length (cm)	47[a]	46[a]	44[a]	44[a]
	No. of flowers per spray	13.4[a]	14.5[a]	12.5[a]	12.9[a]

Note: Figures with a different letter differ significantly based on Duncan's Multiple Range Test at $P = 0.05$. The figures represent mean values obtained from four monopodial orchid hybrids: *Aranda* Kooi Choo, *Holtumara* Loke Tuck Yip, *Aranthera* Beatrice Ng and *Mokara* Chark Kuan. It will be more useful if the data were presented for individual hybrids by the authors.

Redrawn from Zaharah, Saharan & Nuraini (1986).

Table 6.5. Effect of Cytokinin (BAP) on flowering in sympodial orchids.

Spraying date	Parameter	BAP concentration			
		0 ppm	200 ppm	400 ppm	800 ppm
Spray 1 (day 1)	No. of initiated buds per plant	1.1^{c}	1.7^{bc}	1.9^{ab}	2.7^{a}
	No. of developed sprays per plant	$0,8^{b}$	$0,8^{b}$	1.1^{b}	1.6^{a}
	Percentage of buds which developed to sprays	43^{a}	42^{a}	45^{a}	45^{a}
	Average spray length (cm)	41^{ab}	45^{a}	40^{ab}	36^{b}
	No. of flowers per spray	15.5^{a}	14.5^{a}	14.5^{a}	16.0^{a}
Spray II (day 104)	No. of initiated buds per plant	2.4^{b}	7.3^{a}	5.6^{ab}	9.0^{a}
	No. of developed sprays per plant	$0,8^{b}$	2.0^{a}	1.7^{a}	1.9^{a}
	Percentage of buds which developed to sprays	28^{ab}	37^{a}	25^{a}	20^{b}
	Average spray length (cm)	45^{ab}	46^{a}	44^{ab}	41^{b}
	No. of flowers per spray	16.5^{a}	15.8^{ab}	14.4^{ab}	13.0^{b}
Spray III (day 217)	No. of initiated buds per plant	1.9^{c}	3.4^{b}	3.7^{b}	6.3^{a}
	No. of developed sprays per plant	0.2^{c}	1.1^{b}	0.8^{b}	1.7^{a}
	Percentage of buds which developed to sprays	7^{b}	24^{a}	23^{a}	22^{a}
	Average spray length (cm)	58^{a}	49^{b}	58^{a}	46^{b}
	No. of flowers per spray	15.3^{a}	15.1^{a}	17.1^{a}	15.5^{a}

Note: Figures with a different letter differ significantly based on Duncan's Multiple Range Test at $P = 0.05$. The figures represent mean values obtained from four monopodial orchid hybrids: *Dendrobium* Mary Mak, *Dendrobium* Madam Uraiwan, *Dendrobium* Jaquelyn Concert × Jester and *Oncidium* Gower Ramsey. It will be more useful if the data were presented for individual hybrids by the authors.

Redrawn from Zaharah, Saharan & Nuraini (1986).

Generally, GA_3 is ineffective in inducing flowering though it is known to increase spike length and flower size. There are reports that exogenous application of gibberellins accelerates flowering in *Cymbidium*. Abscisic acid (ABA) generally inhibits flowering in tropical orchids. At high concentrations (250–500 mg litre^{-1}), ABA usually causes defoliation in orchids. It has been reported recently that ABA promotes *in vitro* flowering of *Cymbidium ensifolium. The* auxin antagonist TIBA and various growth retardants (maleic hydrazide, MH; chlormequat, CCC; daminozide, B.995) stimulate floral initiation, but many of the initiated buds fail to develop to maturity. Ethephon does not promote flowering and it causes defoliation at high concentrations. Paclobutrazol, a triazole growth regulator inhibiting gibberellin biosynthesis, has been reported to promote early flowering in *Dendrobium* although the plants have reduced growth and flower size. Paclobutrazol (50–400 ppm) and Uniconazole (250–300 µl litre^{-1}) do not affect the flowering date, but effectively restrict inflorescence growth in *Phalaenopsis*. Generally, foliar applied retardant treatments on orchids are less effective than dipping (Hew and Clifford, 1993).

6.6. Bud Drop

An area that deserves more research is the phenomenon of bud drop in orchids. There are frequent reports of bud yellowing or dropping in some popular tropical orchids grown for cut flowers; e.g., *Aranthera* Beatrice Eng and *Dendrobium* Sri Siam. Many farmers in ASEAN have stopped growing these two orchids because of bud drop. Bud drop may be genetical, physiological or pathological in nature. The random occurrence of bud drop in an inflorescence precludes infection as the cause of it. It has been suggested that in some orchids, bud yellowing could be related to death of pollinia at some point. There is a report that 2-napthoxyacetic acid (2-NOA) at 40 mg litre^{-1} prevents bud drop in *Dendrobium bigibbum*. It has been shown that GA_3 at 50 mg litre^{-1} overcomes ethephon-induced bud blasting in a *Cymbidium* hybrid. These studies indicate clearly the potential of using plant hormones to control bud drop in orchids (Hew and Clifford, 1993).

The lack of assimilates may be a possible factor in determining bud drop in orchids. Long-term studies on the flowering of *Dendrobium* Jacquelyn Thomas suggest that there is a significant correlation between the frequency of bud drop and spike length. The spike length of *Dendrobium*, as in other sympodial orchids such as *Oncidium* Goldiana, is dependent on the level of assimilate supply from the leaves. Based on the five-year study, the highest frequency of bud drop occurs in the period when the spike is usually the shortest (Fig. 6.8). The bud drop phenomenon in orchids, or commonly termed 'abortion' of young buds in other plants, may be a response to a lack of assimilates. Competition for available assimilates between the buds is minimal under optimal growing conditions where the leaves are capable of meeting the sink demands of all the actively growing buds. In sub-optimal conditions (e.g., low temperature, cloudy days or water deficit), certain buds may dominate the existing limited supply of assimilates and cause the abortion of the other buds that are poor competitors for assimilates. For example, the abortion of some tomato fruits under source-limiting conditions is due to the competition for available assimilates. The random nature of bud drop and the positive effects of hormones (e.g., increasing sink activity) in preventing bud drop may be adequately explained in terms of assimilate supply and demand. More work is needed to confirm this postulation.

6.7. Controlling Orchid Flower Production

The staggering of flower production can be controlled by the removal of flower buds at the appropriate time. For *Vanda* Miss Joaquim, successful deferment of flowering peak from summer to winter months is achieved by removing young inflorescence during the prescribed period. Substantial increase in *Vanda* flower production is observed two months after bud removal. Staggering of flower production has also been obtained for *Aranda* Christine that has three major flowering peaks: April, June, and December in 1982. By bud removal, one can shift the flower production peak for two weeks (Fig. 6.9). Furthermore, yield increase in *Aranda* Christine is also observed four months following bud removal. This increase in yield compensates for the loss of harvest immediately

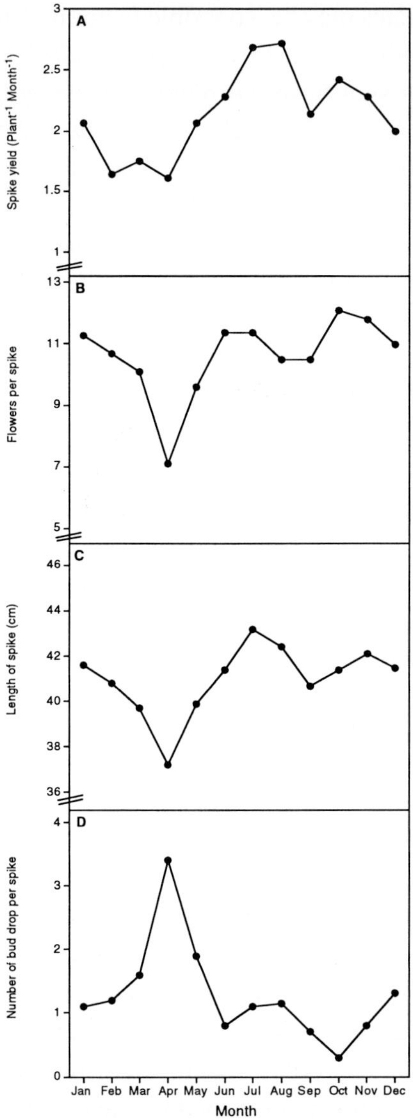

Fig. 6.8. Flowering characteristics of *Dendrobium* Jaquelyn Thomas in Hawaii.
Note: Five-year monthly means of: (A) Spike yield per plant; (B) Flowers per spike; (C) Length of spike; (D) Number of bud drop per spike.

Redrawn from Paull, Leonhardt, Higaki & Imamura (1995).

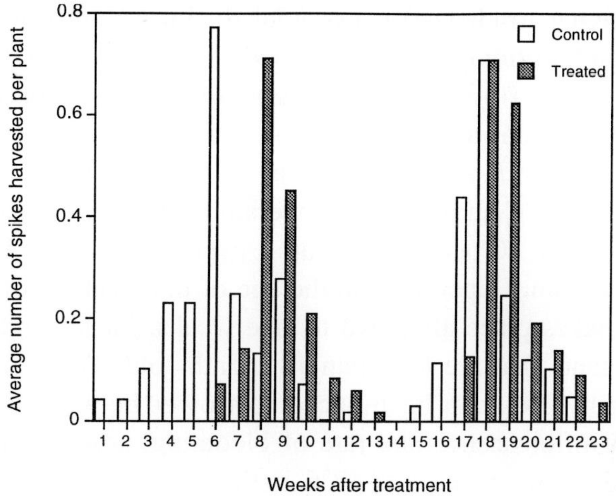

Fig. 6.9. Staggering of flower production in *Aranda* Christine 130 by removing young flower spikes.
Note: For the treated plants, bud removal was done when the spike length is about 5 cm or less.

Redrawn from Hew & Lee (1989).

after bud removal. The market quality and longevity of the control and the treated flowers are comparable. Thus, bud removal is a relatively safe and easy method for staggering flower production in *Aranda* Christine and *Vanda* Miss Joaquim.

From a practical point of view, climatic control of flowering holds considerable potential. Light intensity affects both the growth and flowering of tropical orchids. Members of the *Vanda–Arachnis* group such as *Vanda* Miss Joaquim and *Arachnis* Maggie Oei require extended period of full sunlight to flower while others such as *Oncidium* Goldiana show reduced flowering under high irradiance (Fig. 6.7). It is noteworthy that *Oncidium* Goldiana is a shade plant (see Chap. 3 on Photosynthesis and Fig. 3.5). The required level of irradiance can be regulated by shading. However, information of the optimal light requirement for proper growth and flowering of tropical orchids is lacking. In both the monopodial and sympodial orchids, harvestable yield is source limited and hence increasing the source or photosynthetic capacity (e.g., CO_2

enrichment, higher irradiance) is essential for improving flower production (see Chap. 7 on Partitioning of Assimilates).

The growth and flowering of *Phalaenopsis* have been studied under controlled environment using a phytotron in France. *Phalaenopsis amabilis* and *Phalaenopsis schilleriana* will flower when the night temperature varies between 12°C to 17°C and the day temperature does not exceed 27°C. Night temperature below 12°C or day temperature over 27°–30°C are less favourable for flower induction. Depending on the age of the plant, flower initiation in *Phalaenopsis* takes place after two to five weeks. Flower induction is also observed in plants that are less than 18 months old. There are reports that *Phalaenopsis* hybrids can be induced to flower between 14°C to 25°C. Elongation of inflorescence and rate of blooming, however, require higher temperature (25°C in the day and 17°C in the night).

The fact that flowering can be induced by low temperature has contributed significantly to the large-scale production of *Phalaenopsis* cut-flowers and potted plants in Taiwan and Japan recently. *Phalaenopsis* hybrids are first cultivated in lowlands. When they reach a certain stage of development, they are transported to cooler places in the highlands during the summer for flower induction. The age of the plant, timing of transfer to highlands and their eventual return to the warmer lowlands are critical for flower induction and development. Flowering induction of *Phalaenopsis* hybrids grown under greenhouses can be achieved by cooling the greenhouses, but the high energy requirements make it economically unattractive.

6.8. Concluding Remarks

Considerable progress has been achieved in our understanding of flowering in orchids. Most of the published reports focused on the phenomenon of flower induction especially for *Phalaenopsis* and *Cymbidium*. We still lack information on the nature of flower induction in orchids. More extensive research is needed before we can develop a commercially sound and viable method for the control of flowering in the economically important tropical orchid hybrids such as *Aranda*, *Mokara*, *Oncidium* and *Dendrobium*.

6.9. Summary

1. Orchid inflorescence is either terminal or axillary in nature.
2. The duration of juvenility varies widely among orchids. An orchid may take between one and 13 years to flower. The average time taken by most orchids is between two and three years.
3. Thermoperiodic flower induction in orchids is well-illustrated by *Cymbidium* and *Phalaenopsis*. However, the nature of low temperature induction of flowering remains unclear. There is evidence to indicate that GA_3 can reverse the high temperature inhibition and sugar may play a role in flowering induced by low temperature.
4. Commercially important tropical orchids for cut-flower production such as *Aranda*, *Dendrobium*, *Mokara* and *Oncidium* are all day neutral plants and are indifferent to daylength.
5. Many monopodial orchids such as *Aranda* and *Vanda* exhibit a flowering gradient along the stem.
6. Some of the methods suggested for the control of flowering in selected tropical orchids are (1) decapitation/incision, (2) application of chemical (e.g., BAP), (3) bud removal and (4) environmental control (e.g., temperature and light).

General References

Atherton, J. G., 1987, *Manipulation of Flowering* (Butterworths, London), 438 pp.

Bernier, G., 1988, "The control of floral evocation and morphogenesis," *Annual Review of Plant Physiology and Plant Molecular Biology* **39**: 175–219.

Garner, W. W. and Allard, H. A., 1923, "Further studies in photoperiodism, the response of plants to relative length of day and night," *Journal of Agricultural Research* **23**: 871–920.

Goh, C. J. and Arditti, J., 1985, "Orchidaceae," in *Handbook of Flowering, Vol. 1*, ed. A. H. Halevy (CRC press, Boca Raton), pp. 309–336.

Goh, C. J., Strauss, M. S. and Arditti, J., 1982, "Flower induction and physiology in orchids," *Orchid Biology: Reviews and Perspectives, Vol. VI*, ed. J. Arditti (Cornell University Press, Ithaca, New York), pp. 213–241.

Hamilton, R. M., 1990, "Flowering months of orchid species under cultivation," in *Orchid Biology: Reviews and Perspectives, Vol. V*, ed. J. Arditti (Timber Press, Portland, Oregon), pp. 265–405.

Hew, C. S., 1994, "Orchid cut-flower production in ASEAN countries," in *Orchid Biology: Reviews and Perspectives, Vol. VI*, ed. J. Arditti (John Wiley and Son Inc., New York), pp. 363–401.

Hew, C. S. and Clifford, P. E., 1993, "Plant growth regulators and the orchid cut-flower industry," *Plant Growth Regulation* **13**: 231–239.

Rotor, G. B. Jr., 1952, "Daylength and temperature in relation to growth and flowering of orchids," *Cornell University Agricultural Experimental Station Bulletin* **885**.

Rotor, G. B. Jr., 1959, "The photoperiodic and temperature response of orchids," in *The Orchids: Scientific Studies*, ed. C. L. Withner (Ronald Press, New York), pp. 397–417.

Sanford, W. W., 1974, "The ecology of orchids," in *The Orchids: Scientific Studies*, ed. C. L. Withner (Wiley-Interscience, New York), pp. 1–100.

References

Alphonso, A. G., 1978, "Growing *Paphiopedilum barbatum* under simulated conditions," *Proc. of the Symposium on Orchidology, Singapore*, The Orchid Society of South East Asia, Singapore, pp. 70–76.

Chen, W. S., Liu, H. Y., Liu, Z. H., Yang, L. and Chen, W. H., 1994, "Gibberellin and temperature influence carbohydrate content and flowering in *Phalaenopsis*," *Physiologia Plantarum* **90**: 391–395.

Chia, T. F. and Hew, C. S., 1987, "Effects of floral excision on reversion from reproductive to vegetative development in strawberry," *HortScience* **22**: 672–673.

Chin, T. Y., Chai, B. L. and Hew, C. S., 1989, "Occurrence of abscisic acid-like and gibberellins-like substances in tropical orchid flowers," *Malaysian Orchid Bulletin* **4**: 13–18.

Ding, T. H., Ong, H. T. and Yong, H. C., 1980, "Factors affecting flower development and production of Golden Shower (*Oncidium* Goldiana)," *Proc. of the Third ASEAN Orchid Congress* (Terusan Selatan, Kuala Lumpur, Malaysia), pp. 65–78.

Day, J. S., Loveys, B. R. and Aspinall, D., 1995, "Cytokinin and carbohydrate changes during flowering of *Boronia megastigma*," *Australian Journal of Plant Physiology* **22**: 57–65.

Goh, C. J., 1975, "Flowering gradient along the stem axis in an orchid hybrid *Aranda* Deborah," *Annals of Botany* **39**: 931–934.

Goh, C. J., 1977, "Regulation of floral initiation and development in an orchid hybrid *Aranda* Deborah," *Annals of Botany* **41**: 763–769.

Goh, C. J., 1979, "Hormonal regulation of flowering in a sympodial orchid hybrid *Dendrobium* Louisae," *New Phytologist* **82**: 375–380.

Goh, C. J. and Yang, A. L., 1978, "Effects of growth regulators and decapitation on flowering of *Dendrobium* orchid hybrids," *Plant Science Letters* **12**: 287–292.

Hew, C. S. and Lee, F. Y., 1989, "Control of flowering by floral bud removal in *Aranda* Christine under tropical field conditions," *Journal of the Japanese Society of Horticultural Science* **58**: 691–695.

Higuchi, H., Katano, Y. and Hara, M., 1975, "Temperature and light in relation to flowering of *Dendrobium* Nodoka," *Research Bulletin of the Aichi-Ken Agricultural Research Center (Series B)* **7**: 45–50.

Irawati, 1994, "Blooming season of orchid at Bogor Botanic Garden," *Buletin Kebun Raya Indonesia* (*Indonesian Botanic Garden Bulletin*) **8**: 1–15.

Koay, S. H. and Chua, S. E., 1979, "Evaluations and commercial applications of flowering potential in *Aranda* Peter Edward and *Aranda* Christine 1," *Singapore Journal of Primary Industry* **7**: 51–61.

Koay, S. H. and Chua, S. E., 1981, "Export-oriented orchid productions by chemical regulation of flowering," *Singapore Journal of Primary Industry* **9**: 93–100.

Lee, N. and Lin, G. M., 1987, "Control the flowering of *Phalaenopsis*," in *Proc. Symp. Forcing Culture Hort. Crops*, ed. L. R. Chang, Special Publication 10, Taichung District Agr. Improv. Sta., Taiwan, pp. 27–44.

Murashige, T., Kamemoto, H. and Sheehan, T. J., 1967, "Experiments on the seasonal flowering behavior of *Vanda* 'Miss Joaquim'," *Proc. of the American Society for Horticultural Science* **91**: 672–679.

Ohno, H., 1990, "High temperature-induced flower bud blasting in *Cymbidium*," *Proc. of the Nagoya International Orchid Show* (1990), pp. 125–128.

Ohno, H. and Kako, S.,1991, "Roles of floral organs and phytohormones in flower stalk elongation," *Journal of the Japanese Society of Horticultural Science* **60**: 159–165.

Ong, H. T., 1978, "Climatic influences over the flowering of orchids in Malaysia," *Proc. of the Symposium on Orchidology, Singapore*. The Orchid Society of South East Asia, Singapore, pp. 89–93.

Paull, R. E., Leonhardt, K. W., Higaki, T. and Imamura, J., 1995, "Seasonal flowering of *Dendrobium* 'Jaquelyn Thomas' in Hawaii," *Scientia Horticulturae* **61**: 263–272.

Sakanishi, Y., Imanishi, H. and Ishida, G., 1980, "Effect of temperature on growth and flowering of *Phalaenopsis amabilis*," *Bulletin of the University of Osaka Prefecture* (*Series B*) **32**: 1–9.

Sanford, W. W., 1971, "The flowering time of West African orchids," *Botanical Journal of the Linnean Society* **64**: 163–181.

Sinoda, K., 1994, "Orchid," in *XXIVth International Horticultural Congress*, eds. K. Konishi, S. Iwahori, H. Kitagawa and T. Yakuwa, Horticulture in Japan, Kyoto, 1994 (Asakura Publishing, Tokyo), pp. 161–165.

Tran Thanh Van, K. M., 1974, "Methods of acceleration of growth and flowering in a few species of orchids," *American Orchid Society Bulletin* **43**: 699–707.

Wang, Y. T. and Hsu, T. Y., 1994, "Flowering and growth of *Phalaenopsis* orchids following growth retardant applications," *HortScience* **29**: 285–288.

Wee, S. H., 1971, "Maturation period of pods and time taken for plant to flower," *Malayan Orchid Review* **10**: 42–46.

Zaharah, H., Saharan, H. A. and Nuraini, I., 1986, "Some experiences with BAP as a flower inducing hormone," *Malaysian Orchid Bulletin* **3**: 31–38.

Zhang, N. G, Yong, J. W. H., Hew, C. S. and Zhou, X., 1995, "The production of cytokinin, abscisic acid and auxin by CAM orchid aerial roots," *Journal of Plant Physiology* **147**: 371–377.

Chapter 7

Partitioning of Assimilates

7.1. Introduction

Information on assimilate partitioning between sources (net producers of assimilates, e.g., leaves) and sinks (net importers of assimilates, e.g., flowers) is essential for increasing the harvestable component of economically important plants. The harvestable yield is the result of carbon dioxide fixation and the subsequent allocation of fixed carbon and other assimilates into economically important yield components. For the orchid cut-flower industry, the flower is the harvestable organ. A thorough understanding on how assimilates are allocated among flowers, storage organs (e.g., pseudobulbs) and leaves is useful for the maximisation of harvestable yield in orchids.

This chapter will provide a basic introduction to the source–sink idea of phloem translocation, a general account of the patterns of assimilate partitioning in sympodial and monopodial orchids and to explore possible avenues to increase harvestable yield in orchids.

7.2. The Source–Sink Concept of Phloem Translocation

Plant growth is dependent on water, ions and organic solutes. While water and mineral ions are absorbed from the soil, most organic solutes are autotrophically synthesised either directly or indirectly in the leaves. There is a need to move the assimilates from the leaves to various organs for maintenance and development. Assimilate transport involves the exit of sugars from the chloroplast and entering into the receiving cells. Water is an important medium for transporting ions from the soil to the other plant parts, and to convey assimilates from the leaves to the growing organs. The transport of assimilates in the phloem from a source leaf to a sink is commonly termed 'assimilate translocation.'

Münch hypothesis is the currently accepted view for the movement of assimilates in the phloem. The pressure-driven mass flow hypothesis suggests that translocation through the sieve elements in the phloem is driven by a turgor pressure gradient between source and sink. Loading in the leaf increases the osmotic pressure of the source, while unloading in the sink organs lowers the osmotic pressure at the other end of the phloem. At the sink, the sugars are unloaded into the respective sink organs. The processes of phloem loading at the source and unloading at the sink are believed to produce the driving force for translocation. Translocation through the vascular system to the sink is referred to as long-distance transport (Wardlaw, 1990).

Sources and sinks

The source or sink status of a plant organ is dynamic and may change during development. A plant organ can be described as a source or sink organ, depending on its ability to export or import assimilates. A source may also be defined as an exporter of sugars to the phloem and a sink is an importer of sugar from the phloem. For most plants, healthy fully expanded leaves are the major sources or net producers of photoassimilates. Other green organs (or chlorophyll-containing organs) such as green stems, roots, floral and fruiting organs may also provide additional carbon through photosynthesis for growth. Non-foliar green organs such as pseudobulbs, stems, roots, fruit capsules and flowers are plentiful in orchids and these organs may contribute varying amounts of assimilate for growth. At the extreme of vegetative modification, shootless orchids are known to obtain their only source of carbon from its green photosynthetic roots. The common feature among these non-foliar green organs of leafy orchids is the inability to exhibit net photosynthesis (see Chap. 3 on Photosynthesis). This unique phenomenon could be because these organs solely perform regenerative photosynthesis in the presence of well-developed leaves.

The difference in the photosynthetic capacity of various plant organs can be explained by the relative cost effectiveness of investing scarce resource, in particular nitrogen, for autotrophic functions.

Sinks are present during all phases of a plant life-cycle. These may include shoot and root meristems, young and expanding leaves, flowers, fruits and cambia. Even primary sources like mature leaves are sinks during their early development and expansion. There is usually a priority for assimilate partitioning between different sinks. For example in flowering tomato plants, the priority for assimilate allocation follows this decreasing order: Inflorescence > young leaves > roots. During the fruiting stage, there is a change in the priority order, such that fruit > young leaves > flower > roots. In many plants, it has been found that fruit and seed growth generally dominate the growth of vegetative tissues, but under source-limiting conditions, flowers are poor competitors for assimilates (Wardlaw, 1990). Storage organs, at some point in the life cycle, may be considered as sink organs. For example, storage organs such as bulbs, tubers and corms are sinks during their early development but become sources at a later stage. Orchids have many types of storage organs that are peculiar to their habitat: Pseudobulbs, swollen roots and underground tubers.

Phloem loading

In the photosynthesizing leaf of a C_3 plant, sucrose is synthesised in the cytoplasm from triose-phosphates formed from photosynthesis in the chloroplast. Sucrose then moves from the mesophyll cell to the vicinity of sieve elements in the smallest veins of the leaf (short distance transport). There are two possible routes (symplastic or apoplastic pathway) by which sucrose from the mesophyll cell can be loaded into the sieve elements of the minor leaf veins for export. The specific pathway of phloem loading in both C_3 and CAM orchids is unknown. Sucrose is the main export form of assimilate translocated in the phloem in many higher plants. Recent evidence also indicates that sucrose is the dominant form of reduced carbon transported in thick-leaved orchids.

In most higher plants, the loading of sugar from the mesophyll cells into the sieve elements requires energy. There is also a demand for oxygen during phloem loading. For example, treating source tissues with respiratory inhibitors decreases ATP concentration and inhibits loading of exogenous sugar. Current evidence indicates that a sucrose/H^+ symporter is involved in generating a proton gradient necessary to move sucrose across the plasma membrane. In addition, phloem loading for most species is generally specific and selective in nature, attributed to the involvement of selective carriers for specific solutes. As a result, only certain nonreducing sugars are translocated in the sieve elements and these sugars vary for different species. However, not all substances transported in the phloem are actively loaded into the sieve elements and these include organic acids and plant hormones that probably enter the phloem through diffusion. At the whole plant level, phloem loading does not seem to limit phloem transport, given its capacity to increase over a wide range of sucrose concentrations, and to respond to rapid changes in source/sink ratio (Delrot and Bonnemain, 1985).

Along the path

Transport in the path is considered as a relatively passive process. The precise role of phloem in regulating assimilate partitioning is still unclear. Part of this problem lies in the difficulty in obtaining *in vivo* structural evidence for phloem function. Despite this uncertainty of phloem function, evidence from several studies has shown that the loss of assimilates during long-distance transport from the sieve element-companion cell complex is relatively small compared with the total amount translocated. Loading and unloading processes are responsible for the maintenance of a concentration gradient in the phloem between the source and the sink. This concentration gradient allows the movement of solutes by mass flow from the source to the sink.

Phloem unloading

Unloading begins with the exit of assimilates from sieve tubes in the sink region and is followed by lateral transport to the receiving cells. The pathways of phloem unloading are diverse and dependent on the type of sink tissue. The unloading process is considered to be a critical control mechanism at the whole-plant level and has been the focus of intense research during recent years. As with the phloem loading process, the unloading process may occur apoplastically or through plasmodesmata symplastically into sink cells. Unloading is typically symplastic in growing and respiring sinks such as young leaves and roots. The transported sugar moves through the plasmodesmata to the sink cell where it can be metabolised in the cytosol or vacuole before entering the metabolic pathways associated with growth. Assimilate unloading in most developing seeds occurs across the apoplast because there is no symplastic link between the phloem of the maternal tissues and the young embryo. The transported sugar may be partially metabolised in the apoplast by suitable enzymes in some species (e.g., sugar cane and corn) or remain unchanged while transversing the apoplast for others (e.g., sugar beet root and soybean seeds). Assimilate unloading to its apoplast generally involves some energy input and is influenced by the concentration of sugars in the apoplast as well as by other factors including plant hormones and turgor-sensing mechanism.

7.3. Patterns of Assimilate Movement in Most Higher Plants

Generally, the preferred vascular pathway for transport seems to be the one that offers least resistance to vascular connection. The overall pattern of phloem transport in most higher plants can be described as a source-to-sink movement following five generalizations (Kursanov, 1984; Wardlaw, 1990).

Sources usually supply nearby sinks

The upper mature leaves of a plant usually export assimilates to the growing shoot tip, young expanding leaves and other sinks. Lower leaves predominantly supply the root system whilst intermediate leaves export assimilates in both directions. For very large sinks such as fruits, the subtending leaf usually acts as the main supplier of assimilates for growth and development. A classic example is seen in cereals (such as wheat) where most of the assimilates for the developing ear comes from a single leaf — flag leaf, while the remaining leaves make little or no contribution to ear development but mainly contribute to the other plant parts such as roots.

Assimilate partitioning changes during plant development

Generally, roots and shoot apices are the dominant sinks during the vegetative stages, while flowers and fruits become the main sinks during the reproductive stages, especially for the adjacent and some other nearby leaves. For example, in young soybean plants with two or three leaves, the leaves supply assimilates to both shoot and roots. At this stage, there is no strict specialization of leaves. In older soybean plants before and during flowering, different groups of leaves (such as lower leaves, middle leaves and upper leaves) supply assimilates to specific regions along the stem axis.

Vascular geometry and phyllotaxy can affect partitioning pattern

Vascular geometry (the arrangement of vascular bundles) and phyllotaxy (the arrangement of leaves) affect the pattern of assimilate partitioning. In sugar beet, a mature leaf within the rosette of leaves exports assimilates to a specific segment of the storage root even though all source leaves and the root sinks are of equal distance from each other. In tomato, the basal leaves export assimilates to the upper stem and shoot apex, while the upper leaves export assimilates to the lower stem. This atypical pattern of assimilate partitioning is due to the complex bicollateral phloem system found in tomato. Furthermore, a sink that is near a source may not necessarily receive assimilates from it unless it is connected by vascular tissues. For example, in sunflower plants, the movement of assimilates follows phyllotaxic links between leaves and growing organs.

Fully expanded leaves do not import assimilates

Young, expanding leaves are growing sinks which import assimilates from other sources, while mature or fully expanded leaves do not usually import assimilates from other leaves. At some time during foliar sink to source transition, bi-directional transport within a leaf may be established where there are both influx and efflux of assimilates. The sink-to-source transition is complicated by the fact that maturation of different parts of the leaf blade is staggered in time. Generally, the expanding leaves of dicots and monocots start to export assimilates when they are about 50% and 90% of their final leaf area respectively. The bi-directional movement of assimilates is likely a result of movements in different bundles in the petiole.

Removal of sources can change the general pattern of translocation

Defoliation can alter the general pattern of translocation between sources and sinks. Upper source leaves on a plant can be forced to supply assimilates to the roots by removing the lower source leaves. However, the flexibility of the new pathway is dependent on the availability of vascular interconnections between the new source(s) and sink. For example, in chickpea (*Cicer arietinum*), individual branches on the plant are independent units for assimilate production and utilization. The leaves on the depodded branch cannot transport assimilates to the pods on an adjacent defoliated branch.

7.4. Patterns of Assimilate Movement in Tropical Orchids

Recent findings have indicated that tropical orchids follow a slightly different pattern of assimilate movement when compared to most higher plants. Both rules 1 and 4 are 'broken' by orchids. The patterns of assimilate partitioning in economically important orchids grown for their cut flowers is described briefly for sympodial orchids (e.g., *Dendrobium* and *Oncidium*) and monopodial orchids (e.g., *Aranda*).

Assimilate partitioning in the sympodial orchids

Sympodial orchids exhibit clonal growth where identical shoot or lead is produced sequentially from the bases of stems. As in other clonal plants, the physical connection among the shoots (or ramets) allows physiological integration of resources such as assimilates, water and minerals, although the magnitude and duration of resource sharing vary substantially among species. Each shoot is usually at a different stage of development from another shoot: New basal shoot stage, pseudobulb formation stage, flowering stage and non-flowering stage.

Thin-leaved sympodial orchids

The general patterns of assimilate partitioning in thin-leaved sympodial orchids is based on studies conducted for *Oncidium* Goldiana. The growth habit of *Oncidium* Goldiana is similar to the growth habit of other economically important *Oncidium* hybrids: *Oncidium* Taka and *Oncidium* Gower Ramsey. The complex growth habit of sympodial orchids makes it necessary to study the pattern of assimilate partitioning in a single shoot of a sympodial orchid before proceeding to a more complex system of connected shoots (Fig. 7.1).

Using single shoots of *Oncidium* Goldiana, the time course study indicates that the distribution of ^{14}C-assimilates to the different plant parts is similar for the different time intervals excluding the sixth hour (Fig. 7.2). The total ^{14}C-assimilates recovered from the whole plant (with the same test leaf L2) over seven days is similar, indicating low respiratory loss of ^{14}C-assimilates. This observation is in tandem with the observation that *Oncidium* Goldiana is a shade plant with a low leaf respiration rate of 0.21 $\mu mol\ m^{-2}\ s^{-1}$. A transport time of 33 hours after $^{14}CO_2$ feeding is therefore suitable for studying the pattern of ^{14}C-assimilates within a single shoot of *Oncidium* Goldiana. The percentage of ^{14}C-assimilates exported from test leaves is significantly low (4%) at the sixth hour but the export increases to a higher and constant level of 51% beyond 33 hours. The rate of export of ^{14}C-assimilates in *Oncidium* Goldiana appears to be slower when compared with other C_3 non-orchidaceous plants. Within a single shoot, all test leaves of each growth stage generally

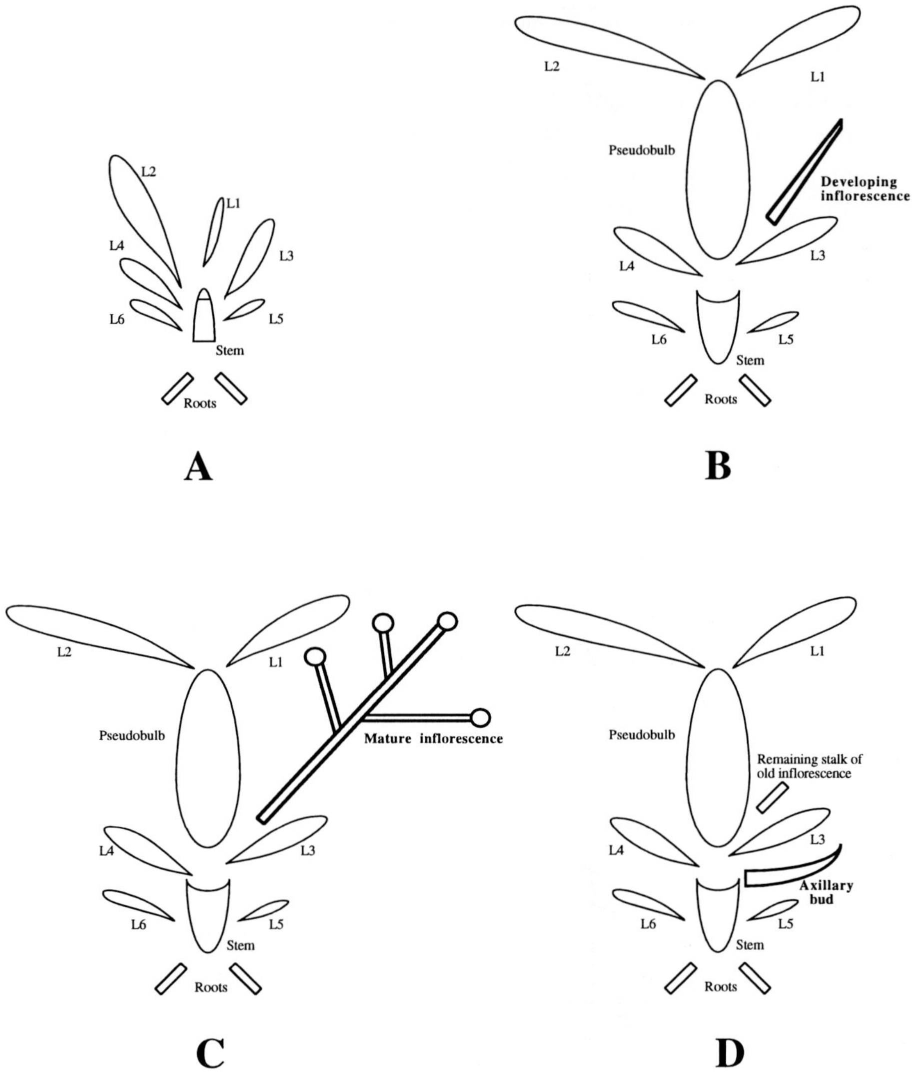

Fig. 7.1. Diagrammatic representation of the four growth stages of a single shoot of *Oncidium* Goldiana.
Note: (A) Stage 1; (B) Stage 2; (C) Stage 3; (D) Stage 4,

Adapted from Yong & Hew (1995a).

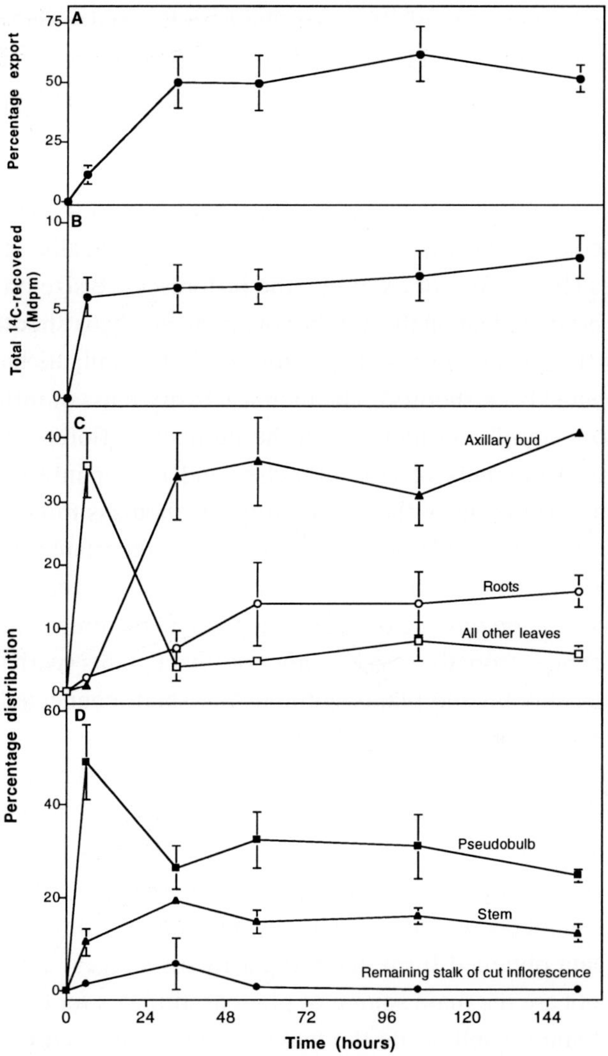

Fig. 7.2. Time course study for the movement of ^{14}C-assimilates within a single shoot of *Oncidium* Goldiana with an axillary bud.

Note: (A) Percentage export of ^{14}C-assimilates to all plant fractions from the test leaf L2; (B) Total recovery of ^{14}C-assimilates from the whole plant; (C) & (D) The pattern of ^{14}C-assimilate partitioning to various plant parts. Mean of 3 or 4 replicates, ± SE.

Redrawn from Yong & Hew (1995a).

contribute similar amounts of ^{14}C-assimilates to the major sinks (Table 7.1). The upper two leaves (L1 and L2) contribute significantly more ^{14}C-assimilates to the pseudobulbs than the lower two leaves, L3 and L4. In comparison to the other three leaves (L1, L2 and L3), the lower leaf L4 supplies significantly more ^{14}C-assimilates to the roots.

Long-term growth studies provide some evidence to the postulation that connected shoots of *Oncidium* Goldiana are physiologically interdependent for assimilates (Fig. 7.3). For example, the inflorescence size on the 'lead' or current shoot is dependent on the number of connected back shoots (Table 7.2). The size of inflorescence increases progressively from plants having no, one and two connected back shoot(s). The increase in dry mass of inflorescences is mainly due to a significant increase in the number of florets and to a lesser extent, increased numbers of side branches and longer inflorescences. The removal of shoot(s) reduces the availability of photoassimilate contribution from leaves of the connected back shoots, which are important for inflorescence development on the current shoot. The study essentially reveals that the size of an inflorescence in *Oncidium* Goldiana is dependent on three connected shoots since inflorescences from the current shoot of intact plants (with three or more connected back shoots) and plants with current shoot connected to two back shoots are similar in size.

The defoliation experiments further demonstrate the relative importance of different leaves on the current shoot, first back shoot and second back shoot as a source of photoassimilates to the inflorescence (Fig. 7.4). In particular, the relative importance of photoassimilate contribution to the inflorescence from the leaves of a shoot decreased with increasing distance from the inflorescence (Table 7.3). Data obtained from these experiments suggest that the leaves of the current shoot are the main source of photoassimilates for the inflorescence on the current shoot itself while the leaves of the connected back shoots are secondary sources of photoassimilates.

Direct evidence to demonstrate that the connected shoots of *Oncidium* Goldiana are physiologically interdependent for assimilates is obtained from $^{14}CO_2$ feeding experiments (Fig. 7.5). Both autoradiographic and quantitative studies indicate that every test leaf on different shoots exports ^{14}C-assimilates to most plant parts within the connected shoots during the vegetative and

Table 7.1. Pattern of ^{14}C-assimilate partitioning within a single shoot of *Oncidium* Goldiana at different growth stages.

	Test leaf (L1, L2, L3 or L4) at different developmental stage was supplied with $^{14}CO_2$		
	Stage 2	Stage 3	Stage 4
A. Percentage of distribution to plant parts			
All other leaves	0.8–3.5	0.7–1.6	1.8–3.8
Pseudobulb	6.1–27.8	2.6–11.0	17.5–42.2
Stem	11.5–20.3	7.2–21.5	11.4–21.4
Inflorescence	51.5–74.6	61.8–85.4	0–0.6
(type)	(Developing inflorescence)	(Mature inflorescence)	(Remaining stalk of old inflorescence)
Axillary bud	absent	absent	35.2–43.8
Roots	1.1–5.1	0.5–4.1	2.1–21.3
B. Sink activity of plant parts			
All other leaves	0.8–3.1	0.6–2.1	2.2–4.8
Pseudobulb	9.6–69.3	4.4–22.5	26.1–69.9
Stem	50.3–56.8	22.5–70.3	45.6–62.2
Inflorescence	318.6–529.5	66.3–84.0	0–9.5
(type)	(Developing inflorescence)	(Mature inflorescence)	(Remaining stalk of old inflorescence)
Axillary bud	absent	absent	477.1–691.1
Roots	6.7–22.4	3.7–17.5	22.1–116.0
Total	408.7–613.3	109.6–178.7	638.7–851.5
C. Assimilation and export			
Total ^{14}C-assimilates recovered from whole plant (Mdpm)	4.46–8.25	3.13–10.08	3.89–9.94
Percentage of ^{14}C-assimilates exported from test leaf	46.4–61.0	38.8–52.6	17.0–46.7

Note: A, Percentage of distribution to plant parts was calculated as the percentage of total ^{14}C-activity exported from a test leaf; and B, sink activity of plant parts was calculated as the percentage of total ^{14}C-activity exported from a test leaf per g dry mass of plant part. Leaves L1 and L2 are above the pseudobulb and leaves L3 and L4 are below the pseudobulb. For each test leaf, values for the percentage of distribution or sink activity are the mean of three or four replicate plants. Transport time = 33 h.

Adapted from Yong and Hew (1995a).

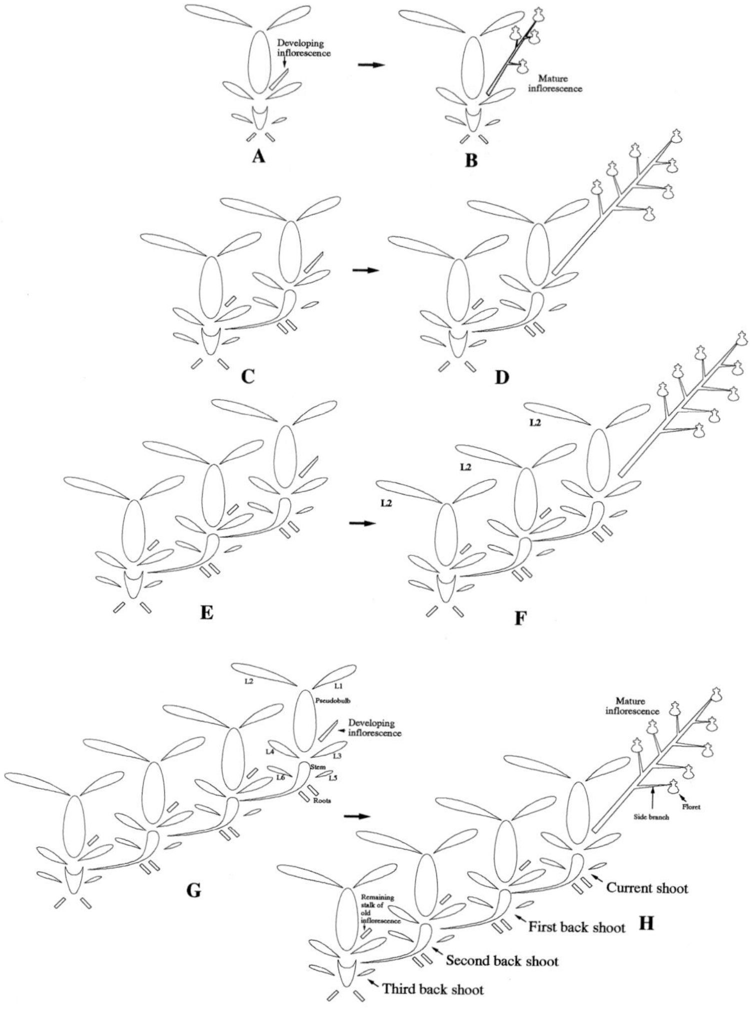

Fig. 7.3. Diagrammatic representations of *Oncidium* Goldiana with current shoots at growth stage 2 or 3 either without or with different number of connected back shoot(s).
Note: (A) Current shoot at stage 2; (B) Current shoot at stage 3; (C) Current shoot at stage 2 connected to one back shoot; (D) Current shoot at stage 3 connected to one back shoot; (E) Current shoot at stage 2 connected to two back shoots; (F) Current shoot at stage 3 connected to two back shoots; (G) Current shoot at stage 2 connected to three or more back shoots; (H) Current shoot at stage 3 connected to three or more back shoots.

Redrawn from Yong & Hew (1995c).

Table 7.2. Inflorescence size in *Oncidium* Goldiana after removal of connected back shoot(s).

	Inflorescence (at stage 3)				
	Dry mass (g)	No. of side branches	No. of florets	Length (cm)	Rate of elongation (cm d^{-1})
Number of connected back shoot(s) (at stage 2)					
none	0.51 ± 0.01^{c}	3 ± 1^{b}	18 ± 1^{c}	56 ± 5^{b}	1.0 ± 0.1^{c}
one	0.77 ± 0.07^{b}	6 ± 1^{a}	41 ± 7^{b}	70 ± 7^{a}	1.2 ± 0.1^{b}
two	1.25 ± 0.04^{a}	7 ± 1^{a}	58 ± 4^{a}	76 ± 2^{a}	1.2 ± 0.1^{b}
Control (three and more)	1.22 ± 0.13^{a}	8 ± 1^{a}	66 ± 4^{a}	74 ± 3^{a}	1.6 ± 0.1^{a}

Note: Data analysis was done using Duncan's Multiple Range Test; $^{a, b, c, d}$ means with the same letter are not significantly different ($\alpha = 0.05$) when compared within a column. Means and SEs of five replicate plants.

Adapted from Yong & Hew (1995c).

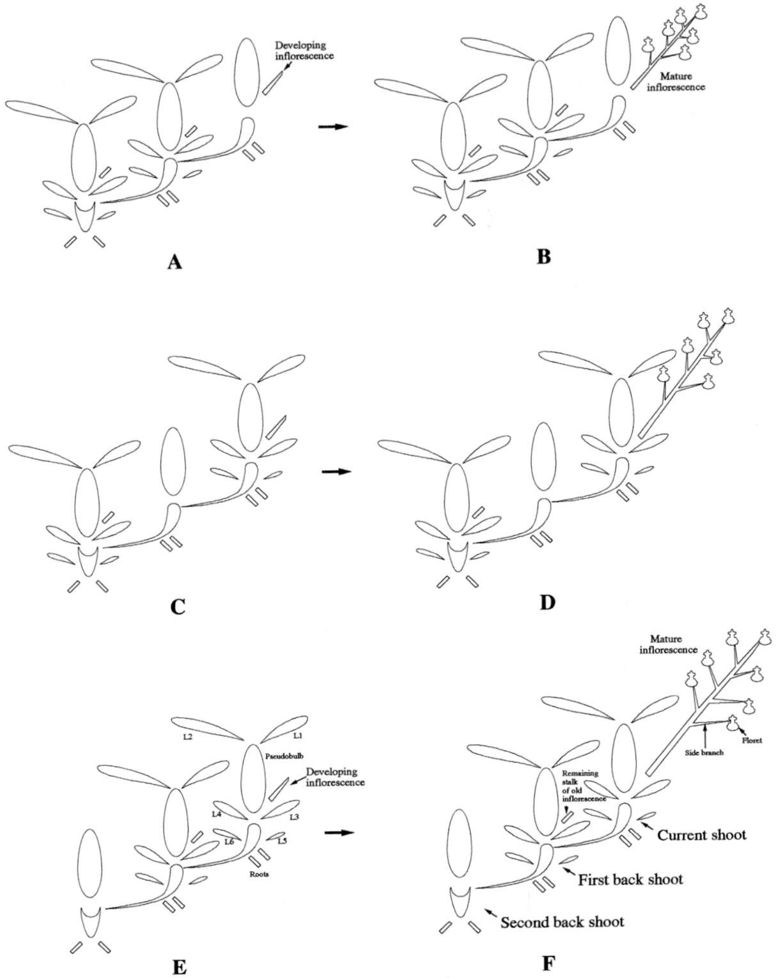

Fig. 7.4. Diagrammatic representation of *Oncidium* Goldiana with the current shoot at growth stage 2 or 3 connected to 2 back shoots for defoliation experiments.

Note: (A) Defoliated current shoot at growth stage 2 connected to two back shoots; (B) Defoliated current shoot at growth stage 3 connected to two back shoots; (C) Current shoot at growth stage 2 connected to the defoliated first back shoot and second back shoot; (D) Current shoot at growth stage 3 connected to the defoliated first back shoot and second back shoot; (E) Current shoot at growth stage 2 connected to the first back shoot and defoliated second back shoot; (F) Current shoot at growth stage 3 connected to the first back shoot and defoliated second back shoot.

Redrawn from Yong & Hew (1995c).

Table 7.3. Inflorescence Size in *Oncidium* Goldiana after selective defoliation on the current shoot or connected back shoots.

	Inflorescence (at stage 3)				
	Dry mass (g)	No. of side branches	No. of florets	Length (cm)	Rate of elongation (cm d^{-1})
Defoliated shoot (at stage 2)					
Current shoot	0.65 ± 0.09^{c}	5 ± 1^{a}	40 ± 6^{b}	59 ± 3^{c}	1.2 ± 0.1^{b}
First back shoot	0.85 ± 0.02^{b}	5 ± 1^{a}	39 ± 5^{b}	67 ± 2^{b}	1.3 ± 0.1^{ab}
Second back shoot	0.96 ± 0.07^{b}	6 ± 1^{a}	50 ± 7^{ab}	74 ± 2^{ab}	1.4 ± 0.1^{a}
Control (All shoots intact)	1.25 ± 0.04^{a}	7 ± 1^{a}	58 ± 4^{a}	76 ± 2^{a}	1.2 ± 0.1^{ab}

Note: Data analysis was done using Duncan's Multiple Range Test; $^{a, b, c, d}$ means with the same letter are not significantly different ($\alpha = 0.05$) when compared within a column. Means and SEs of five replicate plants.

Adapted from Yong & Hew (1995c).

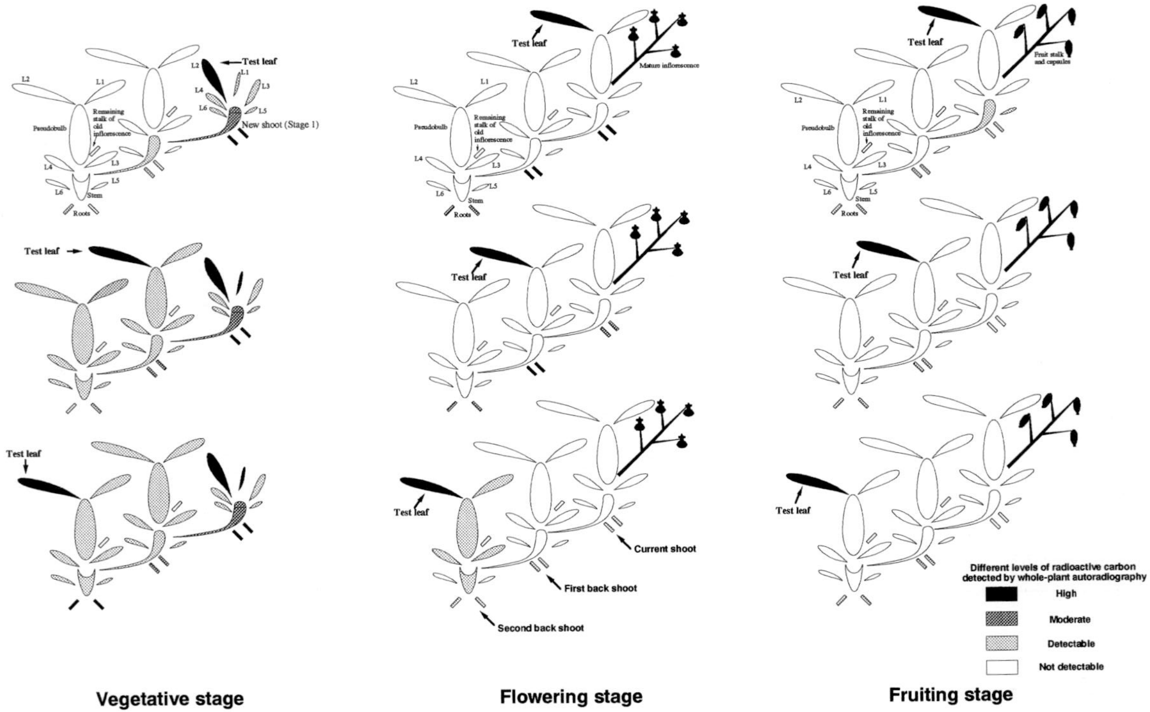

Fig. 7.5. Diagrammatic representation of the distribution pattern of radioactive carbon for the connected shoots of *Oncidium* Goldiana during the vegetative, flowering and fruiting stages.

Note: The plants were harvested after a transport time of 57 hrs. The respective fed leaf is indicated by an arrow. Each diagrammatic representation was based on two replicates.

Redrawn from Yong & Hew (1995b).

reproductive stages. The relative importance of photoassimilate contribution by the leaves on different shoots to the inflorescence is also assessed quantitatively from the different percentages of ^{14}C-assimilates exported by the respective test leaf and the percentage allocation to sink organs. The test leaf on the current shoot exports 70% of the total ^{14}C fixed and a greater proportion (83%) of these assimilates is allocated to the inflorescence (Table 7.4). On the other hand, the test leaves on the first back shoot and second back shoot export 31% to 40% of the total ^{14}C fixed and a lower proportion (73% to 61%) is partitioned to the inflorescence. These experimental data suggest that the leaves on the current shoot are the primary sources of assimilates for the inflorescence, while the leaves on the other connected shoots are secondary sources.

Experimental evidence indicates a polar movement of ^{14}C-assimilates towards the major sink on the current shoot. There is some bi-directional transfer of ^{14}C-assimilates among the current shoot, first back shoot and second back shoot (Table 7.5). The major sinks (new shoot, inflorescence or fruiting structures) on the current shoot dominate the overall supply of ^{14}C-assimilates from all test leaves on the current shoot as well as the leaves on the back shoots of *Oncidium* Goldiana. Competition for ^{14}C-assimilates exported from test leaves does occur between major sinks on different but connected shoots of *Oncidium* Goldiana. The competition for ^{14}C-assimilates exported from leaves of connected shoots is usually between two inflorescences (Fig. 7.6) or a new shoot and an inflorescence (Fig. 7.7) which is situated on another connected shoot. Distance from the source leaf becomes a major factor when two sinks are similar in developmental stage. It is shown that the inflorescence nearer to the fed leaf is able to import more ^{14}C-assimilates than the other competing inflorescence that is further away. The inflorescence appears to be a stronger sink than the new shoot in view of its ability to attract ^{14}C-assimilates from a distant source leaf that is nearer to the new shoot.

Thick-leaved sympodial orchids

The general patterns of assimilate partitioning in thick-leaved sympodial orchids is based on studies conducted for two *Dendrobium* hybrids: *Dendrobium*

Table 7.4. The Pattern of ^{14}C-assimilate partitioning among the connected shoots of *Oncidium* Goldiana at the flowering stage with leaf L2 as the test leaf.

	Test leaf L2 was supplied with $^{14}CO_2$		
	Current shoot	First back shoot	Second back shoot
A. Percentage of distribution to plant parts			
Current shoot			
Leaves	1.1 ± 0.4^{Ac}	0.9 ± 0.3^{Abc}	1.6 ± 0.4^{Ac}
Pseudobulb	2.3 ± 0.7^{Ac}	0.8 ± 0.4^{Abc}	1.4 ± 0.4^{Ac}
Stem	5.2 ± 0.7^{Ab}	2.8 ± 0.5^{Abc}	3.0 ± 1.7^{Ac}
Inflorescence	83.4 ± 2.3^{Aa}	72.6 ± 5.9^{ABa}	61.2 ± 10.6^{Ba}
Roots	2.0 ± 0.6^{Ac}	1.5 ± 0.8^{Abc}	1.4 ± 0.7^{Ac}
First back shoot			
Leaves	1.6 ± 0.4^{Ac}	0.9 ± 0.4^{Abc}	2.8 ± 1.1^{Ac}
Pseudobulb	0.7 ± 0.3^{Ac}	8.1 ± 4.9^{Ab}	2.1 ± 0.5^{Ac}
Stem & remaining stalk of old inflorescence	1.8 ± 0.6^{Ac}	4.9 ± 1.5^{Abc}	4.1 ± 2.4^{Ac}
Roots	0.7 ± 0.2^{Ac}	0.7 ± 0.1^{Abc}	0.7 ± 0.2^{Ac}
Second back shoot			
Leaves	0.4 ± 0.1^{Ac}	5.0 ± 2.8^{Abc}	2.1 ± 0.7^{Ac}
Pseudobulb	0.2 ± 0.1^{Bc}	0.7 ± 0.3^{Bbc}	15.8 ± 8.0^{Ab}
Stem & remaining stalk of old inflorescence	0.4 ± 0.2^{Bc}	0.9 ± 0.4^{Bbc}	3.4 ± 1.1^{Ac}
Roots	0.4 ± 0.2^{Ac}	0.1 ± 0.1^{Ac}	0.5 ± 0.2^{Ac}
B. Assimilation and export			
Total ^{14}C-assimilates recovered from whole plant (Mdpm)	2.7 ± 0.3^{A}	3.6 ± 1.4^{A}	1.8 ± 0.5^{A}
Percentage of ^{14}C-assimilates exported from test leaf	66.9 ± 5.8^{A}	39.5 ± 9.0^{AB}	31.0 ± 14.2^{B}

Note: A, Percentage of distribution to plant parts was calculated as the percentage of total ^{14}C-activity exported from a test leaf. For each test leaf, values for percentage of distribution are the means and SEs of three or four replicate plants. Transport time = 57 h. Data analysis was done using Duncan's Multiple Range Test; $^{A, B}$ means with the same letter are not significantly different ($\alpha = 0.05$) when compared within a row.$^{a, b, c, d}$ means with the same letter are not significantly different ($\alpha = 0.05$) when compared within a column.

Table 7.5. The Pattern of ^{14}C-assimilate partitioning and sink activity FOR an individual shoot within connected shoots of *Oncidium* Goldiana at The flowering stage with leaf L2 as the test leaf.

	Test leaf L2		
	Current shoot	First back shoot	Second back shoot
A. Percentage of distribution to individual shoot			
Current shoot	94.0 ± 1.8^{Aa}	78.5 ± 5.5^{ABa}	68.6 ± 11.1^{Ba}
First back shoot	4.8 ± 1.4^{Ab}	14.5 ± 6.0^{Ab}	9.7 ± 1.5^{Ab}
Second back shoot	1.4 ± 0.4^{Bb}	6.7 ± 2.9^{ABb}	21.7 ± 9.7^{Ab}
B. Sink activity of individual shoot			
Current shoot	106.4 ± 9.2^{Aa}	70.6 ± 4.6^{Ba}	76.5 ± 15.3^{ABa}
First back shoot	8.6 ± 2.4^{Ab}	28.9 ± 10.6^{Ab}	23.7 ± 7.1^{Ab}
Second back shoot	5.0 ± 1.5^{Bb}	11.2 ± 4.0^{Bb}	57.6 ± 18.1^{Aab}
Total	120.0 ± 11.7^{B}	110.7 ± 9.3^{B}	157.8 ± 10.2^{A}

Note: A, Percentage of distribution to plant parts was calculated as the percentage of total ^{14}C-activity exported from a test leaf. Values for each shoot were based on the total sum of ^{14}C-activity exported to leaves (excluding the test leaf), pseudobulb, stem, inflorescence and roots. B, sink activity of plant parts was calculated as the percentage of total ^{14}C-activity exported from a test leaf per g dry mass of plant part. Values for each shoot were based on the total sum of sink activity for leaves (excluding the test leaf), pseudobulb, stem, inflorescence and roots. For each test leaf, values for percentage of distribution and sink activity are the means (with SEs) of three or four replicate plants. Transport time = 57 h. Data analysis was done using Duncan's Multiple Range Test; $^{A, B}$ means with the same letter are not significantly different ($\alpha = 0.05$) when compared within a row.$^{a, b, c, d}$ means with the the same letter are not significantly different ($\alpha = 0.05$) when compared within a column.

Adapted from Yong & Hew (1995c).

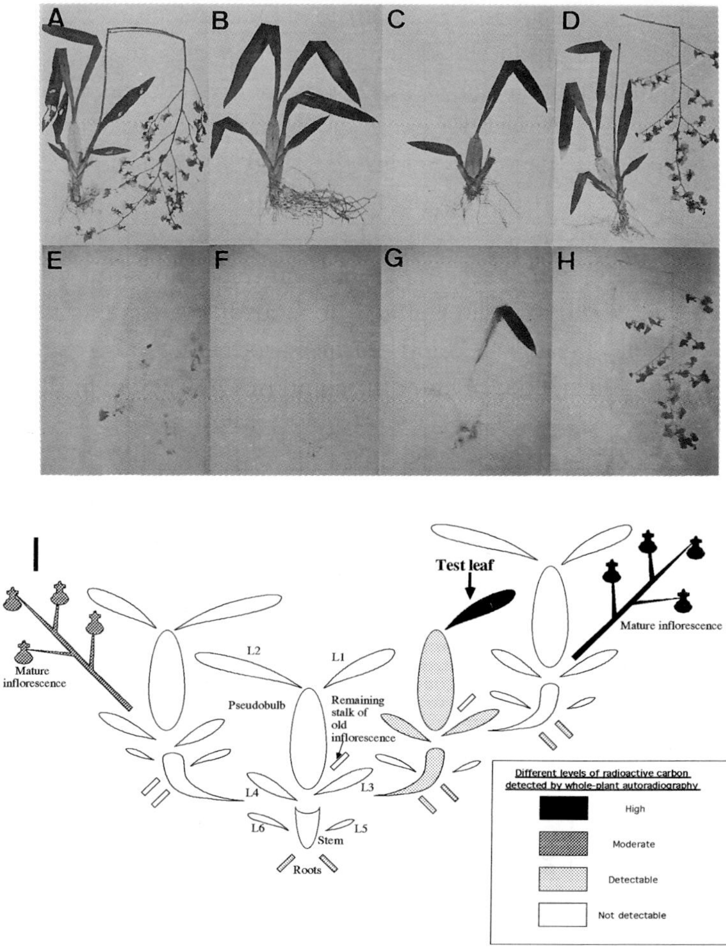

Fig. 7.6. Whole plant autoradiography and diagrammatic representation for the connected shoots of *Oncidium* Goldiana with two mature inflorescences.

Note: A–D, photographs of connected shoots (A, second current shoot with a mature inflorescence; B, second back shoot; C, first back shoot with the test leaf L1; D, current shoot with a mature inflorescence) mounted individually on herbarium sheets. E–H, photographs of x-ray films for the corresponding shoots (A–D respectively) after $^{14}CO_2$ (0.19 MBq) feeding (The fed leaf is indicated by an arrow). I, diagrammatic representation for the connected shoots with four types of shading indicating the different levels of ^{14}C-assimilates (The fed leaf is indicated by an arrow). Note: The plants were harvested after a transport time of 57 h. Each diagrammatic representation was based on two replicates.

Adapted from Yong & Hew (1995b).

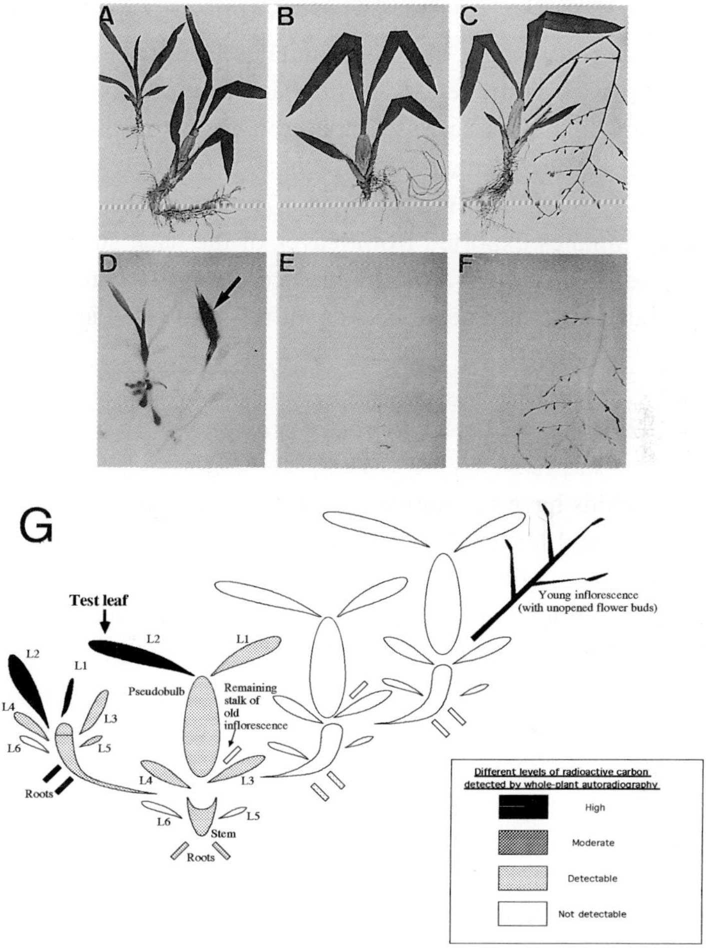

Fig. 7.7. Whole plant autoradiography and diagrammatic representation for the connected shoots of *Oncidium* Goldiana with a young inflorescence and a new shoot.
Note: A–C, photographs of connected shoots (A, new shoot at stage 1 with the second back shoot; B, first back shoot; C, current shoot with a young inflorescence) mounted individually on herbarium sheets; D–F, photographs of x-ray films for the corresponding shoots (A–C respectively) after $^{14}CO_2$ (0.19 MBq) feeding (The fed leaf is indicated by an arrow); G, diagrammatic representation for the connected shoots with four types of shading indicating the different levels of ^{14}C-assimilates (The fed leaf is indicated by an arrow).
Note: The plants were harvested after a transport time of 57 h. Each diagrammatic representation was based on two replicates.

Adapted from Yong & Hew (1995b).

Rong Rong and *Dendrobium* Jashika Pink. The inflorescence of *Dendrobium* dominates the overall supply of ^{14}C-assimilates from all test leaves on the current shoot as well as the leaves on the back shoot (Tables 7.6, 7.7). The *Dendrobium* inflorescence faces competition from the pseudobulb, stem internodes, roots and the vegetative basal shoot when it is present. For example in *Dendrobium* Jashika Pink, the presence of vegetative basal shoot reduces 38% of the assimilates exported from leaves of the back shoot to the inflorescence (Table 7.8). Without an inflorescence, there is a general increase in assimilate allocation to the pseudobulbs and fully expanded leaves.

Assimilate partitioning in the monopodial orchids

Monopodial orchids have vegetative apex that grows indeterminately, giving rise to new stem and leaves. Inflorescences will arise from axillary buds at nodes some distances (five to ten leaves counting from the apex) from the vegetative apex shoot (see Chap. 2 on A Brief Introduction to Orchid Morphology and Nomenclature).

Thick-leaved monopodial orchids

The general patterns of assimilate partitioning in thick-leaved monopodial orchids is based on studies conducted for two *Aranda* hybrids: *Aranda* Tay Swee Eng and *Aranda* Noorah Alsagoff. Selective feeding of $^{14}CO_2$ to different test leaves along the stem axis of *Aranda* Noorah Alsagoff reveals that the inflorescence receives ^{14}C-assimilates from many leaves rather than a few (Table 7.9). The vegetative apical shoot competes with the inflorescence for assimilate supply. The fully expanded leaves also constitute a major sink for assimilates. An increased flux of ^{14}C-assimilates to the inflorescence, especially for the subtending leaf and leaves above the subtending leaf, is obtained by removing the vegetative apical shoot (Table 7.10). Conversely, the removal of the inflorescence consistently leads to an increase in assimilate allocation by all test leaves to the stem internodes and roots, possibly for storage.

Table 7.6. Pattern of ^{14}C-assimilate partitioning in *Dendrobium* Rong Rong.

	Test leaf supplied with $^{14}CO_2$						
	L_u	L_{u-1}	L_{u-2}	L_{u-3}	L_I	L_L	Pooled S.E.
A. Percentage of distribution to plant parts							
Inflorescence	58	59	55	47	32	47	± 5.5
Pseudobulb	4	6	9	10	11	3	± 1.5
Stem internodes	16	14	15	16	16	19	± 2.9
Roots	20	19	20	24	38	29	± 4.8
Fully-expanded leaves	2	2	1	3	3	2	± 0.8
B. Sink activity of plant parts							
Inflorescence	4.8	4.4	3.9	4.3	2.7	2.7	± 0.5
Pseudobulb	0.2	0.3	0.4	0.6	0.7	0.2	± 0.1
Stem internodes	0.5	0.5	0.6	0.7	0.6	0.8	± 0.1
Roots	0.3	0.3	0.3	0.7	1.0	0.9	± 0.2
Fully-expanded leaves	less than 0.1 for all test leaves						± 0.04

Note: Percentage distribution to plant parts was calculated as the percentage of total ethanol-soluble ^{14}C-activity exported from a test leaf to plant parts. Sink activity of plant parts was calculated as the percentage of total ethanol-soluble ^{14}C-activity exported from a test leaf per gram fresh mass of plant part. For each test leaf, values for percentage distribution or sink activity are the mean of seven replicate plants. Mean total ethanol-soluble ^{14}C-activity exported to plant parts ranged between 5.6×10^5 and 7.8×10^5 dpm, but did not differ significantly between test leaves. Vegetative basal shoots are not included as plant part although three of the seven replicate plants possessed a vegetative basal shoot. Fully-expanded leaves on the stem axis are designated as follows: leaf L_U is the uppermost leaf, nearest to the inflorescence; leaf L_{U-2} is the second leaf below the leaf L_U; leaf L_L is the lowermost leaf. The number of leaves below leaf L_{U-4} varies from two to nine and leaf L_I is chosen as the intermediate leaf positioned midway between leaf L_{U-3} and leaf L_L. Transport time = 72 h.

Adapted from Clifford, Neo, Ma & Hew (1994).

Table 7.7. The Pattern of ^{14}C-assimilate partitioning among the connected shoots of *Dendrobium* Jashika Pink at the flowering stage with the basal leaf as the test leaf.

	Test leaf was supplied with $^{14}CO_2$	
	Current shoot	Back shoot 1
Percentage of distribution to plant parts		
Current shoot		
Inflorescence	21.7^{AB}	47.1^{A}
Fully-expanded leaves	8.2^{A}	2.2^{B}
Pseudobulb	6.8^{A}	5.6^{A}
Stem	19.5^{A}	3.2^{B}
Back shoot 1		
Fully-expanded leaves	3.0^{A}	3.2^{A}
Pseudobulb	1.6^{A}	8.4^{A}
Stem	5.1^{A}	5.6^{A}
Back shoot 2		
Fully-expanded leaves	0.5^{A}	2.9^{A}
Pseudobulb and stem	3.7^{A}	2.7^{B}
Other plant parts		
Roots	29.8^{A}	19.1^{B}
Assimilation and export of ^{14}C-assimilates		
Total ^{14}C recovered from the whole plant	15.3^{A}	9.3^{B}
Total ^{14}C exported from test leaf	4.0^{A}	1.4^{B}
Percentage exported from test leaf	26	15

Note: Percentage of distribution to plant parts was calculated as the percentage of total ^{14}C-activity exported from a test leaf. For each test leaf, values for the percentage of distribution are the mean of five replicate plants. Transport time = 48 h. Data analysis with Duncan's Multiple Range Test; $^{A,\,B}$ means with the same letter are not significantly different ($\alpha = 0.05$) when compared within a row.

Adapted from Wadasinghe & Hew (1995).

Table 7.8. Pattern of ^{14}C-assimilate partitioning in *Dendrobium* Jashika Pink plants at different growth stages.

	Test leaf (basal leaf on back shoot 1) was supplied with $^{14}CO_2$ at the following developmental stages		
	Flowering plants with a vegetative shoot	Flowering plants	Non-flowering plants
Percentage of distribution to plant parts			
Current shoot			
Inflorescence	9.6^{B}	47.1^{A}	not present
Fully-expanded leaves	0.9^{B}	2.2^{B}	10.7^{A}
Pseudobulb	2.6^{B}	5.6^{B}	24.8^{A}
Stem	1.6^{B}	3.2^{B}	6.2^{A}
Back shoot 1			
Fully-expanded leaves	2.7^{A}	3.2^{A}	3.6^{A}
Pseudobulb	6.2^{A}	8.4^{A}	4.0^{A}
Stem	4.7^{A}	5.6^{A}	7.3^{A}
Back shoot 2			
Fully-expanded leaves	0.5^{A}	2.9^{A}	2.2^{A}
Pseudobulb and stem	1.3^{B}	2.7^{B}	10.5^{A}
Other plant parts			
New shoot	36.4	not present	not present
Roots	33.5^{A}	19.1^{A}	30.8^{A}

Note: Percentage of distribution to plant parts was calculated as the percentage of total ^{14}C-activity exported from a test leaf. For each test leaf, values for the percentage of distribution are the mean of five replicate plants. Transport time = 48 h. Note: Data analysis with Duncan's Multiple Range Test; $^{A, B}$ means with the the same letter are not significantly different ($\alpha = 0.05$) when compared within a row.

Adapted from Wadasinghe & Hew (1995).

Table 7.9. Pattern of ^{14}C-assimilate partitioning in *Aranda* Noorah Alsagoff.

	Test leaf supplied with $^{14}CO_2$							
	L_S	L_{S+1}	L_{S+2}	L_{S+3}	L_{S+4}	L_{S+5}	L_{S-3}	Pooled S.E.
A. Percentage of distribution to plant parts								
Inflorescence	57	57	58	52	53	31	65	± 5.1
Vegetative apex	17	19	12	17	15	25	10	± 3.9
Stem internodes	15	11	12	14	21	24	12	± 1.8
Roots	4	5	5	10	6	12	7	± 1.4
Fully-expanded leaves	7	8	13	7	5	8	6	± 1.5
B. Sink activity of plant parts								
Inflorescence	3.5	4.4	3.3	2.8	3.4	1.9	3.9	± 0.5
Vegetative apex	3.1	3.1	2.3	1.9	2.1	2.7	1.4	± 0.7
Stem internodes	0.7	0.5	0.5	0.6	0.9	1.2	0.5	± 0.1
Roots	0.1	0.2	0.1	0.3	0.1	0.2	0.2	± 0.1
Fully-expanded leaves	0.1	0.2	0.2	0.1	0.1	0.1	0.1	± 0.1

Note: Percentage distribution to plant parts was calculated as the percentage of total ethanol-soluble ^{14}C-activity exported from a test leaf to plant parts. Sink activity of plant parts was calculated as the percentage of total ethanol-soluble ^{14}C-activity exported from a test leaf per gram fresh mass of plant part. For each test leaf, values for percentage distribution or sink activity are the mean of five replicate plants. Mean total ethanol-soluble ^{14}C-activity exported to plant parts ranged between 1.4×10^5 and 5.7×10^5 dpm, but did not differ significantly between test leaves. L_S is the leaf subtending the inflorescence; L_{S+4} is the fourth leaf above the leaf subtending the inflorescence; L_{S-2} is the second leaf below the leaf subtending the inflorescence. Transport time = 48 h. The growth habit of a monopodial orchid is shown in Fig. 2.2.

Adapted from Clifford, Neo, Ma & Hew (1994).

Table 7.10. Pattern of ^{14}C-assimilate partitioning in *Aranda* Noorah Alsagoff after the removal of sink organ.

	Control plants	Plants with vegetative apical shoot removed two days earlier	Plants with inflorescence removed two days earlier
Percentage of distribution to plant parts			
L_S as the test leaf (subtending leaf of the inflorescence)			
Inflorescence	44	69***	not present
Vegetative apical shoot	11	not present	21**
Stem internodes	23	17	35*
Roots	13	8	22*
Fully expanded leaves	9	6	22**
L_{S+5} as the test leaf (5th leaf above the subtending leaf)			
Inflorescence	43	65**	not present
Vegetative apical shoot	20	not present	22
Stem internodes	16	19	35*
Roots	10	9	25*
Fully expanded leaves	11	7	18
L_{S-5} as the test leaf (5th leaf below the subtending leaf)			
Inflorescence	66	59	not present
Vegetative apical shoot	5	not present	9
Stem internodes	14	21	41***
Roots	11	10	41***
Fully expanded leaves	4	10	9

Note: Percentage of distribution to plant parts was calculated as the percentage of total ^{14}C-activity exported from a test leaf. For each test leaf, values for the percentage of distribution are the mean of six replicate plants. Transport time = 48 h. Note: Differences between treatment means for any test leaf were assessed by one-way ANOVA. * = $P < 0.05$; ** = $P < 0.01$; *** = $P < 0.001$.

Adapted from Clifford, Neo & Hew (1995).

7.5. Import of Assimilates by Mature Orchid Leaves

It is generally accepted that mature leaves do not import and retain photo-assimilates from the other leaves under normal conditions (Wardlaw, 1990). However, the import of photoassimilates by mature leaves has been demonstrated in some plants, for example, tomato. It is reported that mature tomato leaves could import up to 56% of the total ^{14}C-assimilates during early phases of the reproductive period.

At present, it is difficult to provide an explanation for the import of ^{14}C-assimilates by mature leaves of orchids (Table 7.11). For tropical orchids, the percentages of ^{14}C-assimilates imported by mature (or fully expanded) orchid leaves are as follows: 1–4% for *Oncidium* Goldiana; 5–13% for *Aranda* Noorah Alsagoff; 10–41% for *Aranda* Tay Swee Eng; 1–3% for *Dendrobium* Rong Rong; 1–11% for *Dendrobium* Jashika Pink. In *Aranda*, but not in *Dendrobium* and *Oncidium*, the proportion of exported assimilates appears to be more than what would be expected from transpirational import following exchange between phloem and xylem. It is tempting to propose that the mature leaves of monopodial orchids may have a storage function in the absence of storage organs such as pseudobulbs or tubers commonly found in the sympodial orchids. Future experiments should aim to resolve whether the import of assimilates by mature orchid leaves from other leaves is a direct transfer of assimilates between leaves or an indirect import of labelled amino acids through the transpirational stream using exported ^{14}C-assimilates received by the roots. The use of heat-girdling to destroy phloem tissue near the stem–aerial-root junction before $^{14}CO_2$ feeding is one possible way to rule out import into mature leaves through the transpirational stream. Alternatively, we could study the biochemical details of the ^{14}C-labelled products in the mature leaves after $^{14}CO_2$ feeding.

At present, the dual physiological function of mature orchid leaves being a source (predominantly) and a sink organ is attributed to either complex foliar vascular structures or some unique biochemical process occurring within the orchid leaves. The minor transfer of ^{14}C-assimilates between the mature leaves could possibly be due to the complex connections of leaf traces and other vascular bundles at the sheathing bases of leaves in most monocots.

Table 7.11. Patterns of photoassimilate partitioning in tropical epiphytic orchids.

Orchid	Habit	Photosynthetic pathway	Percentages of ^{14}C-assimilates exported from all test leaves: Inflorescence	Pseudobulb*	Stem	Roots	Other plant parts	Mature leaves
***Aranda* Noorah Alsagoff**	Monopodial	CAM						
Plants at flowering stage			31–65	not present	11–24	4–12	10–25 (vegetative shoot apex)	5–13
***Aranda* Tay Swee Eng**	Monopodial	CAM						
Plants at flowering stage			15–65	not present	7–18	7–17	4–18 (vegetative shoot apex)	10–41
***Dendrobium* Rong Rong**	Sympodial	CAM						
Plants at flowering stage			32–59	3–11	14–19	19–38	none	1–3
***Dendrobium* Jashika Pink**	Sympodial	CAM						
Plants at flowering stage			22–47	2–8	3–20	19–30	none	1–8
Plants at vegetative stage			not present	4–25	6–11	31	none	2–11
Plants at both flowering and vegetative stages			9	1–6	1–4	34	36 (vegetative basal shoot)	1–2
***Oncidium* Goldiana**	Sympodial	C_3						
Plants at early flowering stage			52–67	6–28	12–20	1–5	none	1–4
Plants at flowering stage			65–85	3–11	7–22	1–4	none	1–2
Plants at vegetative basal shoot formation stage			not present	18–42	11–21	2–21	35–44 (vegetative basal shoot)	2–4

* Most epiphytic sympodial orchids such as *Oncidium* and *Dendrobium* possess a prominent and enlarged bulbous structure, commonly termed the pseudobulb. In general, the pseudobulb is the enlarged portion of the stem from which leaves and inflorescence may arise. The pseudbobulb is a storage organ for carbohydrates, minerals and water (see Arditti, 1992). Monopodial orchids such as *Aranda* lack pseudobulbs.

Note: Dosing with $^{14}CO_2$ was at 1900 h for CAM hybrids and at 0900 h for C_3 hybrids. Percentage distribution of ^{14}C-assimilates to plant parts was calculated as the percentage of total ^{14}C-activity exported from a test leaf. Transport time = 24 h (*Dendrobium* Jashika Pink), 33 h (*Oncidium* Goldiana), 48 h (*Aranda* Noorah Alsagoff and *Aranda* Tay Swee Eng) and 72 h (*Dendrobium* Rong Rong).

Adapted from Clifford, Neo & Hew (1992); Clifford, Neo, Ma & Hew (1994); Wadasinghe & Hew (1995) and Yong & Hew (1995a).

7.6. The Role of Non-Foliar Green Organs in Assimilate Partitioning

Leaves are the main sources of assimilates for growth, especially in leafy orchids. There are non-foliar green organs in leafy orchids (e.g., pseudobulbs, flowers, fruit capsules and roots) and these organs may potentially contribute to the overall carbon balances. For example, in *Oncidium* Goldiana, it was found that a major portion of the non-foliar photoassimilates is being used within these organs probably for maintenance respiration and other physiological processes, and not exported to other major sinks (Figs. 7.8, 7.9), except for the pseudobulbs (Fig. 7.10). This is unlike the shootless orchids where the roots form more than half of the biomass of the orchid and are responsible for carbon acquisition. It may be concluded that in leafy orchids, the contribution of carbon from non-foliar green sources is generally minimal and certainly not sufficient for growth.

7.7. Improving the Harvestable Yield of Orchids

The highly integrated patterns of assimilate partitioning between sources and sinks exhibited by orchids (Table 7.11) is different from most crop plants. The inflorescence on the current shoot receives assimilates from both nearby leaves and distant leaves on the other connected back shoots. The relatively free movement of assimilates from all the sources suggests that vascular restriction on assimilate movement to the inflorescence is minimal, implying that the potential to divert assimilates for inflorescence growth is high. This is in contrast to plant species in which (a) the inflorescence receives assimilates from a subtending leaf (e.g., cotton) or nearby leaves (e.g., field beans); (b) partitioning is rigidly governed by specific vascular geometry and phyllotaxy patterns (e.g., *Citrus sinensis*).

The growth of the inflorescence may be limited by the supply of assimilate (source-limited) or by the capacity of the inflorescence to import or use that assimilate (sink-limited). Source or sink limitation is discussed only pertaining to the growth of the inflorescence sink. Source or sink limiting situations

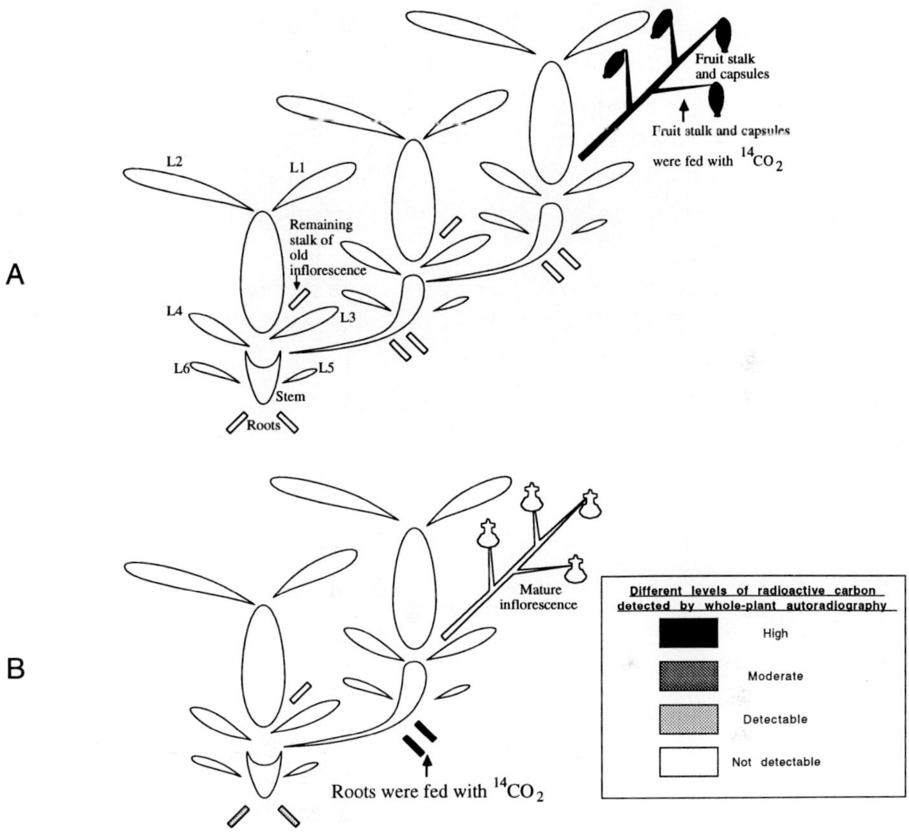

Fig. 7.8. Diagrammatic representation of the distribution pattern of ^{14}C-photoassimilates fixed by non-foliar photosynthetic organs of *Oncidium* Goldiana.

Note: **A**, the fruit stalk and capsules were fed with $^{14}CO_2$ (0.19 MBq); **B**, the epiphytic roots of the current shoot with mature inflorescence were fed with $^{14}CO_2$.

The plants were harvested after a transport time of 57 h. The diagrammatic representation was based on two replicates.

Adapted from Yong & Hew (1995b).

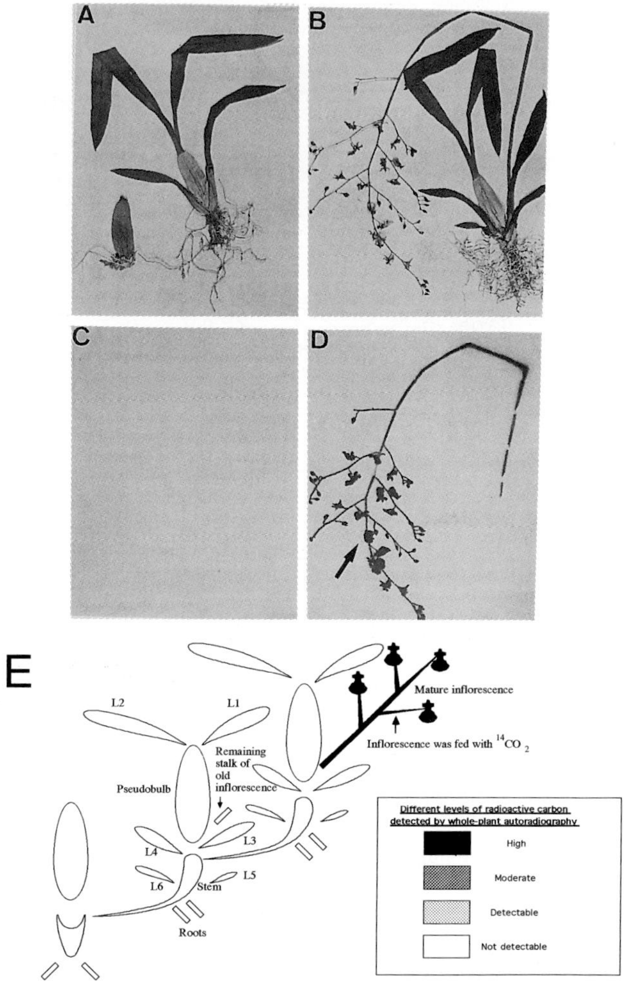

Fig. 7.9. Whole plant autoradiography and diagrammatic representation of the distribution pattern of ^{14}C-photoassimilates fixed by a mature inflorescence of *Oncidium* Goldiana.
Note: A–B, photographs of connected shoots (A, the second back shoot without leaves and first back shoot; B, current shoot with a mature inflorescence) mounted individually on herbarium sheets. C–D, photographs of x-ray films for the corresponding shoots (A and B respectively) after $^{14}CO_2$ (0.19 MBq) feeding and a transport time of 57 h. (The fed inflorescence is indicated by an arrow). E, diagrammatic representation for the connected shoots with four types of shading indicating the different levels of ^{14}C-assimilates (The fed inflorescence is indicated by an arrow).
Each diagrammatic representation was based on two replicates.

Adapted from Yong & Hew (1995b).

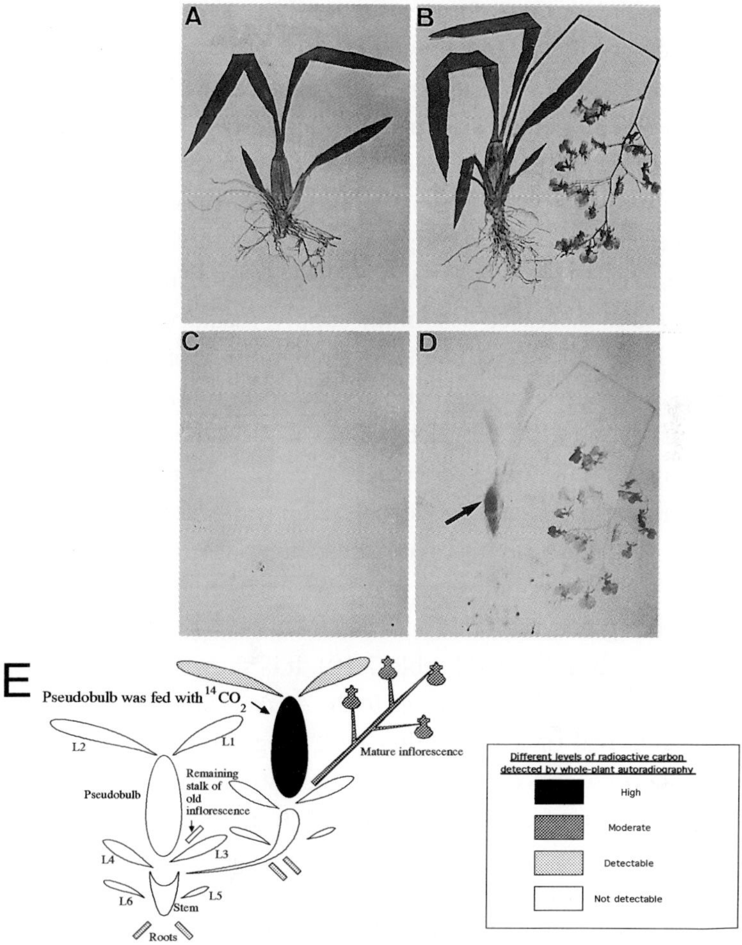

Fig. 7.10. Whole plant autoradiography and diagrammatic representation of the distribution pattern of ^{14}C-photoassimilates fixed by a pseudobulb of *Oncidium* Goldiana.
Note: A–B, photographs of connected shoots (A, first back shoot; B, current shoot with a mature inflorescence) mounted individually on herbarium sheets. C–D, photographs of x-ray films for the corresponding shoots (A and B respectively) after $^{14}CO_2$ (0.19 MBq) feeding and a transport time of 57 h. (The fed pseudobulb is indicated by an arrow and a small portion of cuticle (2 cm by 2 cm) on the pseudobulb was removed prior to $^{14}CO_2$ feeding. E, diagrammatic representation for the connected shoots with four types of shading indicating the different levels of ^{14}C-assimilates (The fed pseudobulb is indicated by an arrow).
The diagrammatic representation was based on two replicates.

Adapted from Yong & Hew (1995b).

invariably change with plant ontogeny as well as environmental influences. It is not easy to decide if harvestable yield is source-limited or sink-limited, but the decision can usually be made by experimentation. The initial step would be to establish whether biomass gain by the harvestable organ (sink) is limited by assimilate supply (i.e., source limited) or saturated by assimilate supply (i.e., sink limited). The harvestable organ is considered source-limited if treatments like elevated CO_2 or the removal of competing sinks increase the growth of the harvestable organ. When the growth of the harvestable organ does not respond to increased assimilate supply, the harvestable organ will be considered to be sink-limited. The least ambiguous approach to study source limitation is to increase leaf photosynthetic rates by elevated CO_2. Manipulations of assimilate supply by removal of competing sinks or changing light levels may be confounded by changes in correlative plant hormone signals.

The increase in dry mass of the *Oncidium* inflorescence under elevated CO_2 indicates that the growth of inflorescence is primarily source-limited (Table 7.12). A similar conclusion is drawn for another study that indicates that inflorescence growth of a thick-leaved monopodial orchid *Aranda* Noorah Alsagoff is also source-limited (Table 7.10). It is shown that an increased flux of ^{14}C-assimilates from the test leaves to the *Aranda* inflorescence is the direct result of the removal of a competing sink (vegetative apical shoot). These findings for *Aranda* and *Oncidium* are interesting and provide exceptional examples to the current consensus that sink limitation is the main factor in controlling harvestable yield in many economically important plants such as soybean, tomato and wheat.

Improvement in the harvestable yield of tropical orchids grown for its cut-flowers should adopt a two-pronged approach that seeks to increase the ability of the inflorescence sink to import assimilates and the photosynthetic capacity of the source leaves. Due consideration must also be given to understand the mineral nutrient requirements (see Chap. 5 on Mineral Nutrition), photosynthetic characteristics of leaves (see Chap. 3 on Photosynthesis) and growth habit of the orchids selected for improvement. Some possible ways for improving the harvestable yield of tropical orchids are suggested.

Table 7.12. Inflorescence size in *Oncidium* Goldiana after growing in elevated CO_2.

	Inflorescence (at stage 3)				
	Dry mass (g)	No. of side branches	No. of florets	Length (cm)	Rate of elongation (cm d^{-1})
Treatments (from stage 2 to stage 3)					
Plants grown under 1% CO_2	1.64 ± 0.11^a	9 ± 1^a	74 ± 5^a	79 ± 3^a	1.5 ± 0.1^b
Plants grown under 10% CO_2	1.76 ± 0.25^a	7 ± 1^{ab}	76 ± 11^a	80 ± 6^a	1.9 ± 0.2^a
Control plants grown under 0.03% CO_2 (ambient levels)	1.13 ± 0.09^b	6 ± 1^b	51 ± 4^b	73 ± 2^a	1.2 ± 0.1^b

Note: Data analysis was done using Duncan's Multiple Range Test; $^{a, b, c, d}$, means with the same letter are not significantly different ($\alpha = 0.05$) when compared within a column. Means and SEs of five replicate plants.

Adapted from Yong (1995).

Increasing the availability of assimilates for flower development by the removal/suppression of competing sinks

The clonal habit of sympodial orchids, such as *Oncidium*, *Dendrobium* and *Cymbidium*, allows interactions between sinks found on the different shoots. Unlike the monopodial orchids, direct competition for assimilates between the growing vegetative apex and the inflorescence is not observed for sympodial orchids. However, competition for available assimilates is observed to occur between sinks growing on different but connected shoots. For example, the *Oncidium* inflorescence on the current shoot competes against the new vegetative shoot growing on the second back shoot for assimilates exported by the leaves on the first back shoot. Similarly, a growing axillary basal shoot of *Dendrobium* is known to cause a 38% reduction of assimilates exported by leaves for the inflorescence.

Studies have shown that the development of a 'normal-size' *Oncidium* inflorescence requires three connected shoots. The frequent occurrence of small inflorescences in *Oncidium* Goldiana plants with simultaneous vegetative and reproductive growth stages in several local orchid farms is probably due to an inadequate supply of assimilates to the inflorescence. The removal of new vegetative shoots (or competing sinks for available assimilates) or the suppression of its growth (possibly by chemical inhibitors) at the time of flowering is likely to enhance flower production in *Oncidium* and *Dendrobium*.

Increasing the photosynthetic capacity of orchid leaves by increasing irradiance

Experimental evidence suggests that harvestable yield improvements for CAM monopodial orchids could be achieved by increasing photosynthetic rates of source leaves by providing higher irradiance under optimal growing conditions. For example, it is shown that flowering plants of *Vanda* Miss Joaquim grown in full sun under optimal conditions produce 30% more flowers than those grown in 30% shade (Table 7.13). Similarly, more racemes (36%) and florets (44%) are produced by flowering plants of *Dendrobium* grown in full sun than those grown in 30% shade.

Table 7.13. Flower production in *Vanda* Miss Joaquim and *Dendrobium* Jaquely Thomas in full sun and 30% shade.

	Plants grown in 30% shade	Plants grown in full sun
***Vanda* Miss Joaquim**		
Number of flowers harvested	14.8 ± 0.2	21.0 ± 0.1*
***Dendrobium* Jaquely Thomas**		
Number of shoots initiated	4.3 ± 0.3	4.9 ± 0.2 n.s.
Number of racemes initiated	2.8 ± 0.3	4.6 ± 0.3*
Number of racemes harvested	2.7 ± 0.2	4.2 ± 0.4*
Number of flowers harvested	27 ± 3	48 ± 4*

Note: Differences between treatment means were assessed by student *t*-test. * = $P < 0.05$; n.s. = not significant. The number of plants used for experiments: $n = 40$ for *Vanda*; $n = 80$ for *Dendrobium*. The field experiments were conducted from February 1989 to December 1989 in Guam, USA.

Adapted from Mcconnell, Guerrero, Leon & Mafnas (1990).

Increasing photosynthetic capacity of orchid leaves by elevated carbon dioxide

Increasing the level of irradiance to enhance assimilate production is not feasible for some shade-loving orchids grown for cut-flowers such as *Oncidium* and *Phalaenopsis*. For example, in *Oncidium* Goldiana, increasing the rate of source photosynthesis by increasing irradiance is not feasible because gas-exchange studies have shown that *Oncidium* Goldiana is a shade plant. Light saturation for *Oncidium* Goldiana leaves (leaf L2, above the pseudobulb; and leaf L3, below the pseudobulb) occurs between 80 $\mu mol\ m^{-2}s^{-1}$ and 100 $\mu mol\ m^{-2}s^{-1}$ for all the different stages of development. Any further increase in irradiance (beyond 700 $\mu mol\ m^{-2}\ s^{-1}$) may lead to photoinhibition of the leaves. Moreover, results obtained from a long-term field study on *Oncidium* Goldiana indicate that there is a reduction in flower production when the annual total number of sunshine hours is high (see Chap. 6 on Control of Flowering and Fig. 6.7).

The use of elevated CO_2 as an alternative to increase photosynthetic rates of source leaves appears to be a logical solution since further increase in irradiance may result in photoinhibition of the leaves of this shade-loving orchid.

There is an average of 50% increase in inflorescence dry mass and 94% increase in dry matter accumulation in the pseudobulbs of current shoot and first back shoot for *Oncidium* Goldiana plants grown in elevated CO_2 (Tables 7.12, 7.14). In addition, the higher dry matter content in these pseudobulbs after CO_2 enrichment may provide more assimilates for growth and development of the next shoot. The data obtained from the elevated CO_2 experiments also indicate that the growth of inflorescence in *Oncidium* Goldiana is primarily source-limited. The increase in dry mass of inflorescence under elevated CO_2 is attributed to more florets produced on the stalk. Since the inflorescence growth of *Oncidium* Goldiana is primarily source-limited, the use of elevated CO_2 in improving the harvestable yield of *Oncidium* Goldiana is therefore justifiable.

More work is urgently needed to study the effects of elevated CO_2 on flower production in tropical orchids, especially monopodial orchids, under field conditions. There is a recent report on the positive effects of elevated CO_2 on the growth and photosynthetic capacity of CAM monopodial orchid plantlets, suggesting that adult plants may respond positively to elevated CO_2 under field conditions.

Increasing assimilate availability for flower development by selecting specific cultivars with more leaves

There is additional evidence to support the idea that increasing source capacity could increase the availability of assimilates for inflorescence growth in some orchids such as *Oncidium*. A cultivar ('Seven-leaf' cultivar) of *Oncidium* Goldiana is discovered recently within a population of mericloned *Oncidium* Goldiana plants which could produce two inflorescences sequentially on a single shoot (Fig. 7.11). The first inflorescence is subtended by leaf L3 and the second inflorescence is subtended by leaf L4. It is likely that the 'Seven-leaf' cultivar is a somaclone arising from tissue culture. The 'Seven-leaf' cultivar has an additional leaf (termed leaf L0) above the pseudobulb. Since experiments on single shoots of *Oncidium* Goldiana have shown that all test leaves contribute similar amounts of ^{14}C-assimilates to the major sink of the growth stage, the

Table 7.14. Dry mass and water content of *Oncidium* Goldiana pseudobulbs after growing in elevated CO_2.

	Pseudobulbs (at stage 3)					
	Dry mass (g)			Water content (%)		
Treatments (from stage 2 to stage 3)	Current shoot	First back shoot	Second back shoot	Current shoot	First back shoot	Second back shoot
Plants grown under 1% CO_2	0.93 ± 0.07^{Aab}	0.89 ± 0.12^{Aab}	0.53 ± 0.08^{Ba}	95.1 ± 0.2^{Aa}	93.8 ± 0.4^{Aa}	91.5 ± 0.6^{Ba}
Plants grown under 10% CO_2	1.24 ± 0.16^{Aa}	1.08 ± 0.16^{Aa}	0.96 ± 0.23^{Aa}	94.0 ± 0.5^{Aab}	92.0 ± 0.8^{Bab}	90.5 ± 0.5^{Ba}
Control plants grown under 0.03% CO_2 (ambient levels)	0.64 ± 0.03^{Ab}	0.56 ± 0.05^{Ab}	0.62 ± 0.09^{Aa}	93.6 ± 0.3^{Ab}	91.1 ± 0.6^{ABb}	88.7 ± 2.4^{Ba}

Note: Data analysis was done using Duncan's Multiple Range Test; $^{A, B}$ means with the the same letter are not significantly different ($\alpha = 0.05$) when compared within a row.$^{a, b, c, d}$ means with the same letter are not significantly different ($\alpha = 0.05$) when compared within a column. Means and SEs of five replicate plants.

Adapted from Yong (1995).

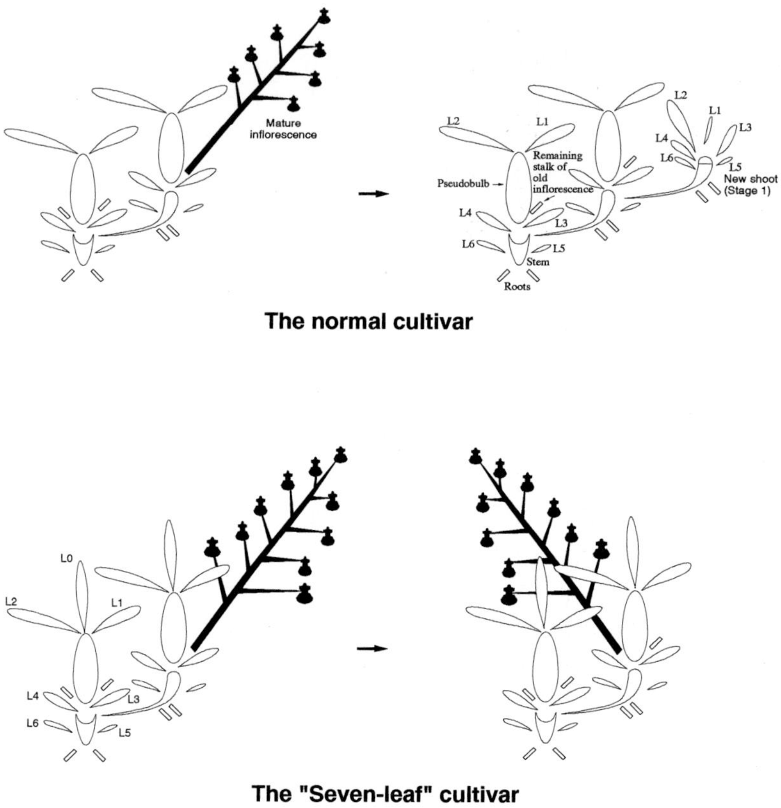

Fig. 7.11. The development of a second inflorescence on a shoot of the "Seven-leaf" cultivar of *Oncidium* Goldiana.

leaf L0 on the current shoot bearing the inflorescence will also produce assimilates and this will certainly increase the total amount of assimilates for flower development. Preliminary observations indicate that two inflorescences are produced in sequence. Each of these inflorescences is much larger in terms of dry mass, length, number of florets and side branches than the inflorescences produced by other *Oncidium* Goldiana plants. The physiological basis for the production of the second inflorescence is not known, but it is likely that the presence of an extra leaf may affect the correlative hormonal signals within the orchid. More work is needed to substantiate this hypothesis. Currently, many efforts have been undertaken to propagate the 'Seven-leaf' cultivar for future experiments and field trials. The selection of this cultivar for future replanting and mericloning is a practical alternative to increase flower production since the flowering period for each shoot is extended by two folds.

7.8. Concluding Remarks

Over the years, the selection of new orchid hybrids has only emphasised the aesthetic value of the inflorescence and no consideration is given to hybrids that exhibit greater partitioning of assimilates to the inflorescence. This suggests that there is still a great potential in increasing harvestable yield of tropical orchids. Although strategies favouring assimilate transfer to the harvestable component and increasing total biomass production hold great potential, it is important to note that increase in yield is never the result of a single factor. There exists a ceiling to which assimilates may be partitioned to the harvestable portion without affecting the capacity of the plant to support the yield component structurally and nutritionally.

Commercial orchid growers should consider the possibility of using elevated carbon dioxide to control and improve orchid flower production in view of the positive results obtained for the orchids tested so far and for many non-orchidaceous plants. Careful and well-planned usage of elevated carbon dioxide, coupled with the other physiological 'tools' such as plant hormones and the appropriate fertiliser, should allow the growers to 'speed-up' vegetative growth or to extend the flowering period in tandem with market demands. More work is still needed to study the effects of elevated carbon dioxide on the pattern of assimilate partitioning and how we can channel more carbon for flower development.

Considerable advancements have been achieved in understanding the patterns of assimilate partitioning in tropical orchids in the last five years. Much remains to be understood in the mechanisms of some partial processes along the source–path–sink system (e.g., phloem loading) and the regulation of photoassimilate partitioning between sources and sinks in orchids at the whole plant level.

7.9. Summary

1. Tropical orchids have a highly integrated pattern of assimilate partitioning in which both major sinks (inflorescence and vegetative apex) and minor sinks (leaves, stems and roots) receive ^{14}C-assimilates from nearby and distant leaves.
2. The relatively unrestricted assimilate movement between sources and sinks within an orchid suggests the high potential in diverting additional assimilates for inflorescence growth.
3. Inflorescence growth of orchids is primarily source-limited and larger inflorescences could be obtained by increasing source capacity through the usage of elevated CO_2 treatments, removal of competing sinks or possibly, by selecting a specific cultivar with additional source leaves.
4. Improvements in the harvestable yield of orchids grown for its cut-flowers should adopt a two-pronged approach which seeks to increase both the photosynthetic capacity of the source leaves and the ability of the inflorescence sink to import assimilates.

General References

Baker, D. A. and Milburn, J. A., 1989, *Transport of Photoassimilates* (Longman, Harlow), 384 pp.

Crafts, A. S. and Crisp, C. E., 1971, *Phloem Transport in Plants* (W. H. Freeman and Co., San Francisco), 481 pp.

Cronshaw, J., Lucas, W. J. and Giaquinta, R. T., 1986, *Phloem Transport* (Liss, New York), 650 pp.

Delrot, S. and Bonnemain, J. L., 1985, "Mechanism and control of phloem transport," *Physiologie Végétale* **23**: 199–220.

Farrar, J. F., 1992, "The whole plant: Carbon partitioning during development," in *Carbon Partitioning Within and Between Organisms*, eds. C. J. Pollock, J. F. Farrar and A. J. Gordon (BIOS Scientific Publishers, Oxford), pp. 163–179.

Geiger, D. R. and Fondy, B. R., 1991, "Regulation of carbon allocation and partitioning: Status and research agenda," in *Recent Advances in Phloem Transport and Assimilate Compartmentation*, eds. J. L. Bonnemain, S. Delrot, W. J. Lucas and J. Dainty (Ouest Editions, Nantes, France), pp. 1–10.

Gifford, R. M. and Evans, L. T., 1981, "Photosynthesis, carbon partitioning and yield," *Annual Review of Plant Physiology* **32**: 485–509.

Halevy, A. H., 1987, "Assimilate allocation and flower development," in *Manipulation of Flowering*, ed. J. G. Atherton (Butterworths, London), pp. 363–378.

Hew, C. S., Clifford, P. E. and Yong, J. W. H., 1997, "Aspects of carbon partitioning in tropical orchids," *Journal of Orchid Society of India* **10** (in press).

Ho, L. C., 1988, "Metabolism and compartmentation of imported sugars in sink organs in relation to sink strength," *Annual Review of Plant Physiology and Plant Molecular Biology* **39**: 355–378.

Kursanov, A. L., 1984, *Assimilate Transport in Plants*, translated from Russian by V. Vopian. (Elsevier, Amsterdam, New York and Oxford), 660 pp.

Marshall, C., 1990, "Source–sink relations of interconnected ramets," in *Clonal Growth in Plants: Regulation and Function*, eds. J. van Groenendael and H. de Kroon (SPB Academic Publishing, The Hague, Netherlands), pp. 23–41.

Nelson, C. D., 1963, "Effect of climate on the distribution and translocation of assimilates," in *Environmental Control of Plant Growth*, ed. L. T. Evans (Academic Press, New York), pp. 149–174.

Patrick, J. W., 1988, "Assimilate partitioning in relation to crop productivity," *HortScience* **23**: 33–40.

Sachs, R. M., 1987, "Roles of photosynthesis and assimilate partitioning in flower initiation," in *Manipulation of Flowering*, ed. J. G. Atherton (Butterworths, London), pp. 317–340.

Turgeon, R., 1989, "The sink–source transition in leaves," *Annual Review of Plant Physiology and Plant Molecular Biology* **40**: 119–138.

Van Bel, A. J. E., 1993, "Strategies of phloem loading," *Annual Review of Plant Physiology and Plant Molecular Biology* **44**: 253–281.

Wardlaw, I. F., 1990, "The control of carbon partitioning in plants," *New Phytologist* **116**: 341–381.

Wareing, P. F. and Patrick, J. W., 1975, "Source–sink relations and the partition of assimilates in the plant," in *Photosynthesis and Productivity in Different Environment*, ed. J. P. Cooper (Cambridge University Press, Cambridge), pp. 481–499.

References

Clifford, P. E., Neo, H. H. and Hew, C. S., 1992, "Partitioning of ^{14}C-assimilate between sources and sinks in the monopodial orchid *Aranda* Tay Swee Eng," *Annals of Botany* **69**: 209–212.

Clifford, P. E., Neo, H. H, Ma, C. W. and Hew, C. S., 1994, "Photosynthate partitioning in tropical orchids," *Singapore Journal of Primary Industry* **22**: 1–7.

Clifford, P. E., Neo, H. H. and Hew, C. S., 1995, "Regulation of assimilate partitioning in flowering plants of the monopodial orchid *Aranda* Noorah Alsagoff," *New Phytologist* **130**: 381–389.

Hew, C. S. and Lee, F. Y., 1989, "Control of flowering by floral bud removal in *Aranda* Christine under tropical field conditions," *Journal of the Japanese Society of Horticultural Science* **58**: 691–695.

Hew, C. S., Hin, S. E., Yong, J. W. H., Gouk, S. S. and Tanaka, M., 1995, "*In vitro* CO_2 enrichment of CAM orchid plantlets," *Journal of Horticultural Science* **70**: 721–736.

Kimball, B. A., 1983, "Carbon dioxide and agricultural yield: An assemblage and analysis of 430 prior observations," *Agronomy Journal* **75**: 779–789.

Mcconnell, J., Guerrero, Leon, R. and Mafnas, J., 1990, " Environmental factors affecting flowering in some vandas and dendrobiums in the tropics," *Proc. of the Nagoya International Orchid Show* (1990), pp. 174–175.

Neo, H. H., Clifford, P. E. and Hew, C. S., 1991, "Partitioning of ^{14}C-photosynthates between sources and sinks in monopodial orchids," *Singapore Journal of Primary Industry* **19**: 94–103.

Neo, H. H., 1993, "Photosynthate partitioning in orchids." M.Sc. Dissertation. Department of Botany, The National University of Singapore, 98 pp.

Paull, R. E., Leonhardt, K. W., Higaki, T. and Imamura, J., 1995, "Seasonal flowering of *Dendrobium* 'Jaquelyn Thomas' in Hawaii," *Scientia Horticulturae* **61**: 263–272.

Pitelka, L. F. and Ashmun, J. W., 1985, "Physiology and integration of ramets in clonal plants," in *Population Biology and Evolution of Clonal Organisms*, eds. J. B. C. Jackson, L. W. Buss and R. E. Cook (Yale University Press, New Haven and London), pp. 399–435.

Rogers, H. H. and Dahlman, R. C., 1993, "Crop responses to CO_2 enrichment," in *CO_2 and Biosphere*, eds. J. Rozema, H. Lambers, S. C. Van De Geijn and M. L. Cambridge (Kluwer Academic Publishers, Dordrecht), pp. 117–131.

Singh, B. K. and Pandey, R. K., 1980, "Production and distribution of assimilate in chickpea (*Cicer arietinum* L.)," *Australian Journal of Plant Physiology* **7**: 727–735.

Thrower, S. L. and Thrower, L. B., 1980, "Translocation into mature leaves — the pathway of assimilate movement," *New Phytologist* **86**: 145–154.

Wadasinghe, S. and Hew, C. S., 1995, "The importance of back shoots as a source of photoassimilates for growth and development in *Dendrobium* Jashika Pink (Orchidaceae)," *Journal of Horticultural Science* **70**: 207–214.

Yong, J. W. H., 1995, "Photoassimilate partitioning in the sympodial thin-leaved orchid *Oncidium* Goldiana," M.Sc. Dissertation. Department of Botany, The National University of Singapore, 132 pp.

Yong, J. W. H. and Hew, C. S., 1995a, "Partitioning of ^{14}C-assimilates between sources and sinks during different growth stages in the sympodial thin-leaved orchid *Oncidium* Goldiana," *International Journal of Plant Sciences* **156**: 188–196.

Yong, J. W. H. and Hew, C. S., 1995b, "The patterns of photoassimilate partitioning within connected shoots for the thin-leaved sympodial orchid *Oncidium* Goldiana during different growth stages," *Lindleyana* **10**: 92–108.

Yong, J. W. H. and Hew, C. S., 1995c, "The importance of photoassimilate contribution from the current shoot and connected back shoots to inflorescence size in the thin-leaved sympodial orchid *Oncidium* Goldiana," *International Journal of Plant Sciences* **156**: 450–459.

Zimmerman, J. K., 1990, "Role of pseudobulbs in growth and flowering of *Catasetum viridiflavum* (Orchidaceae)," *American Journal of Botany* **77**: 533–542.

Chapter 8

Flower Senescence and Postharvest Physiology

8.1. Introduction

The process of senescence is an important developmental program in plants. A basic understanding of the physiology of flower senescence is crucial to the development of postharvest technology. To the cut-flower industry, the ability to retard flower senescence or to prolong the vase-life of cut-flowers is vital. Chemical solutions to extend vaselife in cut-flowers are experimentally formulated to inhibit certain physiological processes along the complex pathway of senescence.

This chapter aims at understanding the basic physiology and biochemistry associated with orchid flower senescence. Emphasis will be given to the efforts made in developing appropriate postharvest technology for tropical orchid cut-flower industry.

8.2. Senescence in Plants

Plants senesce in many different ways according to their habit of growth. In the first type, the whole plant senesces and dies at one time (e.g., annuals). Next, we have the progressive senescence of plant parts as the whole plant ages. Usually, the oldest parts (e.g., older leaves) senesce and die while other parts remain alive and active. Third, there may be a simultaneous or sequential senescence of a part of the plant (e.g., the top of a biennial or perennial) while the rest remains active. Lastly, certain cells senesce and die (e.g., xylem) while the whole plant is actively growing.

Therefore, senescence is distinct from aging. Senescence may be simply defined as those changes that lead eventually to the death of an organism or some part of it. Aging refers to processes accruing maturity with the passage of time. To avoid confusion, the usage of the three terms — maturation, aging and senescence — as defined by Avadhani *et al.* (1994) is adopted:

(1) Maturation will be used to denote the gradual changes that result from the genetic program of an individual. For orchid flowers, maturation consists of events that occur within a short period after anthesis.
(2) Aging refers to the changes in time without reference to death. In the case of orchid flowers, this would be changes in all segments, essential and non-essential, and occurs gradually as a degradative process.
(3) Senescence is that phase (or the final phase) of the aging process that leads to death. The term has been used to describe the changes that occur in the shedding of leaves in deciduous plants or during the ripening of fruits.

Information on the ultrastructural, biochemical and physiological changes associated with flower senescence is obtained mainly through studies on morning glory, carnation, rose, chrysanthemum and others. Although, there is a fair amount of information on pollination-induced senescence in orchid flowers (Avadhani *et al.*, 1994), by contrast, we know little about the senescence of unpollinated orchid flowers. The ultrastructural, biochemical and physiological changes associated with flower senescence have been reviewed by Halevy and Mayak (1979). Only a brief account of flower senescence will be described here.

The ultrastructural changes in petal senescence are of two types: For petal without plastids, invagination of the tonoplast is the first observed sign of senescence. Autophagic activity of the vacuole is the next event. The destruction of the vacuole and the subsequent release of digestive enzymes result in the death of the cells. For petals with plastids, the first microscopically visible change is the breakdown of plastids. The disappearance of tonoplast and plasmalemma will set in later.

An increase in free space and membrane permeability is observed during flower senescence. The enhanced permeability of the plasma membrane resulting from a decrease in phospholipids causes cell leakage. Increase in respiration and enhanced hydrolysis of cellular components are two major biochemical and physiological changes associated with petal senescence. Breakdown of macromolecular components such as starch, protein and nucleic acid has been observed during senescence.

Discolouration or fading of colour is a common feature associated closely with senescence. The major classes of pigments responsible for flower colour are carotenoids and anthocyanins. Co-pigmentation with other flavonoids and related compounds often determine the intensity of colour in most flowers and the degree of co-pigmentation is greatly influenced by pH.

Blueing is often observed in red flowers (e.g., rose and morning glory) during senescence and it is also influenced by changes in pH. An increase in free ammonia resulting from the breakdown of protein causes the cytoplasmic pH to rise. The loss of fresh mass (i.e., wilting) in petal is the final stage of senescence.

Among the five natural plant hormones, ethylene has been implicated to play an important role in regulating senescence of some flowers. The effects of ethylene in flowers are as follows:

(1) The in-rolling of carnation petals, often termed 'sleepiness.'
(2) Fading and in rolling of corolla of *Ipomea* flowers.
(3) Fading and wilting of sepal tips in orchid labella.
(4) Induction of anthocyanins formation in orchid labella.

8.3. Growth and Development of Orchid Flower and Inflorescence

For carnation and rose, the number of flowers is usually not more than three or four, and each flower is subtended by green leaves. The number of flowers (or florets) per stalk in an orchid inflorescence, for example, in *Oncidium* Goldiana, may be up to 70 and there are no leaves on the inflorescence. The flowers along an inflorescence may open at various times making the studies of flower senescence difficult. Two approaches are usually adopted for studies when examining the physiological factors regulating the senescence of the orchid blooms. By using detached flowers, the senescence itself can be studied more precisely. Alternatively, the growth and development of flowers along the axis of an inflorescence can be studied.

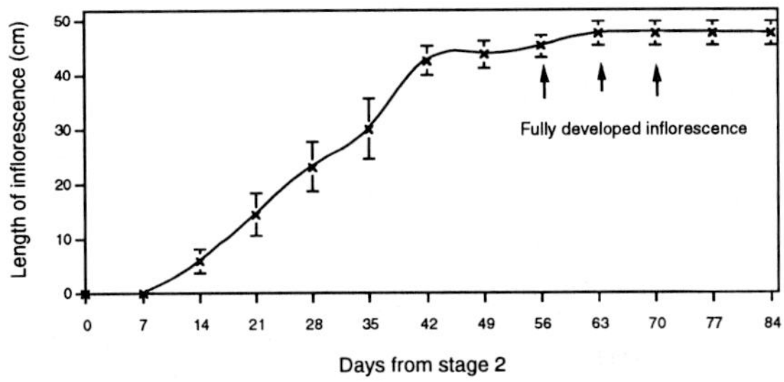

Fig. 8.1. The rate of growth of an *Oncidium* inflorescence.

Redrawn from Hew & Yong (1994).

The growth of an orchid inflorescence varies considerably. For example, in *Oncidium* Goldiana, the average time taken for an inflorescence to develop is about 56–70 days (Fig. 8.1). Evidently, the rate of growth depends on cultural and environmental conditions. For *Dendrobium* Pompadour, the inflorescence may bear as much as 30 flowers with varying stalk length. Buds open acropetally along the stalk when they reach a size of 2.8 cm in length and 4.8 cm in width. The time taken for full floral anthesis is about 16.5 h. For *Aranda* Wendy Scott and *Aranda* Christine 1, the fresh mass, dry mass, anthocyanin content, water potential and protein content are lower in the bud stage than in the fully opened flowers (Figs. 8.2, 8.3). Osmotic concentration and sugar content, on the other hand, are higher in the buds and decrease rapidly during the development of the flowers (Fig. 8.3). The dry matter of a bud is usually about one-third to one-fourth of the fully opened flower. A similar pattern has also been observed for *Dendrobium* Multico. Fully opened detached flowers of *Dendrobium* Pompadour pass through four visually distinct floral stages during aging and senescence (Fig. 8.4).

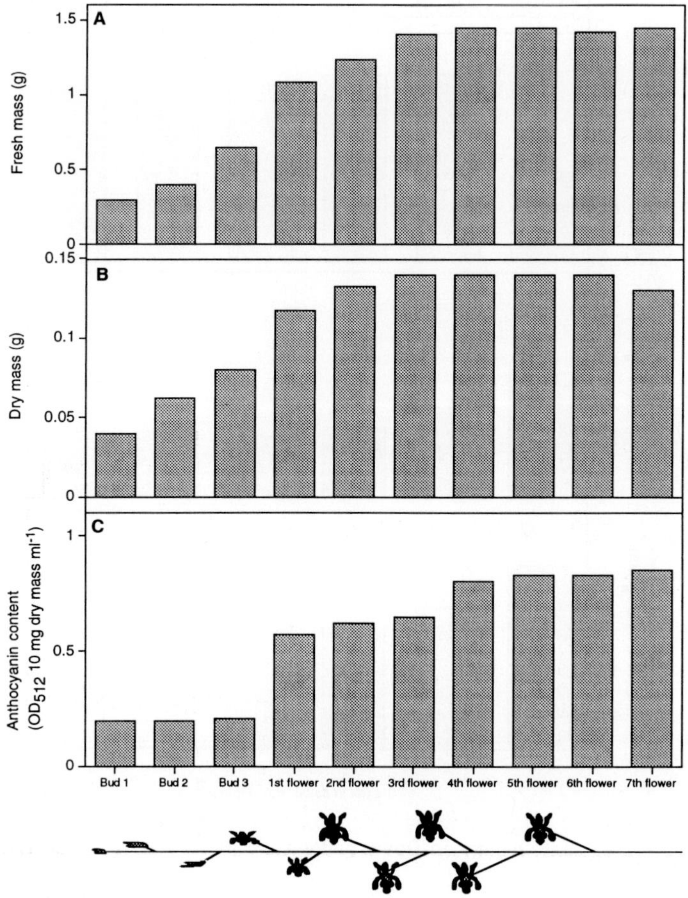

Fig. 8.2. Developmental changes in buds and flowers along the axis of an inflorescence of *Aranda* Wendy Scott.
Note: Changes in (A) fresh mass, (B) dry mass and (C) anthocyanin content.

Redrawn from Hew (1980).

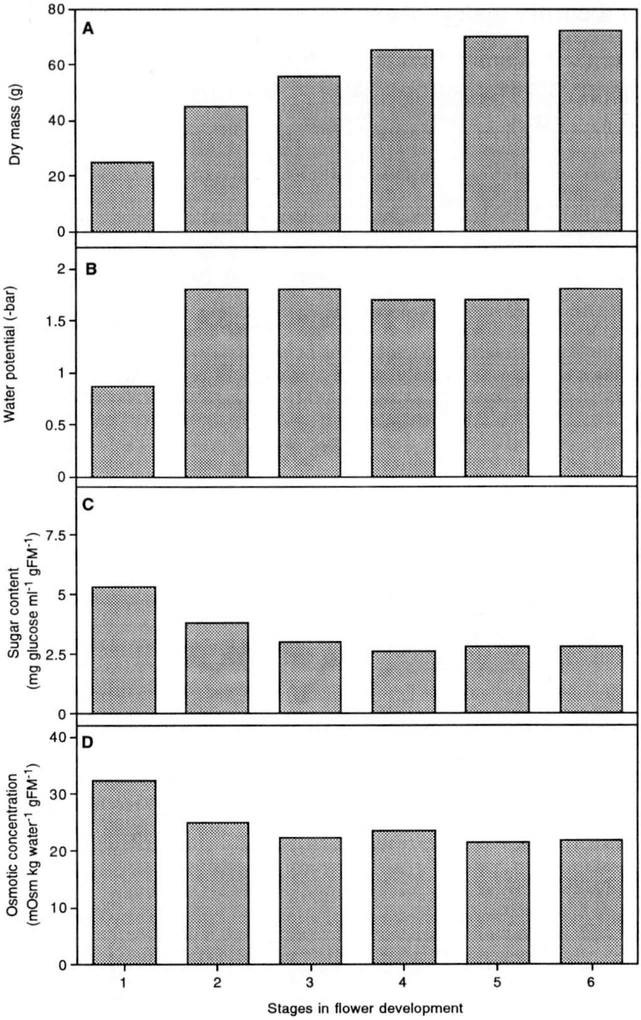

Fig. 8.3. Developmental changes in flowers along the axis of an inflorescence of *Aranda* Christine 1.

Note: Changes in (A) dry mass, (B) water potential, (C) sugar content and (D) osmotic concentration of flowers.

Adapted from Hew, Wee, Wong, Ong & Lee (1989).

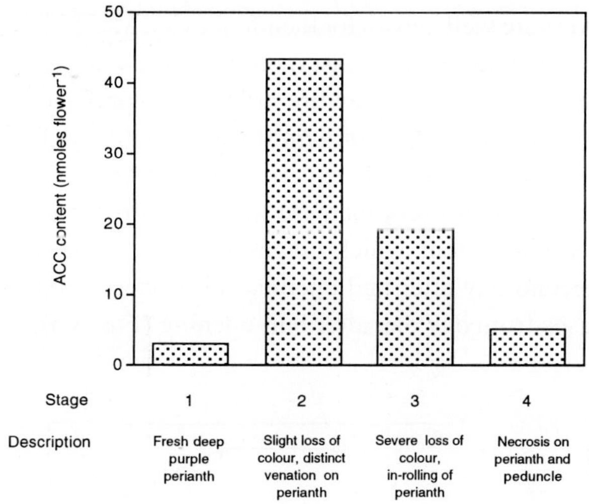

Fig. 8.4. Visual changes and ACC content of *Dendrobium* Pompadour flowers at various stages of senescence.

Redrawn from Nair & Tung (1987).

Table 8.1. Life span of some orchid flowers.

Orchid	Lifespan (days)
Aranda Wendy Scott	24–32
Arundina graminifolia	5
Dendrobium crumenatum	1
Dendrobium Rose Marie	30
Dendrobium Jacquelyn Thomas	16
Dendrobium Louisae Dark	44–45
Paphiopedilum villosum	70
Phalaenopsis violacea	30
Vanda suavis	60
Vanda Tan Chay Yan	28

Adapted from Arditti (1992).

Orchid flowers are well known for their longevity. Many of the economically important tropical orchid flowers last more than a few weeks (Table 8.1). For scientific study, flowers of shorter life are usually preferred. The relatively short life span of *Arundina graminifolia* flowers (5–6 days from bud opening to senescence) makes it a suitable experimental material. The fresh mass and dry mass of *Arundina* flowers increase after bud opening. There are signs, such as anthocyanin content, which suggest that senescence might have taken place on the second day after bud opening. Phosphorus content in *Arundina* flowers decreases from day one after bud opening (Fig. 8.5).

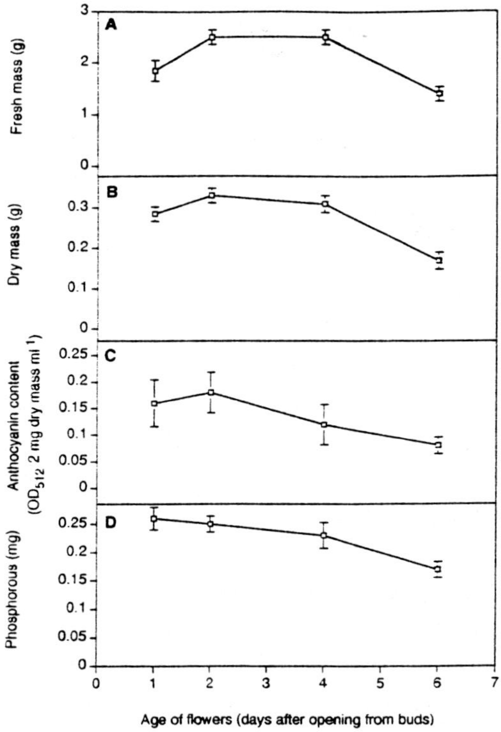

Fig. 8.5. Developmental changes in *Arundina graminifolia* flowers.
Note: Changes in (A) fresh mass, (B) dry mass, (C) anthocyanin content and (D) phosphorous content.

Adapted from Lim, Chin & Hew (1975).

The buds of *Cymbidium* and *Aranda* generally contain the highest level of sugar. For the *Cymbidium* flowers, the sugar level in the perianth decreases with age. With the exception of aspartate, the other amino acids increase in the *Cymbidium* flower during development.

There are no significant changes in the activities of peroxidase and acid phosphatase in the first four days after bud opening for the *Arundina* flowers. The enzyme activities increase markedly thereafter. Multiple forms of peroxidase and acid phosphatase are obtained at different stages of flower development. However, there is no positive correlation between the increase of total enzyme activity and the number of isoenzyme bands. The peroxidase activity also increases gradually at different stages of development for the *Cymbidium* flowers. Increase in peroxidase and acid phosphatase activity during flower development has also been reported in other flowers.

It is well-documented that endogenous hormonal levels in roses and other flowers change with growth and development. Generally, young flowers have high cytokinin and gibberellin levels but are low in ABA content (Halevy and Mayak, 1979). Gibberellin-like activity is higher in buds than in mature flowers of *Arachnis* Maggie Oei. Both *Cymbidium faberi* and *Phalaenopsis aphrodite* have higher GA_3 content in the newly opened flowers than in those undergoing senescence. As for the other flowers, GA may also play a role in the aging and senescence of orchid flowers.

In roses, ABA level is higher in mature flowers and its role in regulating flower senescence is well-established. ABA-like substances have been detected in extracts of buds and mature flowers of *Arachnis* Maggie Oei and *Oncidium* Goldiana. Newly opened flowers of *Cymbidium faberi* contain less ABA than those undergoing senescence.

High levels of cytokinins are detected in developing inflorescence of *Cymbidium*. Cytokinins are also present in all parts of a *Cymbidium* flower. Their levels decrease following pollination. Zeatin is present in the newly opened flowers of *Cymbidium faberi* and *Phalaenopsis aphrodite* at a concentration of 0.43 and 0.51 mg per kilogram fresh mass, respectively. At present, we need more information about the changing pattern and levels of endogenous cytokinins during orchid flower development. Such information is important for the effective control of flowering by exogenous application of cytokinins (see Chap. 6 on Control of Flowering) and the extension of vase-life.

In relation to orchid flower senescence, the most extensively studied plant hormone is ethylene because many orchid flowers are sensitive to ethylene. The biosynthesis and regulatory role of ethylene will be discussed separately.

8.4. Flower Senescence in Orchids

Post-pollinated phenomena

As would be expected, there exists wide-ranging variation in the longevity of individual orchid flower. The duration of flower opening in some flowers persists for a few hours while others last few days. Many orchid cut-flowers persist for several weeks under normal cultural conditions. It is not easy to study the senescence of orchid flowers, particularly those for cut-flowers that have been selected for their longevity. Furthermore, it is also difficult to identify the specific signal for senescence or the onset of senescence during natural senescence of flowers. However, flower senescence can be induced and events leading to senescence can then be followed. Senescence can be induced following cutting, emasculation (removal of pollen) and pollination. Cutting of flowers (i.e., harvesting) affect senescence less dramatically. Emasculated flowers *in situ* or after being harvested, senesce and die faster than those that have been cut. Pollination generally hastens the senescence process. The post-pollination phenomenon has been extensively studied by Arditti and his associates. Readers are recommended to read the review articles in these areas (Arditti, 1992). The available experimental evidence suggests that the senescence of undisturbed, cut, emasculated and pollinated flowers are essentially the same. The differences lie only in the timing, rate and intensity of the physiological processes.

Post-pollination phenomena can be induced by a number of factors. Pollination will of course induce all of these phenomena under natural or horticultural conditions. Auxins can bring about many but not all the changes associated with post-pollination phenomena. Abscisic acid and gibberellins can initiate some of the events. Interactions between hormones and other substances can also affect the post-pollination phenomena.

The biochemical and metabolic changes in orchid flowers induced by pollination are listed below (Arditti, 1992; Avadhani *et al*., 1994):

(1) Increase in respiration.
(2) Increase in RNA synthesis or the production of new RNA, or both.
(3) Production of new proteins, increased synthesis of existing ones, or both.
(4) Activation and/or synthesis of several enzymes, transport of organic and inorganic substances from perianth segments into the gynostemium and ovary.
(5) Anthocyanin synthesis or destruction.
(6) Chlorophyll production or destruction.
(7) Cessation of scent production.
(8) Hydrolysis of storage and structural molecules.
(9) Appearance of yellow pigments.
(10) Starch accumulation in ovaries.
(11) Ethylene evolution.

Additional changes induced by pollination are as follows:

(1) Swelling of ovaries.
(2) Changes in pedicel curvature.
(3) Closure of stigmas.
(4) De-resupination.
(5) Swelling and loss of curvature by the gynostemium.
(6) Ovule development.
(7) Senescence of some or all perianth parts.
(8) Re-differentiation of some floral segments.
(9) Nastic movements, such as hyponasty of sepals and petals.

The advantage of reducing flower longevity following pollination has been discussed by Stead (1992). He believes that the advantage may be twofold.

First, a shorter flower longevity would ensure that no excessive amount of pollen will be deposited upon the stigma for a full seed set. Any further deposition of pollen is deemed wasteful as growth of excessive pollen tubes competes for a limited pool of resources. Second, the maintenance of elaborate floral structures is a costly process in terms of water and energy. To achieve cost-effectiveness, the strategies taken by plants following pollination are:

(1) Reduction or modification of nectar composition.
(2) Structural modification (e.g., corolla wilting) and modification of colour (e.g., fading).
(3) Abscission of all or part of the corolla.

Pollination appears to affect floral longevity of long-lived flowers but not the short-lived ones. Orchid flowers are long-lived and as described earlier, their senescence is significantly affected by pollination.

Ethylene and senescence

Orchid flowers are particularly sensitive to ethylene. Davidson (1949) is probably the first scientist to give a full account of the injury of orchid flowers caused by ethylene. He found that concentration as low as 0.002 μl $litre^{-1}$ for 24 h or 0.1 μl $litre^{-1}$ for 8 h can damage the sepals of *Cattleya* flowers that have started to open. The injury to orchid flowers is characterised by a progressive drying and bleaching of the sepals beginning at the tips and extending towards the bases. Abnormalities in the sepals become apparent as the bloom reaches maturity. The dry-sepal injury is a major cause of flower loss for producers in areas with poor air quality.

Ethylene is produced by orchid flowers. For *Oncidium* Goldiana flowers, ethylene production starts after a lag period of 100 h after harvest (Fig. 8.6). It shows a climacteric-like burst that peaks at the 265th h. For *Dendrobium* Pompadour flowers, no ethylene evolution is detectable even after one week of excision. Flowers detached among 10 and 19 days after floral opening produce negligible amount of ethylene. Ethylene begins to be evolved at stage 2 (Fig. 8.7) and reaches a peak at stage 3. This coincides with the folding-in of the perianth as the flower senesces. The timing of an upsurge in ethylene production follows closely the changes in ACC in the tissues (Fig. 8.4). Exogenous ACC application accelerates senescence and abscission of fully mature flowers while the immature flowers are unaffected by ACC. Immature floral tissues appear to lack the ability to convert ACC to ethylene, as has been reported for pre-climacteric fruit. This interesting finding suggests that ACC synthase system only becomes fully operational with full anthesis in *Dendrobium* flowers.

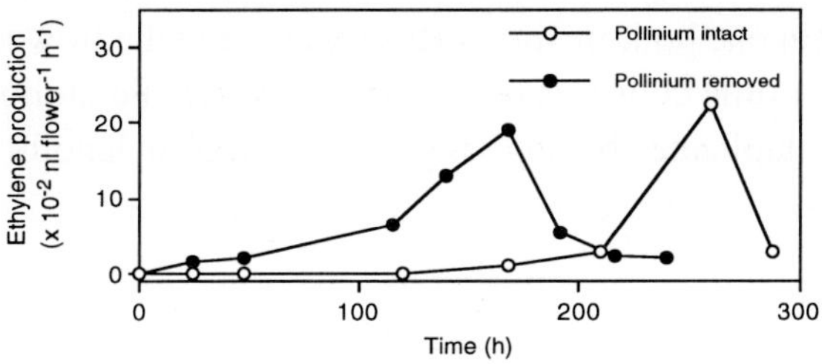

Fig. 8.6. Production of ethylene by emasculated (removal of pollinium) and control flowers of *Oncidium* Goldiana.

Adapted from Nair (1984).

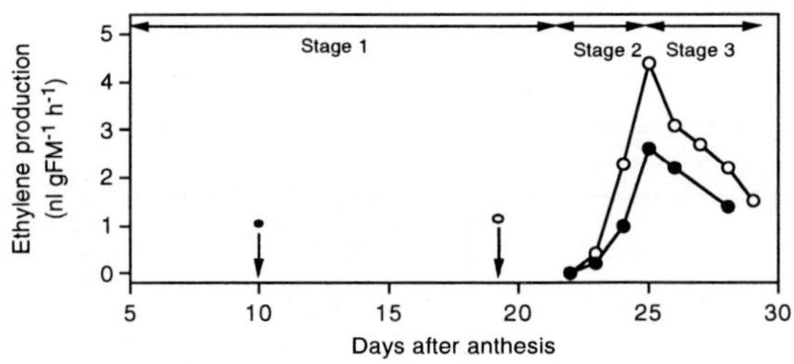

Fig. 8.7. Ethylene evolution by flowers of *Dendrobium* Pompadour.
Note: The flowers were harvested at 10 and 19 days from full anthesis (opening). Arrows indicated on the figure refer to the day of harvest.

Adapted from Nair (1984).

The pathway of ethylene biosynthesis has been well worked out (Fig. 8.8). Methionine is first converted to S-adenosylmethionine (SAM). ACC synthase then catalyses the conversion of SAM to 1-aminocyclopropane-1-carboxylic acid (ACC). Since this enzyme requires pyridoxal phosphate for maximum activity, it is inhibited by aminoethoxyvinylglycine (AVG) and aminooxyacetic acid (AOA) which are two well-known inhibitors for the pyridoxal phosphase enzymes. In air, ACC is oxidised by ACC oxidase to ethylene. The functional nature of the gene responsible for encoding ACC synthase is well-established, but not for ACC oxidase. Work on ACC oxidase is actively being pursued and we will soon know the nature and genetic control of this enzyme.

Pollination induces ethylene production. Self pollination induces ethylene production within one hour in *Vanda* Rose Marie and the flower fades within 8 to 10 h. A similar time course is observed with *Vanda* Petamboeran (Fig. 8.9). This response is duplicated by applying 5 mM of IAA in lanolin paste. Ethylene

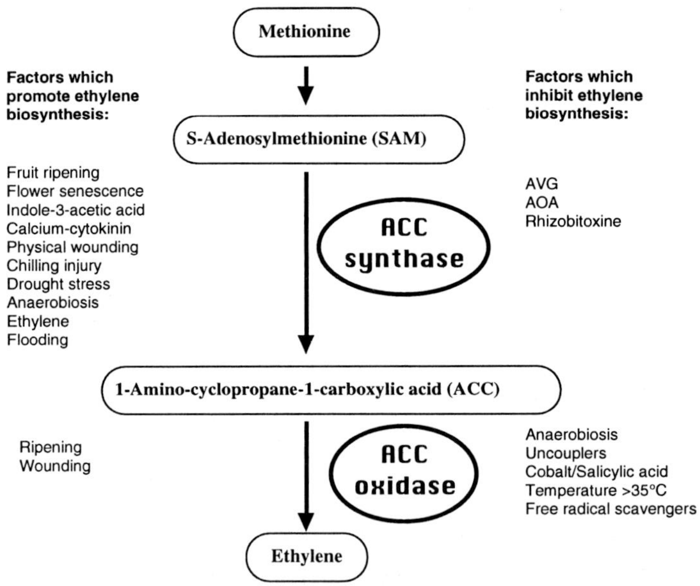

Fig. 8.8. Ethylene biosynthesis and its regulation in higher plants.

Redrawn from Abeles, Morgan & Saltveit (1992), Yang & Hoffman (1984) and Mathooko (1996).

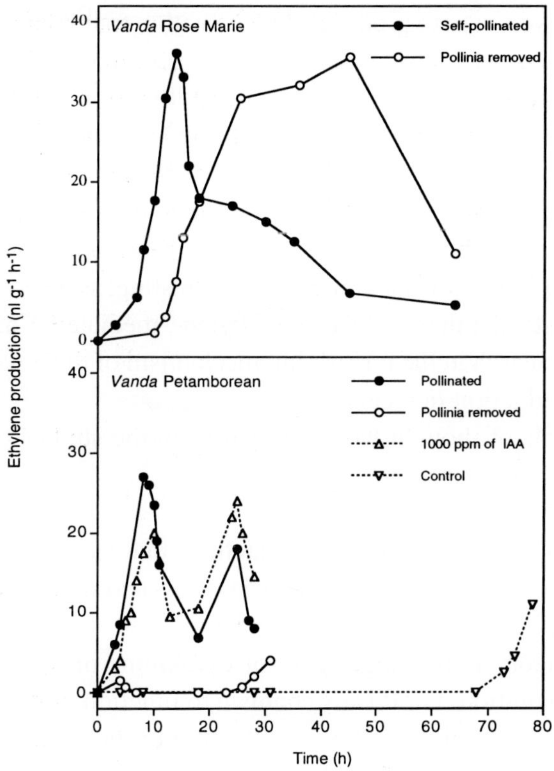

Fig. 8.9. Ethylene production by *Vanda* flowers.

Note: (A) Production of ethylene following self-pollination and emasculation (the removal of pollinia) of *Vanda* Rose Marie. (B) Evolution of ethylene by *Vanda* Petamborean after pollination (of flowers with pollinia intact), application of 1000 ppm (or 5 mM) IAA in lanolin to stigmas and removal of pollinia (emasculation).

Redrawn from Burg & Dijkman (1967) and Dijkman & Burg (1970).

is evolved primarily by the column and lip and, to a lesser extent, by the perianth. Earlier experiments suggest that pollination causes a transfer of auxin from the pollen to the stigma, resulting in the spread of a growth hormone to the column and lip, and the induction of ethylene formation in these tissues. Later experiments suggest a possible role for ACC as the inter-organ translocation signal in ethylene production following pollination.

Cattleya flowers start to produce ethylene within four hours after pollination. Ethylene production by *Cymbidium* flowers starts after two hours of treatment and becomes noticeable within 4–12 h after pollination. For the other flowers such as *Phalaenopsis* and *Arachnis*, ethylene is induced 10–12 h after pollination. Emasculation (removal of pollinia) also induces ethylene production but there is a longer lag period before ethylene evolution (Fig. 8.9). The findings that emasculation or the mere dislodgment of the anther cap caused the onset of several post-pollination phenomena are interesting, as the available evidence strongly suggests that these effects are ethylene-mediated. This has promoted considerable research in the underlying mechanism of the emasculation effect on ethylene production in recent years.

Three schools of thought have evolved from the studies on emasculation (Avadhani *et al.*, 1994):

(1) Removal of pollinia injures the rostellum that starts to produce what is probably wound-induced ethylene. That stress and wound-induced ethylene production is well-documented in plant system.
(2) Another theory is that high level of cytokinins in the pollinia prevents ethylene evolution. For this mechanism to function, it would be necessary for cytokinins from the pollen to diffuse into the rostellum. Evidence in supporting this theory is based on the diffusion of auxin from the pollen to the stigma.
(3) It has been shown more recently that desiccation following emasculation plays a major role in the induction of ethylene evolution in *Cymbidium* and *Phalaenopsis* flowers. Under conditions of high relative humidity (100%) and covering the rostellum with water insoluble grease, the normal response to emasculation (i.e., increased ethylene production, lip colouration and wilting) is absent (Table 8.2). However, under the only condition of high relative humidity, this response can be restored by the addition of ACC to the rostellar surface. Under conditions of low relative humidity, the response is inhibited by AVG and the inhibition could be partially reversed by addition of ACC (Fig. 8.10). Changes in lip colouration following emasculation in *Cymbidium* are visible within 36 h. This has led to the investigation of the signal responsible for the rapid senescence

occurring at short distance (several centimetres) from the site of desiccation. It has been suggested that following emasculation, ACC is synthesised and transported from the central column to the other floral parts where it is converted to ethylene. More information is needed pertaining to the signal that causes ACC to accumulate following emasculation. It has been suggested that ethylene could be the signal.

When all available evidences are considered, the three views (mechanical injury, cytokinin effects and desiccation) are complementary rather than contradictory in nature (Avadhani *et al.*, 1994). Both the injury and the desiccation processes create stress that is expected to lead to ethylene production, especially in the absence of cytokinins. More recently, attempts have been made to differentiate between pollination and emasculation in relation to ethylene

Table 8.2. Effects of various treatments on lip colouration in *Cymbidium* and on wilting in *Phalaenopsis* at low and high relative humidity.

Treatment	Time to lip coloration in *Cymbidium* (day)	Time to wilting in *Phalaenopsis* (day)
Low humidity		
Intact	10	6.5
Emasculation (E)	1.5	2.5
E + water	1.5	2.5
E + AVG (10.0 nmol)	18	5
E + AVG (10.0 nmol) + ACC (2.0 nmol)	1	5
E + grease (water-insoluble)	12	5.5
E + wet capillary tube	7	6
High humidity		
Intact	11	10
Emasculation	12	7
E + water	10	8
E + ACC (2.0 nmol)	2	2
E + wet capillary tube	3	3

Note: Low RH = 60 %; High RH = 100 %; n = 5.

Redrawn from Woltering & Harren (1989).

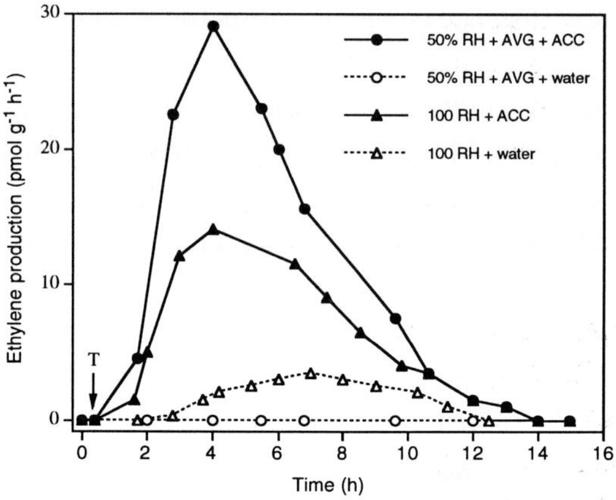

Fig. 8.10. The effects of ACC and AVG on ethylene production in emasculated *Cymbidium* flowers.

Note: Flowers were placed under conditions of high (100%) relative humidity and the rostellum was subsequently treated with water (1.0 mm^3) or with ACC (1.0 nmol in 1.0 mm^3). Similarly, flowers treated previously with AVG (50.0 nmol in 10.0 mm^3 applied to the rostellum) were placed in low humidity (50%) and subsequently treated with water (2.0 mm^3) or with ACC (2.0 nmol in 2.0 mm^3). T = start of experiment.

Adapted from Woltering & Harren (1989).

production. Evidence shows that the effect of pollination on petal senescence is not similar to the effect of senescence caused by emasculation. Pollination induces an increase in ethylene synthesis and tissue sensitivity to ethylene.

The suggestion that ACC may act as an inter-organ translocation signal for ethylene production following pollination has received considerable interest. All these arise from the findings that the complete ethylene biosynthetic pathway is not active in all the different flower parts. High ACC synthase and ACC oxidase activities are present in the columns. The perianth, however, has only ACC oxidase activity but not ACC synthase activity. Consequently, the ethylene biosynthesis in the perianth is dependent on the translocation of ACC from other flower parts such as column and pollen. ACC is present in fair amounts in the pollen and it is translocated to the other floral parts following pollination.

By examining the spatial and temporal location of ethylene biosynthesis within the orchid flower following pollination, we now have a better understanding of how the process of ethylene biosynthesis is regulated. These regulatory factors influence the expression of genes responsible for the encoding of the two major enzymes (ACC synthase and ACC oxidase) in the ethylene biosynthesis pathway. The model for the regulation of pollination-induced ethylene production that caused the senescence of orchid flowers, as suggested by O' Neill and her associates, is summarised in Fig. 8.11. In the orchid system, the pollen-derived auxin induces the expression of genes encoding enzymes involved in ethylene biosynthesis. Ethylene, however, is required for the gene expression of both auxin-induced ACC synthase and ACC oxidase and the full spectrum of pollination induced developmental events. They believe that ACC produced by the column, and not pollen-derived ACC, is translocated to the

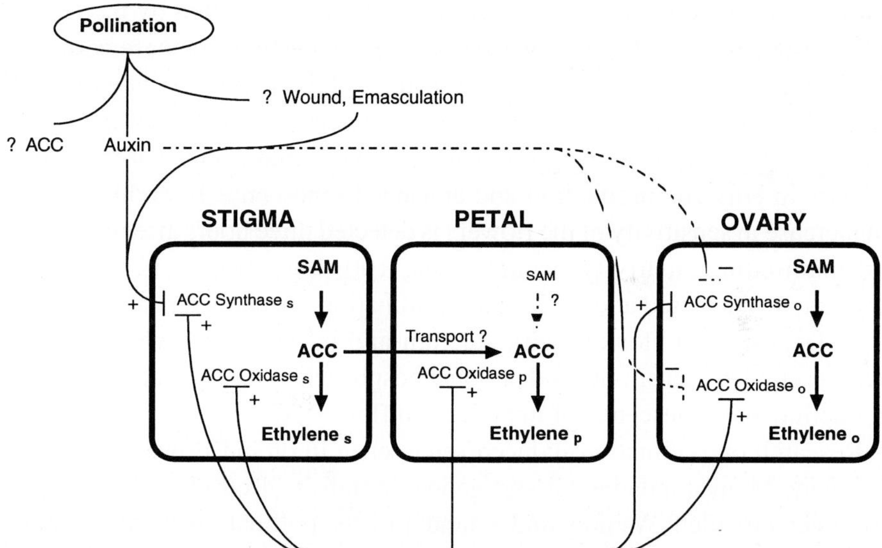

Fig. 8.11. A model for the regulation of pollination-induced ethylene production that causes senescence of *Phalaenopsis* flowers.

Note: stigma (s), petal (p) and ovary (o). The regulation of pollination-induced ethylene production that brings about senescence of the perianth and the other developmental events of post-pollination syndrome is proposed in this scheme.

Redrawn from Nadeau, Bui, Zhang & O'Neill (1993).

perianth (petal). This ACC stimulates ethylene production by the perianth which in turn regulates their senescence. They also believe that the transmitted signal, which acts to propagate the pollination response throughout the flower is ethylene produced by the perianth. On the contrary, results obtained recently by Woltering and his associates using ^{14}C-ACC as a tracer on *Cymbidium* flowers does not confirm the presumed role of ACC as a signal in interorgan communication during orchid flower senescence. In these flowers, ethylene produced in the stigmatic region following pollination or emasculation serves as a mobile factor responsible for senescence symptoms observed in the other flower parts.

This phenomenon of pollination-induced flower senescence has been observed in many, but not all, flowers. The main event that occurs following pollination is an increase in ethylene production. It is important to note that many processes in plant development are not only controlled by the level of plant hormones *per se*, but also by the sensitivity of the tissue to these hormones. On the basis of the research work carried out by Halevy and associates working on *Phalaenopsis* flowers, there is strong evidence to indicate that increase in ethylene sensitivity following pollination is the initial event that triggers an increase in ethylene production and enhanced senescence in orchid flowers. An increase in sensitivity of the flowers is detected three hours after pollination and maximum sensitivity occurs at the 10th hour (Fig. 8.12). Ethylene production is detected about 12 h after pollination, reaching a peak at the 30th hour. Treatment of flowers with calcium and its inophore, which serve to increase ethylene sensitivity and protein phosphorylation, also leads to ethylene production and senescence of unpollinated flower (Table 8.3).

Research on pollination-induced senescence in orchid flower has come a long way. Many fundamental and genetic controls of ethylene biosynthesis have been unveiled. We now understand how the pollination event is initially perceived by the flower and how the signal is communicated to the other parts of the flower to cause the developmental events leading to petal senescence. More work is needed to identify the sensitivity factor. Ethylene sensitivity is known to be a major factor in determining flower longevity (Fig. 8.13). The same study carried out on *Phalaenopsis* should also be extended to *Aranda*, *Oncidium*, *Mokara* and *Dendrobium* flowers that form the backbone of orchid cut-flower production in the tropics (Hew, 1994).

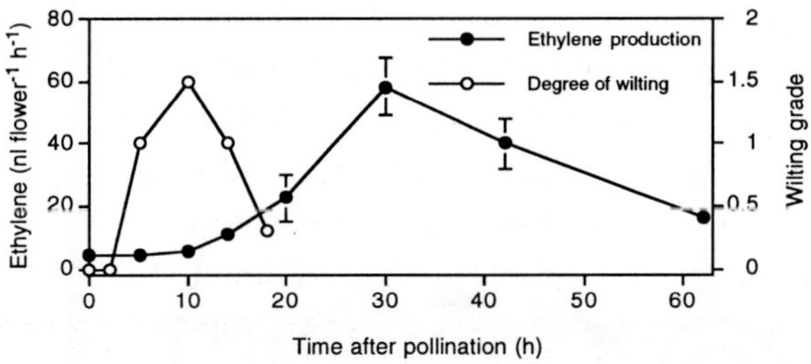

Fig. 8.12. Time course of changes in ethylene production and sensitivity to ethylene following pollination of *Phalaenopsis* flowers.
Note: The changes in ethylene sensitivity are represented by the increase in wilting grades following exposure to 4 μl l^{-1} of ethylene for 4 h at different periods after pollination as compared to pollinated flowers not exposed to ethylene. The values were measured 12 h after the ethylene treatment (n = 4, ± SE).

Redrawn from Porat, Halevy, Serek & Borochov (1995).

Table 8.3. Effects of calcium ions on the sensitivity of pollinated *Phalaenopsis* flowers to ethylene.

Treatment	Time to 50 % wilting (h)	Percent of control (%)
Unpollinated	76 ± 4	–
Pollinated	66 ± 4	100
+ Ca^{2+} and A23187	40 ± 4	66
+ EGTA	84 ± 6	127
+ $LaCl_3$	84 ± 6	127

Note: Flowers were held in 0.5 mM AOA together with either 1 mM $CaCl_2$ and 2 μM A23187 or 10 mM EGTA or 10 mM $LaCl_3$ and then pollinated and exposed to 0.5 μl l^{-1} ethylene for 12 h. Values are means of 4 replicates ± SE.
AOA = (Aminooxy)acetic acid; A23187 = ionophore; $LaCl_3$ = Lanthanum chloride; EGTA is a calcium chelator.

Redrawn from Porat, Halevy, Serek & Borochov (1995).

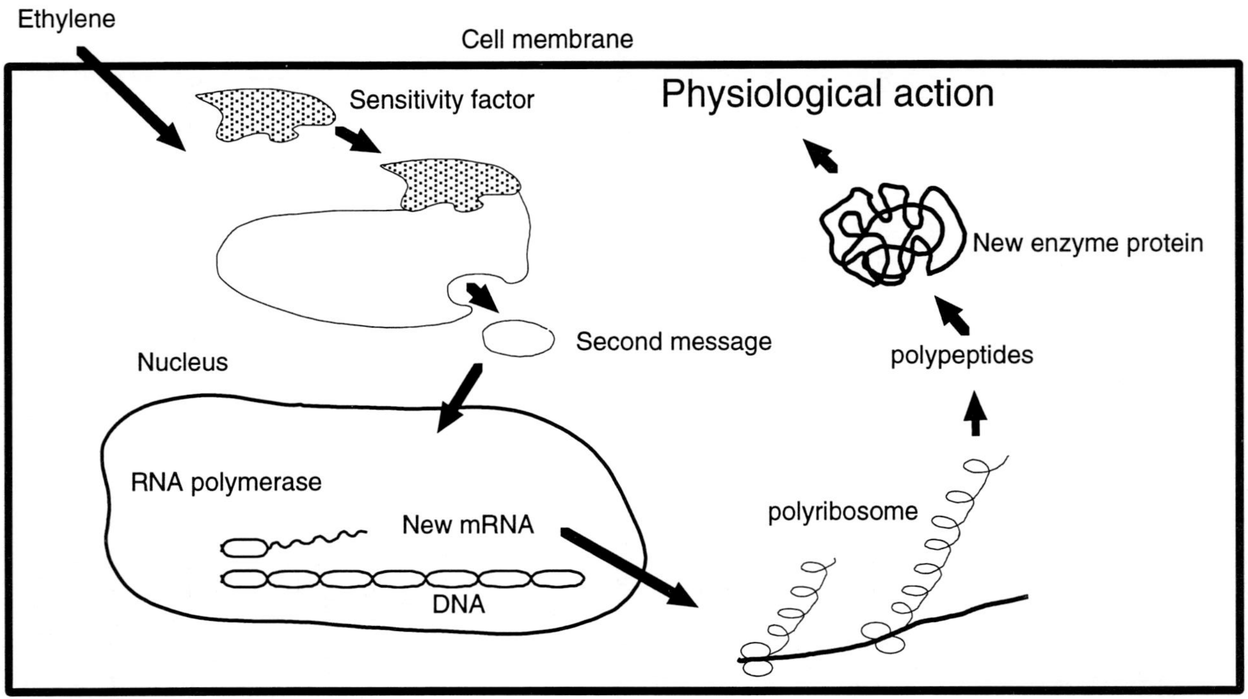

Fig. 8.13. Hypothetical scheme for the action of ethylene in inducing flower senescence.

Note: This scheme suggests a membrane-based binding site that is activated or repressed by a "sensitivity factor." The ethylene molecule binds to a site where the inhibitors of ethylene action, silver ions and 2,5-norbornadiene (NBD), can also bind. When the binding site is sensitized and ethylene binds to it, a second message is generated which interacts with the 5′ (promotor) regions of genes involved in ethylene-regulated senescence, inducing transcription of the genes, and synthesis of the proteins encoded by these genes. A considerable amount of evidence has accumulated from studies with cut-flowers that is consistent with this scheme.

Adapted from Reid & Wu (1991).

8.5. Postharvest Handling of Cut-Flowers

When an inflorescence is excised from the plant, a number of physiological processes are affected. These include the supply of water, depletion of respiratory substrates and ethylene production. Excessive water loss is closely associated with the termination of vase-life of cut-flowers. The failure of water uptake as a result of stem blockage is often the major cause for wilting. Stem blockage can be due to air blockage, microbial growth or physiological plugging.

In leaves, the bulk of transpiration occurs through stomata (stomatal transpiration) whereas a small amount of water vapour is lost through the cuticle (cuticular transpiration). The role of stomata in flowers is unclear and their involvement in flower transpiration remains debatable. There is evidence to indicate that cuticular transpiration plays an important role in the water loss of orchid flowers. The transpiration rate of tropical orchid flowers ranges from 0.15 to 0.17 mg water cm^{-2} h^{-1} or 0.4 to 1.9 g of water per inflorescence per day, depending on the total floral surface area (Table 8.4). Water loss in orchid flowers is considerably lower than that reported for roses and carnations due to the absence of supporting leaves in orchid sprays. Removing the leaves in roses, for example, will cause a tenfold reduction in the transpiration rate of a flower stem.

Since water loss in orchid flowers is not high, uptake of water through the cut ends of the stalk need not be massive to maintain an adequate level of water in the tissues. Water loss by an orchid inflorescence can be estimated easily if the number of flowers and flower size are known. As transpiration in orchid flower occurs primarily through the cuticular surface of the flower, the contribution of water loss through the stomata along the inflorescence axis may not be excessive.

Once a flower is cut, the continuous supply of respiratory substrate from the leaves and storage organs is eliminated. It has been noted that carbohydrate levels in mature flowers are lower than the levels in the tight buds (Fig. 8.3). Moreover, the levels of carbohydrates in the flowers decrease markedly with time after harvest, as reflected in the decreasing rate of respiration. However, the problem can be partially relieved by the exogenous supply of sucrose. The physiological and environmental factors that affect respiration in relation to floral senescence have already been discussed (see Chap. 4 on Respiration).

Table 8.4. The rate of water loss in orchid flowers.

Orchid	No. of flowers per inflorescence	Surface area of flower (cm^2)	Rate of water loss		
			$mg\ cm^{-2}\ h^{-1}$	$g\ flower\ d^{-1}$	$g\ inflorescence\ d^{-1}$
Vanda Tan Chay Yan	9	46.4	0.17	0.189	1.7* (Calculated) 1.9** (Determined experimentally)
Aranda Wendy Scott	8	16.3	0.15	0.058	0.4* (Calculated)

Note: * Calculated from data as measured using a differential psychometer; ** Measured using the weighing method.

Redrawn from Lee & Hew (1985).

Table 8.5. The vase-life of *Dendrobium* Pompadour flowers harvested during different months.

Month	Vase-life (days)	Bud opening (%)	Temperature (°C)	Relative humidity (%)
January	15^{ab}	34.1^{ns}	29.5	62.5
February	13^{bc}	23.9	31.7	55.1
March	12^{bc}	15.4	33.4	54.7
April	10^{c}	16.6	31.0	66.8
May	15^{ab}	22.8	30.6	73.1
June	17^{a}	19.1	30.4	71.0

Note: Figures with a different letter differ significantly based on Duncan's Multiple Range Test at $P = 0.05$.

Redrawn from Ketsa & Amutiratana (1986a).

Another important factor affecting the keeping quality of orchid cut-flower is ethylene production. This is because orchid flowers are sensitive to ethylene and senescing flowers produce fair amounts of ethylene. As ethylene production in orchid flowers is an autocatalytic process, more ethylene will be produced. Many chemicals have been used to reduce ethylene damage to flower either by blocking its biosynthesis or its action. Attempts have also been made to remove the ethylene immediately after it has been formed.

Preharvest conditions

In postharvest handling of orchid cut-flowers, we have often ignored the importance of preharvest conditions to the keeping quality of the flowers, particularly the cultural and environmental condition. It has been claimed that 30–70% of the keeping potential of many floral crops is pre-determined at harvest. The many reports of injury to orchid flowers by ethylene-polluted air are typical examples illustrating the importance of environmental factor in maintaining good quality orchid flowers.

Differences in the keeping quality of *Aranda* Christine cut-flowers supplied by the same nursery in Singapore at different times of the year have been observed. The vase-life is variable and it ranges from 18 to 28 days. Plant mineral nutrition may be an important contributing factor to the differences as light and temperature remain fairly constant throughout the year in Singapore. Local nurseries often stay away from a heavy fertiliser program when the demand for orchid cut flower is low.

A seasonal variation in the vase-life of *Dendrobium* Pompadour is observed from the same grower in Bangkok. The vase-life of flowers ranged from 10 to 17 days between January to June 1984 (Table 8.5). In this case, it is not clearly known how the variability in the vase-life of *Dendrobium* flowers is influenced by environmental conditions such as light intensity, temperature and relative humidity, and mineral nutrition. An important environmental factor which affects the postharvest behaviour in most flowers is the available total light energy. Light affects carbohydrate levels in flowers before harvest, which in turn influence the keeping quality. It is well-documented that high sugar levels improve the water status of cut-flowers.

Longevity of *Dendrobium* Pompadour flowers is significantly correlated to inflorescence size and total water uptake, but not to the number of stomata. The latter can be explained by the fact that either floral stomata of orchids are practically non-functional or the water loss through floral stomata is minimum. Flower size is presumably dependent on the availability of assimilates from the leaves (see Chap. 7 on Partitioning of Assimilates). Hence, the positive correlation between water uptake ability and longevity of *Dendrobium* Pompadour flower is probably due to the higher sugar levels present in the flowers.

An important consideration must be given to the time of harvest. *Aranda* and *Dendrobium* inflorescences, which are harvested early in the morning, generally last longer than those harvested in the late morning. There is a correlation between the longevity of cut-flowers and relative water content. It is possible that the lower water content in orchid flowers harvested in the late morning is attributed to a lower sugar level. *Aranda* and *Dendrobium* are thick-leaved orchids which exhibit typical CAM activities. As stomata in CAM plants are open at night, it would favour greater water uptake during the night. Watering the plants in the late afternoon, prior to a harvesting session in the following morning, should improve the keeping quality of the flowers. At present, data on the water potential of orchid flowers and the rate of water absorption at different times of the day are not available.

In Singapore and Malaysia, orchid flowers in nurseries are known to open faster during rainy season. However, these flowers do not last long. This may be attributed to low sugar levels as a result of a decrease in photosynthesis. There is no information pertaining to the effects of fertiliser, particularly nitrogen sources, on the keeping quality of orchid cut-flowers. The same applies to the effects of pesticides used in orchid nurseries on the vase-life of orchid flowers.

Extension of vase-life

Different terms have been used to evaluate the postharvest quality of cut-flowers: Shelf-life, vase-life, bench-life, longevity and keeping quality. Like in the case for the evaluation of other cut-flowers, the difficulty lies not in the terminology, but with the criteria and conditions adopted for measurement. These are often ill-defined and a proper comparison cannot be made for different studies.

Various criteria have also been used to evaluate vase-life in orchid cut-flowers. A common criterion is the percentage (%) of flowers dropped. However, the percentage of flowers dropped used by different workers varies from 30% to 100%. In addition, temperature, light conditions, relative humidity and developmental stages of inflorescences used for evaluating the vase-life of cut-flowers are often not well-defined. The manner in which the peduncle is cut and the frequency of changing the holding solution are also known to affect the vase-life of orchid flowers. There is certainly a need to establish a standard procedure for critical evaluation of vase-life of orchid cut-flowers.

Formulation of various solutions

There are four major solutions for the postharvest handling of cut-flowers: Conditioning, pulsing, holding and bud-opening solutions. In Malaysia, Singapore and Thailand, orchid flowers are harvested in groups and placed in uncovered boxes located at convenient points in the field for a considerable period of time before they are transported to the packing station. Flowers could experience water stress under these conditions. Conditioning of the flowers in water or a solution containing preservatives to restore tissue turgidity is necessary. No special treatment of water is carried out in the conditioning of orchid cut-flowers in ASEAN countries. Research on water quality and its relation to vase-life in Hawaii has shown that water quality does affect shelf life of cut *Dendrobium* flowers.

Pulsing is a short-term pre-shipment treatment to keep the cut flowers fresh during shipment. Temperate flowers are often pulsed in 2–20% sucrose for various periods before shipment. Shelf-life of *Oncidium* Goldiana flowers increases with silver nitrate pre-treatment for 30 min. For *Aranda* flowers, pulsing with 4 mM solution of silver thiosulphate for 10 min extends the vase-life significantly. Treating *Dendrobium* Pompadour flowers with silver nitrate (25 ppm) and sodium thiosulphate (135 ppm) for 30 min extends its vase-life from 23 days to 36 days (Table 8.6).

Table 8.6. Effects of silver-containing compounds on vase-life of some orchid flowers.

Orchid	Treatment*	Results
Aranda Christine	4 mM solution of STS pulsed for:	
	0 min	100% wilting[a]
	2 min	32% wilting[a]
	10 min	5% wilting[a]
	30 min	7% wilting[a]
	10 min pulse with:	
	0.5 mM solution of STS	100% wilting[a]
	1 mM solution of STS	64% wilting[a]
	2 mM solution of STS	48% wilting[a]
	4 mM solution of STS	4% wilting[a]
Cymbidium hybrid	Dip flowers for 10 min in 4 mmol $AgNO_3$ and 16 mmol $Na_2S_2O_3$	Senescence delayed
Dendrobium hybrid	Flowers treated for 4 or 8 h with 2 mM STS	No effect
Dendrobium Pompadour	Water, 30 min before placing the flowers in 400 ppm 8-HQC plus 5% sucrose.	Vase-life = 23 days
	400 ppm 8-HQC plus 5% sucrose	Vase-life = 32 days
	30 min in 25 ppm $AgNO_3$ plus 135 ppm $Na_2S_2O_3.5H_20$ before placing the inflorescence in 400 ppm 8-HQC plus 5% sucrose.	Vase-life = 36 days

Note: * 8-HQC, 8-hydroxyquinoline citrate; STS, sodium thiosulphate.
[a], percent wilting after 28 days.

Redrawn from Arditti & Hew (1994).

The beneficial effect of the silver thiosulphate (STS) complex on the extension of *Aranda* and *Dendrobium* flowers is in agreement with other reports published for temperate flowers. The relatively low concentration and short immersion time needed for treatment indicate that STS is highly mobile in plants (Table 8.7). Silver ions are known to be an effective ethylene antagonist. By blocking the receptor site for ethylene, silver ions prevent the autocatalytic increase in ethylene production. The fact that STS does not bring about changes in the fresh mass, dry mass and water potential patterns of *Aranda* flowers after harvest further supports the suggestion that ethylene plays a major role in controlling the senescence of orchid flowers. Whether STS can extend the vase-life of other tropical orchid flowers awaits further research because different degrees of sensitivity to ethylene have been observed in some orchid flowers.

Table 8.7. Distribution of silver ions in *Dendrobium* sprays after treatment with silver thiosulphate.

Duration of STS pulsing (h)	Silver content (ng g fresh mass^{-1})					
	Flowers			Stem		
	Top	Middle	Bottom	Top	Middle	Bottom
0	0	0	0	0	0	0
4	0.03	0.06	0.07	0.12	0.20	0.21
8	0.07	0.14	0.18	0.23	0.38	0.29

Note: Limit of detection: Stem tissue = 0.002 ng g fresh mass^{-1}, flowers = 0.001 ng g fresh mass^{-1}. The sprays were treated for 4 or 8 h with 2 mM silver thiosulphate, then placed in deionized water for an additional 20 or 16 h respectively, before silver analysis.

Redrawn from Dai & Paull (1991).

Recently, a volatile ethylene action inhibitor, 1-MCP (1-Methyl-cyclopropene), has been used to extend the vase-life of cut-flowers. The effects of 1-MCP on flower abscission are comparable to that of a pulse treatment with STS. 1-MCP is an odourless and non-toxic cyclic olefin gas which binds to the cellular ethylene receptors. This compound has been shown to be effective in inhibiting the ethylene responses in cut-flowers and potted flowering plants. In future, 1-MCP may serve as an alternative to the commercial treatment of cut-flowers with STS, the latter being an environmental hazard. At present, there is no report on the use of 1-MCP in extending the vase-life of cut orchid flowers.

Considerable efforts have been made to formulate appropriate holding solutions for the extension of the vase-life of tropical orchid cut-flowers. The ingredients of holding solutions include minerals, sugar, bactericides and plant hormones. The beneficial action of the various ingredients in holding solution has been studied singly or in combination. Minerals such as aluminum chloride, boric acid, ammonium molybdenate and silver nitrate give varying results. For example, the effects of silver nitrate on the extension of vase-life is rather variable. There are reports that 10–30 ppm of silver nitrate could extend the life of *Dendrobium* Pompadour. However, others observed that silver nitrate (10–400 ppm) shortens vase-life.

Hydroxylquinoline sulphate (HQS) and hydroxylquinoline citrate (HQC) are bactericides commonly incorporated into the holding solutions. HQS, together with sucrose, extends the vase-life of some tropical orchid flowers. HQS (50–100 ppm) extends the vase-life of *Dendrobium* Pompadour. The same result is obtained with *Oncidium* Goldiana and *Dendrobium* Youppadeewan. However, an extension of the vase-life of *Oncidium* flowers is observed only when HQS is used with sucrose and not when it is added alone. The reason remains unclear. By comparison, the beneficial effects of HQS plus sucrose on the extension of vase-life and flower opening of several temperate flowers are well-documented.

Physan, a quaternary ammonium compound, has been used with sucrose in pulsing, bud opening and holding solutions for flowers such as carnations, chrysanthemums and gypsophilas. Physan alone (100–200 ppm) or in combination with sucrose (200 ppm and 4% sucrose) prolongs the vase life of *Dendrobium* Pompadour flowers.

Reports on the effects of sugar on the vase-life of tropical orchid cut-flowers are conflicting. Sucrose (2–10%) reduces bud opening and the vase-life of *Dendrobium* Pompadour and *Oncidium* Goldiana flowers. However, there are reports which do not agree with this. It is possible that the detrimental effect of sucrose supplied alone is due to microbial occlusion which may develop in the vascular system when the inflorescences are kept in unchanged solution for a long duration. It has also been reported that glucose is better than sucrose as a carbon source for bud opening and the extension of vase-life of *Dendrobium* flowers. Inconsistent effects of sugar on vase-life of the other temperate cut-flowers have also been reported (Halevy and Mayak, 1979).

At present, there are many non-commercial and commercial holding or bud-opening solutions for cut-flowers. Some of the better known non-commercial preparations are Cornell, Cornell Modified, Davis, Ottawa, Marusky, Kagawa and Washington solution (Table 8.8). Commercial solutions available in the market are Chrysal, Floralife, Proflovit, Everbloom and Florever. Many of the commercial solutions are recommended for general use, whereas the non-commercial ones are for specific flowers. The vase-life of *Dendrobium* Pompadour flowers held in distilled water, Cornell, Cornell Modified, Davies, Kagawa, Washington, Chrysal and Florever solutions have been compared. Cut orchid flowers held in Cornell solution gave the best results. The cut-flowers also take up more water and show more bud opening.

Table 8.8. Composition of solutions used in extending the vase-life of orchid cut-flowers.

Vase solution	Composition	
Cornell	8-hydroxyquinoline sulphate	200 mg l^{-1}
	Silver nitrate	50 mg l^{-1}
	Sucrose	5%
Cornell Modified	8-hydroxyquinoline sulphate	200 mg l^{-1}
	Silver nitrate	25 mg l^{-1}
	Aluminium sulphate	50 mg l^{-1}
	Sucrose	5%
Davis	Silver nitrate	25 mg l^{-1}
	Citric acid	75 mg l^{-1}
	Sucrose	10%
Kagawa	Alar	700 mg l^{-1}
	8-hydroxyquinoline sulphate	400 mg l^{-1}
	Sucrose	6%
Washington	Alar	300 mg l^{-1}
	8-hydroxyquinoline sulphate	400 mg l^{-1}
	Sucrose	3%
Chrysal®	unknown — commercial product	
Florever®	unknown — commercial product	

Bud opening

Harvesting orchid flowers in the bud stage is an attractive concept with considerable commercial potential. This has been shown to be feasible for gladiolus, lilac, snapdragon and chrysanthemums. An important consideration in harvesting flowers at the bud stage is the availability of an appropriate holding solution which can allow the flowers to open normally.

Pollinia of some orchid flowers can be easily dislodged and emasculation stimulates ethylene production, which accelerates the senescence of flowers. As ethylene production by senescing flower is autocatalytic, this may adversely affect packed blooms. Cutting flowers at the bud stage may alleviate such a problem.

Acetylsalicyclic acid, when combined with sucrose, have a beneficial influence on the opening of *Oncidium* flowers. The percentage of opened flowers and flower size in acetylsalicylic acid plus sucrose (80%) solution are comparable to those of uncut (i.e., *in situ*) inflorescences. Conversely, only 50% of the flowers open in the solution containing sucrose. HQS and silver nitrate have also been reported to have a beneficial effect on bud opening of *Oncidium* flowers.

Bud opening of *Dendrobium* Pompadour is affected by the same factors which influence vase-life, cutting method and the frequency of changing holding solution. HQS or silver nitrate ($AgNO_3$) at 50 ppm gives a high percentage of bud opening. Physan alone does not increase the percentage of bud opening but bud opening is enhanced when it is used in combination with sucrose. The success for the formulation of a bud-opening solution has made it practical to harvest *Oncidium* Goldiana and *Dendrobium* Pompadour flowers at the bud stage.

8.6. Storage and Transport

Although freshly harvested orchid flowers in ASEAN countries are shipped overseas on the same or the next day, storage technology of flowers has to be developed in cases where the need may arise. Orchid flowers are packed in paper cartons and these are stored in air-conditioned rooms at a temperature of 20–21°C. We have yet to establish the optimal conditions for short term storage and transport of orchid cut-flowers.

There are three major technologies available for the storage of fruits, vegetables and cut-flowers. By comparison, storage of cut-flowers presents more problems.

Low-temperature storage

Lowering the temperature reduces respiration and other biochemical activities. For temperate flowers, including temperate orchids, they can be stored at 5–7°C for 10–14 days. Some *Cymbidium* hybrids store well at –5°C. In general, tropical flowers are damaged by chilling at 10–15°C. Mature *Dendrobium* Pompadour flowers can be stored for four days between 10–25°C. However, a four-day storage at 28°C is found to be unsuitable. For a longer storage period of eight days, it is best to store the flowers at 10°C. *Dendrobium* Pompadour flowers display chilling injury at 4°C under a storage of four or eight days. Chilling injury is a common disorder in plant tissues of tropical and subtropical origin when subjected to low temperatures.

Hypobaric storage/controlled storage

Vanda Miss Joaquim has been successfully stored under reduced atmospheric pressure (hypobaric storage) conditions and/or low temperature for more than two-weeks-days (Table 8.9). The possibility of using controlled atmosphere and reduced pressure storage for delaying the fading of *Vanda* Miss Joaquim during simulated shipping has been explored. Static exposure to 1.5–2.0% carbon dioxide or 1.0–2.6% oxygen modified with nitrogen or reduced pressure of 125 mm Hg for two to three days effectively delays the fading of *Vanda* flowers for several days. The beneficial effects of controlled atmosphere and sub-atmospheric pressure are residual in that they delay the fading process even after the flowers have been removed from storage. For *Dendrobium*

Pompadour flowers, the best condition for storage is at 10% carbon dioxide. Storage of flowers at higher levels of carbon dioxide, such as 20%, is harmful.

Table 8.9. Storage of some cut-flowers under hypobaric conditions and cold storage.

Plant	Cold storage temperature (°C)	Storage life (days)		LPS pressure (mm Hg)
		Cold storage	LPS storage*	
Carnation	0–2	21–28	63	40
Rose	0	7–14	42	40
Vanda Miss Joaquim	12	16	41	40

Note: * The maximum storage life in Low Pressure Storage (LPS) has not yet been determined.

Redrawn from Burg (1973).

Premature fading resulting from exposure to ethylene is a problem during the shipping of orchid flowers. Brominated charcoal and potassium permanganate impregnated materials are effective in controlling the fading of *Vanda* flowers under simulated transport conditions (Tables 8.10, 8.11).

From a commercial viewpoint, the development of an appropriate storage technology which is practical and simple would be a major consideration. In controlled atmosphere storage (CA), the cut-flowers are kept under a modified atmosphere in which low oxygen and/or high carbon dioxide levels prevail. Generally, this method requires relatively sophisticated storage structures. Shelf-life of many vegetables and fruits is increased and the quality is maintained by a controlled atmosphere. But most floral crops have not responded well to CA storage. Results obtained with roses and carnations are inconclusive although *Vanda* and *Dendrobium* flowers seem to store well under CA.

Hypobaric storage works on the principle of storing flowers under a controlled sub-atmospheric pressure. However, maintaining a sub-atmospheric pressure during storage is costly. The commercial utilisation of such a storage technology will depend on its cost effectiveness. Cold storage is by far the

Table 8.10. Effect of Purafil and Brominated activated charcoal in controlling fading of *Vanda* flowers.

Treatment	Percentage of normal flowers faded on day: 1	2	3–8	9	10–19
5 g Purafil	0	0	0	0	0
5 g brominated activated charcoal	0	0	9.1	100	
Control	0	100			

Note: Purafil is a commercially available product containing activated alumina pellets impregnated with potassium permanganate. There were 11 normal flowers with one ethylene-generating flower sealed in a 6.3-litre glass jar. The ethylene-generating flower faded in one day in all treatments. The experiments was terminated on day 19 when the control flowers started to decay.

Redrawn from Akamine & Goo (1981a).

Table 8.11. Effect of potassium permanganate impregnated in inert supports on fading of *Vanda* flowers.

$KMnO_4$ (%)	Days for normal flowers to fade: Perlite	Vermiculite
0	1.8 ± 0.1	1.8 ± 0.1
0.75	21.9 ± 1.4	22.0 ± 1.0
1.5	20.7 ± 1.8	23.0 ± 1.4
3.0	20.4 ± 2.0	21.4 ± 1.4

Note: There were 12 normal flowers with one ethylene-generating flower sealed in a 6.3-litre glass jar. The ethylene-generating flower faded in one day in all treatments. Mean ± SE.

Redrawn from Akamine & Goo (1981a).

most common method employed for the storage of cut-flowers. It is relatively simple, practical and less costly. However, many plant species of tropical origins are sensitive to low but non-freezing temperature and such storage may cause chilling injury. Therefore, the sensitivity of tropical orchid flowers to chilling injury poses serious postharvest problems under cold storage. Success would depend on finding ways to alleviate the chilling injury. Manipulation of storage environments and programs, use of chemicals and genetic application are areas which deserve future research.

8.7. Concluding Remarks

There have been significant advances in the physiology of senescence in orchid flowers in recent years. Many of the fundamental aspects and genetic control of pollination-induced senescence and ethylene biosynthesis have been unveiled. One of the important findings is that an increase in ethylene sensitivity following pollination is the initial event which triggers an increase in ethylene production and enhances the senescence of orchid flowers. The identification of the sensitivity factor will certainly improve the postharvest handling of orchid cut-flowers as ethylene sensitivity is known to be a major factor in determining flower longevity. The progress made in the postharvest physiology and handling of orchid cut-flowers has been impressive but much remains to be done. There are many conflicting reports on the effects of chemicals used in promoting the vase-life of cut-flowers. A set of standardised protocols and approaches may be needed for the proper and critical evaluation of vase-life of orchid cut-flowers. We have not found a simple and an effective technology for storing orchid cut-flowers. All these shortcomings have made the development of an appropriate postharvest technology and management of tropical orchid cut-flowers difficult. Evidently, more extensive research is needed in areas related to the postharvest handling of orchid cut-flowers if we wish to have good and marketable orchid cut-flowers.

8.8. Summary

1. Orchid flowers are well-known for their longevity. Many tropical orchid cut-flowers may last for a few weeks.
2. As in the other non-orchidaceous flowers, the endogenous hormonal levels (e.g. cytokinins, gibberellins, abscisic acid) in orchid flowers change during

development. Ethylene is the most extensively studied plant hormone in relation to orchid flower senescence.

3. Orchid flowers are particularly sensitive to ethylene. The injury caused by ethylene to most orchid flowers is characterised by the progressive drying and fading of sepals. Ethylene is produced by orchid flowers and its production is enhanced following pollination or emasculation (removal of pollinia).
4. Significant advances have been made in the physiology and molecular biology of pollination/emasculation-induced senescence in orchid flowers.
5. When an inflorescence is excised from the plant, a number of physiological processes are affected. These include the supply of water, depletion of respiratory substrates and ethylene production. The keeping quality of orchid cut-flowers can be extended by ensuring a positive water balance and an adequate supply of sugar. Attempts must also be made to block any ethylene biosynthesis, remove the ethylene formed and prevent ethylene from interacting with the orchid tissues.
6. The importance of formulating various solutions for postharvest handling of orchid cut-flowers has been discussed. The four major solutions are conditioning, pulsing, holding and bud-opening solutions.
7. The three important approaches under investigation for the storage of orchid cut-flowers are: (1) Low-temperature or cold storage; (2) Sub-atmospheric (Hypobaric) storage; (3) Controlled atmosphere storage. Orchid flowers seem to store well under sub-atmospheric and controlled atmosphere storage conditions. Cold storage is simple and more cost-effective to implement; however, the sensitivity of tropical orchid flowers to chilling injury poses serious impediment. Success would depend on finding ways to alleviate the chilling injury.

General References

Abeles, F. B., Morgan, P. W. and Saltveit, M. E., 1992, *Ethylene in Plant Biology*, 2nd ed. (Academic Press, San Diego), 414 pp.

Arditti, J., 1992, *Fundamentals of Orchid Biology* (John Wiley and Sons, New York), 691 pp.

Avadhani, P. N., Nair, H., Arditti, J. and Hew, C. S., 1994, "Physiology of orchid flowers," in *Orchid Biology: Reviews and Perspectives, Vol. VI*, ed. J. Arditti (John Wiley and Sons, New York), pp. 189–358.

Halevy, A. H. and Mayak, S., 1979, "Senescence and postharvest physiology of cut flowers, Part 1," in *Horticultural Reviews 1*, ed. J. Janick (AVI Publishing, West Point, Conn.), pp. 204–236.

Hew, C. S., 1987, "Respiration in orchids," in *Orchid Biology: Reviews and Perspectives, Vol. IV*, ed. J. Arditti (Cornell Univ. Press, Ithaca), pp. 229–259.

Hew, C. S., 1994, "Orchid cut-flower production in ASEAN countries," in *Orchid Biology: Reviews and Perspectives, Vol. VI*, ed. J. Arditti (John Wiley and Sons, Inc., New York), pp. 363–401.

Hew, C. S. and Clifford, P. E., 1993, "Plant growth regulators and the orchid cut-flower industry," *Plant Growth Regulation* **13**: 231–239.

Mathooko, F. M., 1996, "Review: Regulation of ethylene biosynthesis in higher plants by carbon dioxide," *Postharvest Biology and Technology* **7**: 1–26.

Mayak, S. and Halevy, A. H., 1980, "Flower senescence," in *Senescence in Plants*, ed. K. V. Thimann (CRC Press, Boca Raton), pp. 131–156.

O'Neill, S. D., Nadeau, J. A., Zhang, X. S., Bui, A. Q. and Halevy, A. H., 1993, "Interorgan regulation of ethylene biosynthetic genes by pollination," *The Plant Cell* **5**: 419–432.

Stead, A. D., 1992, "Pollination-induced flower senescence: A review," *Plant Growth Regulation* **11**: 13–20.

Trewavas, A. J., 1982, "Growth substance sensitivity: The limiting factor in plant development," *Physiologia Plantarum* **55**: 60–72.

Yang, S. F. and Hoffman, N. E., 1984, "Ethylene biosynthesis and its regulation in higher plants," *Annual Reviews of Plant Physiology* **35**: 155–189.

References

Akamine, E. K., 1976, "Postharvest handling of tropical ornamental cut crops in Hawaii," *HortScience* **11**: 125–126.

Akamine, E. K. and Goo, T., 1981a, "Controlling premature fading in *Vanda* Miss Joaquim flowers with potassium permanganate." Research Series no. 7. Coll. of Tropical Agriculture and Human Resources, University of Hawaii.

Akamine, E. K. and Goo, T., 1981b, "Effects of static controlled atmosphere and reduced pressure storage on fading of *Vanda* Miss Joaquim flowers." Research Series no. 8. Coll. of Tropical Agriculture and Human Resources, University of Hawaii.

Arditti, J. and Hew, C. S., 1994, "Extending life of cut orchid flower by silver thiosulphate," *Malayan Orchid Review* **28**: 48–50.

Burg, S. P., 1973, "Hypobaric storage of cut flowers," *HortScience* **8**: 202–205.

Burg, S. P. and Dijkman, M. J., 1967,"Ethylene and auxin participation in pollen induced fading of *Vanda* orchid blossoms," *Plant Physiology* **42**: 1648–1650.

Chin, T. Y., Chai, B. L. and Hew, C. S., 1989, "Occurrence of abscisic acid-like and gibberellins-like substances in tropical orchid flowers," *Malaysian Orchid Bulletin* **4**: 13–18.

Dai, J and Paull, R. E., 1991, "Effect of water status on *Dendrobium* flower spray postharvest life," *Journal of the American Society for Horticultural Science* **116**: 491–496.

Davidson, O. W., 1949, "Effects of ethylene on orchid flowers," in *Proc. of the American Society of Horticultural Science* **53**: 440–446.

Dijkman, M. J. and Burg, S. P., 1970, "Auxin-induced spoiling of *Vanda* blossoms," *American Orchid Society Bulletin* **39**: 799–804.

Hew, C. S., 1980, "Respiration of tropical orchid flowers," in *Proc. 9th World Orchid Conference*, Bangkok (1978), ed. M. R. Sukshom Kashemsanta, pp. 191–195.

Hew, C. S., 1985, "The effects of 8-hydroxylquinoline sulphate, acetylsalicylic acid and sucrose on bud opening of *Oncidium* flowers," *Journal of Horticultural Science* **62**: 75–78.

Hew, C. S., 1986, "Effects of storage temperature on bud opening of *Oncidium* flowers," *Malaysian Orchid Bulletin* **3**: 39–41.

Hew, C. S., 1989, "Chilling injury and cold storage of orchid cut flowers," *Malayan Orchid Review* **23**: 44–47.

Hew, C. S., Lee, G. L. and Wong, S. C., 1980, "Occurrence of non-functional stomata in the flowers of tropical orchids," *Annals of Botany* **46**: 195–201.

Hew, C. S. and Veltkemp, C. J., 1985, "Orchid floral stomata under the scanning electron microscope," *Malayan Orchid Review* **19**: 26–32.

Hew, C. S. and Ong, T. K., 1987, "*Vanda* Miss Joaquim under scanning electron microscope," *Malayan Orchid Review* **21**: 36–41.

Hew, C. S., Wee, K. H. and Lee, F. Y., 1987, "Factors affecting the longevity of cut *Aranda* flowers," *Acta Horticulturae* **205**: 195–202.

Hew, C. S., Wee, K. H., Wong, S. M., Ong, T. K. and Lee, F. Y., 1989, "Water relation and longevity of orchid cut flowers," *Malayan Orchid Review* **23**: 36–43.

Hew, C. S. and Yong, J. W. H., 1994, "Growth and photosynthesis of *Oncidium* Goldiana," *Journal of Horticultural Science* **69**: 809–819.

Ketsa, S., 1986, "Effect of peduncle length, cutting method of peduncle and change of water on water uptake of *Dendrobium* Pompadour flowers," in *Proc. 6th ASEAN Orchid Congress*, Bangkok, Thailand, pp. 116–119.

Ketsa, S., 1986, "Effect of physan-20 and sucrose on vase life of *Dendrobium* Pompadour flowers," *Proc. 6th ASEAN Orchid Congress*, Bangkok, Thailand, pp. 120–123.

Ketsa, S., 1986, "A comparative study of vase solutions for *Dendrobium* Pompadour flowers," in *Proc. 6th ASEAN Orchid Congress*, Bangkok, Thailand, pp. 130–134.

Ketsa, S. and Amutiratana, D., 1986a, "Relationship between the vaselife and some anatomical, morphological and physiological aspects of *Dendrobium* Pompadour flowers," in *Proc. 6th ASEAN Orchid Congress*, Bangkok, Thailand, pp. 113–115.

Ketsa, S. and Amutiratana, D., 1986b, "Effect of sucrose, silver nitrate and 8-hydroxylquinoline sulphate on postharvest behaviour of *Dendrobium* Pompadour flowers," in *Proc. 6th ASEAN Orchid Congress*, Bangkok, Thailand, pp. 124–129.

Ketsa, S. and Boonrote, A., 1990, "Holding solutions for maximizing bud opening and vase-life of *Dendrobium* Youppadeewan flowers," *Journal of Horticultural Science* **65**: 41–47.

Lee, F. Y. and Hew, C. S., 1985, "Water loss by tropical orchid flowers," *Proc. 4th ASEAN Orchid Congress*, Los Banos, Philippines, pp. 109–117.

Lim, S. L., Chin, T. Y. and Hew, C. S., 1975, "Biochemical changes accompanying the senescence of *Arundina* flowers," in *Biology in Society, Proc. of Seminar*. Singapore Institute of Biology and Singapore National Academy of Science, Singapore, pp. 18–26.

Nadeau, J, A., Bui, A. Q., Zhang, X. S. and O'Neill, S. D., 1993, "Interorgan regulation of post-pollination events in orchid flowers," in *Cellular and Molecular Aspects of the Plant Hormone Ethylene: Proc. of the International Symposium on Cellular and Molecular Aspects of Biosynthesis and Action of the Plant Hormone Ethylene*, France (1992), eds. J. C. Pech, A. Latche and C. Balague (Dordrecht , Kluwer Academic Press), pp. 304–309.

Nadeau, J, A., Zhang, X. S., Nair, H. and O'Neill, S. D., 1993, "Temporal and spatial regulation of 1-aminocyclopropane-oxidase-1-carboxylate in the pollination induced senescence of orchid flowers," *Plant Physiology* **103**: 31–39.

Nair, H., 1984, "Postharvest physiology and handling of orchids," *Malayan Orchid Review* **18**: 62–68.

Nair, H. and Tung, H. F., 1980, "Investigations on cut flowers longevity of *Oncidium flexuosum* × *Oncidium spacelatum*," in *Proc. 3rd ASEAN Orchid Congress*, Malaysia (1980), pp. 85–95.

Nair, H. and Tung, H. F., 1987, "Ethylene production and 1-aminocyclopropane-1-carboxylic acid levels in detached orchid flowers of *Dendrobium* Pompadour," *Scientia Horticulturae* **32**: 145–151.

Nair, H., Idris, Z. M. and Arditti, J., 1991, "Effects of 1-aminocyclopropane-1-carboxylic acid on ethylene evolution and senescence of *Dendrobium* (Orchidaceae) flowers," *Lindleyana* **6**: 49–58.

Ong, H. T., 1982, "Use of solutions with trace elements to influence the flowering and shelf life of flowers of *Oncidium* Goldiana," *Orchid Review* **90**: 264–266.

Ong, H. T., Ding, T. H. and Yang, H. C., 1980, "Effects of some trace elements and chemicals on shelf-life of flowers of Golden Shower (*Oncidium* Goldiana)," *Proc. 3rd ASEAN Orchid Congress*, Malaysia (1980), pp. 79–84.

Ong, H. T. and Lim, L. L., 1983, "Use of silver nitrate and citric acid to improve shelf life of *Oncidium* Golden Shower flowers," *Orchid Review* **91**: 141–144.

Porat, R., 1994, "Comparison of emasculation and pollination of *Phalaenopsis* flowers and their effects on flower longevity, ethylene production and sensitivity to ethylene," *Lindleyana* **9**: 85–92.

Porat, R., Borochov, A., Halevy, A. H. and O'Neill, S. D., 1994, "Pollination-induced senescence of *Phalaenopsis* petals. The wilting process, ethylene production and sensitivity to ethylene," *Plant Growth Regulation* **15**: 129–136.

Porat, R., Borochov, A. and Halevy, A. H., 1994, "Pollination-induced changes in ethylene production and sensitivity to ethylene in cut *Dendrobium* orchid flowers," *Scientia Horticulturae* **58**: 215–221.

Porat, R., Halevy, A. H., Serek, M. and Borochov, A., 1995, "An increase in ethylene sensitivity following pollination is the initial event triggering an increase in ethylene production and enhanced senescence of *Phalaenopsis* orchid flowers," *Physiologia Plantarum* **93**: 778–784.

Porat, R., Shlomo, E., Serek, M., Sisler, E. C. and Borochov, A., 1995, "1-Methylcyclopropene inhibits ethylene action in cut phlox flowers," *Postharvest Biology and Technology* **6**: 313–319.

Porat, R., Reiss, N., Atzorn, R., Halevy, A. H. and Borochov, A., 1995, "Examination of the possible involvement of lipooxygenase and jasmonates in pollination-induced senescence of *Phalaenopsis* and *Dendrobium* orchid flowers," *Physiologia Plantarum* **94**: 205–210.

Reid, M. S. and Wu, M. J., 1991, "Ethylene in flower development and senescence," in *The Plant Hormone Ethylene*, eds. A. K. Mattoo and J. C. Suttle (CRC Press, Boca Raton, Florida), pp. 215–234.

Serek, M., Sisler, E. C. and Reid, M. S., 1995, "Effects of 1-MCP on the vase-life and ethylene response of cut-flowers," *Plant Growth Regulation* **16**: 93–97.

Sheehan, T. J., 1954, "Orchid flower storage," *American Orchid Society Bulletin* **23**: 579–584.

Vergano, P. J. and Pertuit, A. J. Jr., 1993, "Effects of modified atmosphere packaging on the longevity of *Phalaenopsis* florets," *HortTechnology* **3**: 423–427.

Wee, K. H. and Hew, C. S., 1986, "Effect of silver thiosulphate on the longevity of cut *Aranda* orchid flowers," *Malaysian Orchid Bulletin* **3**: 25–28.

Wen, Z. Q., Lee, Y. W., Pan, R. C. and Hew, C. S., 1990, "Biochemical and physiological changes associated with the development of *Cymbidium sinense* flower," *Journal of Singapore National Academy of Science* **18/19**: 100–103.

Woltering, E. J., 1990, "Inter-organ translocation of 1-aminocyclopropane-1-carboxylic acid coordinates senescence in emasculated *Cymbidium* flowers," *Plant Physiology* **92**: 837–845.

Woltering, E. J., 1990, "Interrelationship between the different flower parts during emasculation-induced senescence in *Cymbidium* flowers," *Journal of Experimental Botany* **41**: 1021–1029.

Woltering, E. J. and Harren, F., 1989, "Role of rostellum desiccation in emasculation-induced phenomena in orchid flowers," *Journal of Experimental Botany* **40**: 907–912.

Woltering, E. J., Somhorst, D. and Van Der Veer, Pieter., 1995, "The role ethylene in interorgan signaling during flower senescence," *Plant Physiology* **109**: 1219–1225.

Yip, K. C. and Hew, C. S., 1988, "Ethylene production by young *Aranda* orchid flowers and buds," *Plant Growth Regulation* **7**: 217–222.

Zainudin, R. and Nair, H., 1992, "Pre- and post-harvest investigations with blossoms of *Oncidium* Golden Shower. II Floral stomata and transpiration by detached flowers," *Malaysian Orchid Bulletin* **6**: 49–58.

Chapter 9

Recent Advances in Orchid Tissue Culture

9.1. Introduction

The development of Knudson's asymbiotic method has vastly improved the germination of orchid seeds and paved the way for orchid tissue culture. To date, improved tissue culture methods using orchid roots, leaves, flower buds, stems and inflorescences have been adopted, making orchid cultivation faster and easier (see Vajrabhaya [1977]; Arditti and Ernst [1993] for details in media composition). There is an active market for micropropagated orchid plantlets (see Chap. 1 on The Relevance of Orchid Physiology to the Industry). However, there are also many problems associated with the commercial production orchid plantlets: Slow growth of orchid plantlets, low multiplication rate, vitrification, poor rooting and high mortality during acclimatisation. Among others, the shortage of high quality planting materials further constrains the full expansion of the orchid industry (Hew, 1994). It is therefore important to formulate economically viable strategies to improve the quality and production rate of micropropagated orchid plantlets.

This chapter focuses on the recent findings in understanding the physiology of orchids under the artificial environment of a culture vessel. There has been much debate in recent years on the question of whether the established culture protocols, involving agar and sugars, should continue to be used. Many scientists are now in favour of the idea that plants would grow better in high light and low carbohydrate system. This widely held opinion is based on the observation that *in vitro* cultures expose the plants to unnaturally high humidity and sugar, thus suppresses the need and opportunity for photosynthesis. With new culture protocols and innovative design of culture vessels, plantlets cultured *in vitro* could derive a considerable portion of their carbon requirements from photosynthesis. In this chapter, some of the problems relating to the conventional means of micropropagation in orchids are highlighted and possible solutions arc suggested.

9.2. Factors Affecting Orchid Growth *in Vitro*

Most factors affecting growth of excised plant organs, tissues and cells *in vitro* are similar to those limiting the growth of whole plants *in vivo*. These factors include carbohydrate and mineral nutrition, plant hormones, photosynthetic active radiation (or simply, light), temperature, medium pH, humidity, gas exchange and the presence of microorganisms such as fungi and bacteria.

More recently, there is a renewed interest in improving *in vitro* culture conditions by the optimisation of environmental factors that includes light, gaseous environment, temperature and humidity (Fig. 9.1) (Buddendorf-Joosten and Woltering, 1994). The atmosphere in which most plants grow contains nitrogen (78%), oxygen (21%), carbon dioxide (0.035%) and other trace gases. In contrast, the gaseous composition inside *in vitro* culture vessels is often different. This is due to the restriction of gas exchange between culture vessels and the surroundings as there is a need to protect the aseptic culture from microbial contamination. Many different types of culture vessels and sealings are used in scientific and commercial practice. Culture vessels are usually made of glass, polypropylene and polyvinyglycine with a wide range of volumes. Sealing materials, such as cotton plugs, screw caps, aluminium foil, transparent film and many others, have different gas permeability and light transmittance. The following factors are assessed when considering the feasibility and practicality in understanding and improving orchid cultures:

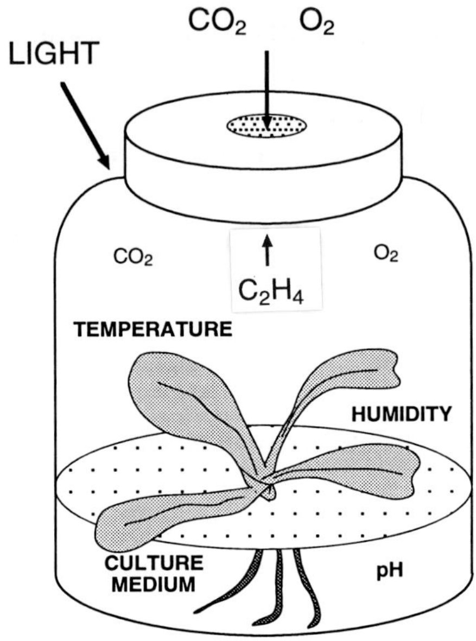

Fig. 9.1. Factors affecting the growth of orchids under *in vitro* culture conditions.

Sugar

Explants, shoots and plantlets *in vitro* (in tissue culture containers) have been considered to have little or low photosynthetic ability to attain a positive carbon balance. Therefore, there is a need to provide an exogenous source of carbon (in the form of sugars) for growth. Direct evidence to show that orchid plantlets are heterotrophic under culture is substantiated by experiments involving the use of C_3 or C_4 sugar as the carbon source (Table 9.1). The $\delta^{13}C$ values of *Dendrobium* plantlets after three months are similar to the $\delta^{13}C$ values of the exogenously supplied sugars, indicating that the orchid plantlets are dependent on the medium for carbon and could not achieve net carbon gain using its own photosynthesis (see Chap. 3 on Photosynthesis for an explanation of $\delta^{13}C$ values).

Considerable attention has been paid to the effect of sugars on orchid tissue cultures (Arditti, 1977). Many media for orchid tissue culture contain sucrose as the carbon source. The effects of other sugars, such as glucose and fructose

Table 9.1. $\delta^{13}C$ Values of *Dendrobium* plantlets after growing in different concentrations of cane sugar and beet sugar.

	4 weeks ‰	12 weeks ‰
Beet sugar (C_3)		
0.1%	−24.10 ± 0.40	−22.13 ± 0.72
1.0%	−25.20 ± 0.52	−23.37 ± 0.53
Cane sugar (C_4)		
0.1%	−21.80 ± 0.49	−22.35 ± 0.05
1.0%	−19.33 ± 0.83	−14.83 ± 0.60

Note: Control plantlets = −20.10 ± 0.28‰.

Adapted from Lim, Hew, Wong & Hew (1992).

in culture media, have been studied with varying results. For example, *Cymbidium* grows better on sucrose than on maltose, glucose, or fructose. Glucose is reported to inhibit the multiplication of *Cymbidium* protocorms. In contrast, *Vanda* tissues proliferate best in sugar-free basal medium containing coconut water, possibly indicating that *Vanda* is more sensitive to high sugar levels. Both *Dendrobium* and *Aranda* tissues have a strong affinity for fructose relative to glucose and sucrose. When sucrose is included in the culture medium as the sole carbon source, it is hydrolyzed into glucose and fructose. Glucose accumulated in the medium is then taken up only after all the fructose has been consumed. For *Dendrobium* tissues, the relative growth rate increases with increasing sugar concentration in the media and this observation is most marked with fructose as the carbon source. It is noteworthy that although fructose appears to be a better carbon source, it cannot be autoclaved (due to chemical decomposition) with the rest of the culture media unlike glucose and sucrose, thus making it unsuitable for large-scale implementation.

The process of sugar uptake by *Aranda* and *Dendrobium* tissues follows linear kinetics and is a function of the initial sugar concentration according to the Monod relation. Studies have shown that the peripheral layers of cells in orchid callus tissues are involved in sugar uptake, which agrees with Morel's (1974) observation on the growth of *Cymbidium* protocorm-like bodies. The overall rate of sugar uptake by orchid callus tissues is determined to a large extent by their surface area-to-volume ratio. The importance of sugar uptake and its subsequent utilisation for growth in orchid callus tissues is demonstrated in sugar uptake kinetics studies. *Aranda* callus shows a high specific rate of glucose uptake when grown in glucose-containing media. Depending on its concentration, the rate of glucose uptake is 10–100 times higher than the specific biomass growth rate. This suggests that glucose accumulates in the cells more rapidly than it can be used for growth. This hypothesis is substantiated by the frequent observation that *Aranda* callus turns brown and dies after being transferred to fresh culture medium. Furthermore, the comparatively faster growing *Dendrobium* callus cultures appear to be less sensitive to sugar stress during transfers to fresh medium (i.e., less browning of tissues occurs in high sugar media).

Orchid tissues of different genera show different affinities for the various sugars and this makes the formulation of a standard solution difficult. A culture medium with a lower sugar concentration may prevent excessive intracellular accumulation of sugar. Therefore a carbon-limited continuous flow culture system could be advantageous.

The presence of sugar in culture medium strongly encourages rapid growth of bacteria and fungi. Hence, sterile and air-tight vessels containing the sugar-rich medium must be handled with care to prevent any possible contamination. The seriousness of the contamination problem is acknowledged by the industry on the whole. Therefore, to prevent the sudden loss of plantlets due to rapid growth of contaminants, small vessels of 100 to 500 ml headspace are used.

Carbon dioxide

In conventional closed system of orchid culture, gas exchange is very much restricted and there is usually a decrease in CO_2 concentration in the culture vessel during photoperiod. Evidence has shown that the increased growth of most plants under carbon dioxide enrichment (CDE) is due to the suppression of photorespiratory loss of carbon. Some efforts have been made in studying the effects of CDE on growth of orchid plantlets *in vitro*. CDE involves the constant supply and maintenance of elevated CO_2 to plantlets, thereby ensuring that plant growth is not limited by the level of CO_2.

Orchids with C_3 photosynthesis

In vitro CDE studies have shown that growth of most non-orchidaceous C_3 plants can be enhanced (Buddendorf-Joosten and Woltering, 1994). Kozai and coworkers have reported that the growth of a C_3 orchid *Cymbidium* Reporsa could be increased using CDE. The photosynthetic response curves as a function of CO_2, photosynthetic active radiation (PAR) and temperature for intact *Cymbidium* plantlets cultured *in vitro* on Hyponex medium (Fig. 9.2) are based on the elegant work of Kozai and his coworkers. Generally, the light response curves of *in vitro Cymbidium* plantlets are similar to those of the shade plants.

Orchids with Crassulacean Acid Metabolism

For CAM orchid plantlets, research has shown that the mode of photosynthesis changes during ontogeny (see Chap. 3 on Photosynthesis). Young protocorms of CAM orchids exhibit a considerable portion of C_3 photosynthesis and have low CAM activity. As the orchids grow older, the proportion of CAM photosynthesis increases with age. This observation is important since short-term CDE is more effective in promoting the growth of C_3 plants than that of CAM plants. Recently, increased growth for CAM orchid plantlets (M*okara* White) is achieved using CO_2 enrichment (Fig. 9.3).

Ethylene

Gaseous composition changes with the growth of orchid plantlets in conventional closed system due to the restriction of gas exchange with the external environment. Gases like CO_2 and O_2 are depleted rapidly while ethylene accumulates in the headspace. Besides the release of ethylene from

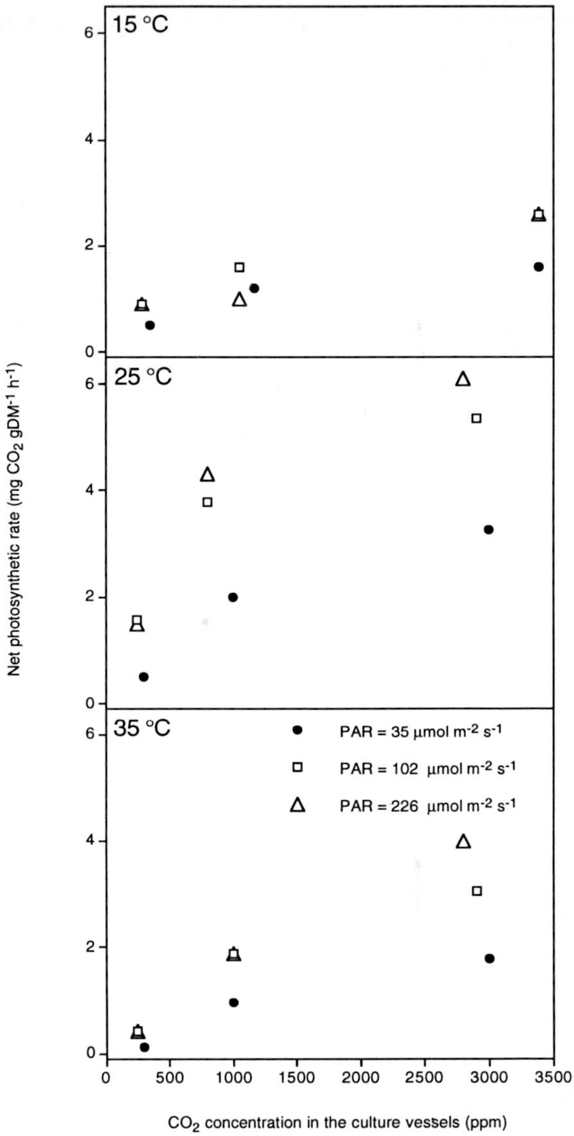

Fig. 9.2. Net photosynthetic curves of *Cymbidium* plantlets under *in vitro* conditions.

Redrawn from Kozai, Oki & Fujiwara (1990).

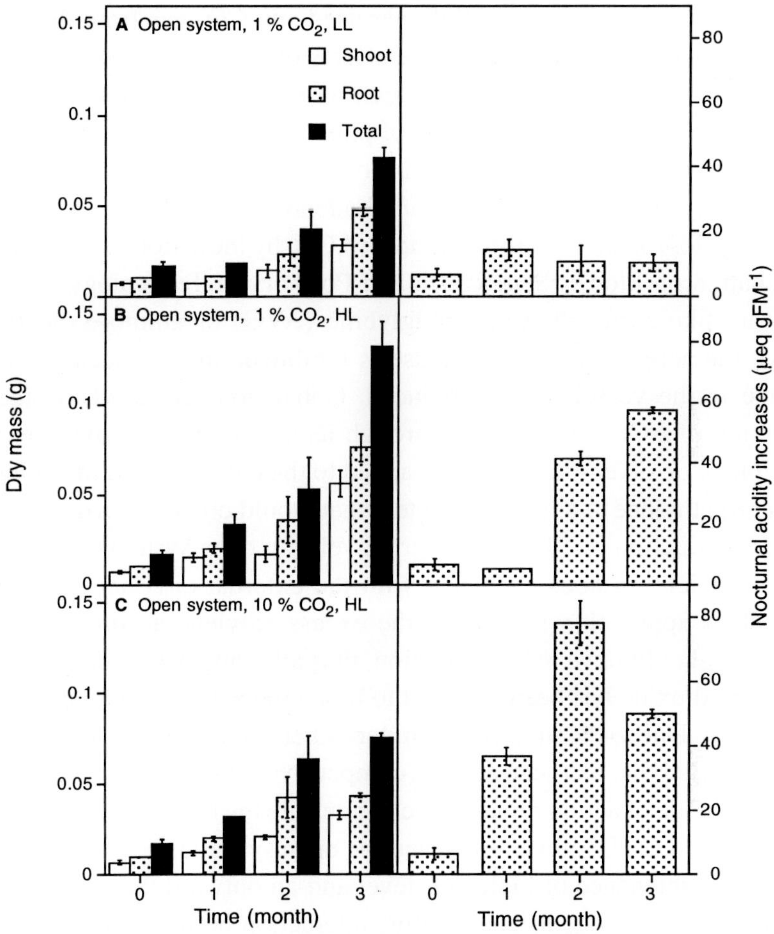

Fig. 9.3. Dry mass changes and nocturnal acidity increases in *Mokara* White plantlets grown in open systems without sucrose at two carbon dioxide concentrations and light intensities. Note: Low light (LL) = 80 µmol m^{-2} s^{-1}; High light (HL) = 200 µmol m^{-2} s^{-1}. External carbon dioxide concentration = 1% and 10% (Mean, ± SE).

Redrawn from Hew, Hin, Yong, Gouk & Tanaka (1995).

plant materials, ethylene accumulation is also contributed by the type of sealing in the culture system, the brand of agar used and CO_2 gas from the cylinder used for CO_2 enrichment studies. The effects of ethylene on plant tissues are very diverse, with both positive and negative results. The prominent negative effects of ethylene are the inhibition of plant growth and the enhancement of senescence (Buddendorf-Joosten and Woltering, 1994).

In plants, ethylene is produced from S-adenosylmethionine (SAM) through 1-aminocyclopropane-1-carboxylic acid (ACC) by the action of ACC synthase (see Chap. 8 on Flower Senescence and Postharvest Technology). By using inhibitors like aminoethyoxylvinylglycine (AVG) or aminooxyacetic acid (AOA), the activity of ACC synthase is inhibited, and the accumulation of ethylene in the vessels can be reduced. Cobalt ions are used to block the conversion of ACC into ethylene through its action on ACC oxidase. Silver thiosulphate or silver nitrate may be added to the culture medium to block the binding of ethylene to a receptor protein that would otherwise trigger a whole cascade production of ethylene. Alternatively, an open system can be used to allow sufficient gaseous exchange with the external environment (Tanaka, 1991). This approach will remove the excess ethylene accumulated in the headspace of culture vessel by diffusion, thus allowing better plantlet growth.

Carbon dioxide has been reported to block the action of ethylene (Abeles *et al.*, 1992). Due to the molecular similarity between CO_2 and ethylene, it has been suggested that CO_2 can act as a competitive inhibitor of ethylene action. The exact mode of action remains unclear despite the known beneficial effects of CO_2 against C_2H_4 (ethylene). When the open system is coupled with CDE, there is a maintenance of high CO_2 level and an outward diffusion of excess ethylene from the system. The positive interaction between the open system and CDE may serve to enhance growth of both C_3 and CAM orchid plantlets *in vitro*.

Nitrogen sources

Numerous studies have shown the preferential uptake of ammonium ions in comparison to nitrate ions by *Dendrobium* tissues and plantlets. It was also observed that *Cattleya* embryos during germination and early stages are unable to utilise nitrate ions. They are only able to utilise these ions after 60 days, which coincides with the appearance of nitrate reductase activity. At present, the mechanism involved in the preferential uptake of ammonium ions over nitrate ions by orchids remains unclear. It is hypothesised that the epiphytic origin of most orchids may require them to exploit whatever nitrogen source available. Chemical analysis of the composition of stemflow in forest canopies shows that nitrogen in the form of ammonium is 40 times higher than that of nitrate. Therefore, the preferential uptake of ammonium by orchids would enable it to obtain nitrogen in the mineral-scarce epiphytic habitat.

Light

Light intensity

Due to the limiting CO_2 concentration in conventional closed system, light (or photosynthetic active radiation, PAR) is kept low at about 60–65 $\mu mol\ m^{-2}\ s^{-1}$. It is known that CDE for plantlets *in vitro* would promote photosynthesis at relatively high PAR of 80–200 $\mu mol\ m^{-2}\ s^{-1}$. Growth of numerous plant species during acclimatisation and multiplication stage is also promoted at high PAR along with CDE. For CAM orchid plantlets, higher plant dry matter accumulation, nocturnal acidity increases, relative growth rates and nitrate uptake are recorded for plantlets grown under open systems (enriched with 1% CO_2) at 200 $\mu mol\ m^{-2}\ s^{-1}$ than those at 80 $\mu mol\ m^{-2}\ s^{-1}$ (Fig. 9.4).

Lighting direction

There is no data available on the effect of lighting direction on orchid growth. Nonetheless, the numerous advantages of this approach warrant a brief discussion. A sidewide (or lateral) lighting system has been developed by Kozai and coworkers to promote growth of plantlets and to control plant height. An attractive feature of this system is that the vessels (Magenta, GA7) and the

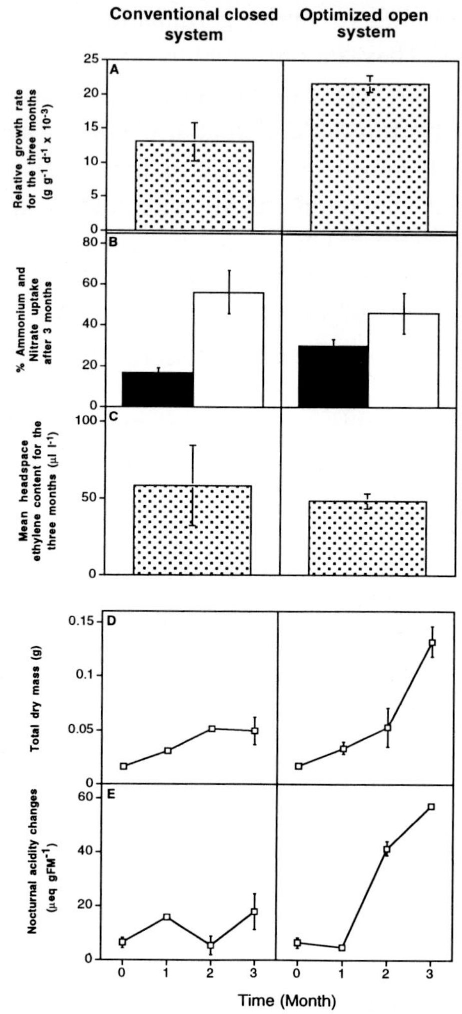

Fig. 9.4. A comparison between an optimized open system of culture and the conventional closed system.

Note: The growth parameters recorded for *Mokara* 'White' plantlets are: (A) relative growth rate; (B) percentage ammonium and nitrate uptake; (C) mean headspace ethylene accumulation; (D) total dry mass; (E) nocturnal acidity increases. Closed system conditions: +2% sucrose, light intensity (PAR) of 80 µmol m^{-2} s^{-1}. Open system conditions: without sucrose, 1% CO_2 (external; internal CO_2 of *ca.* 0.327% within culture vessels), light intensity (PAR) = 200 µmol m^{-2} s^{-1}) (mean, ±SE).

Redrawn from Hew, Hin, Yong, Gouk & Tanaka (1995).

fluorescent tubes are stacked vertically to maximise space usage, without reducing the levels of PAR received by the plantlets. This design allows the plantlets to receive more light energy from the existing light source. Furthermore, the system design gives significant saving in electricity consumption since lesser lighting and cooling facilities are needed. Unlike the conventional system where light is provided vertically, plants in this system can receive light uniformly at all levels along the shoot. In comparison to the conventional downward lighting treatment, favourable growth results (e.g., 1.8 times increase in dry matter and leaf area, shorter plants with larger leaves at the bottom and smaller leaves at the top) are obtained for potato grown under sideward lighting. An improvement to the sideward lighting system can be implemented by using either diffusive optical fibres (string light source) or light emitting diodes (LEDs, point light source). These light sources produce less conducive heat and emit less longwave radiant energy than fluorescent lamps, thus allowing them to be placed closer to the culture vessels. Gallium-aluminum-arsenide LEDs, with high output in the red region of photosynthetic absorption and action spectra, offer a tremendous technical advantage over conventional light sources for plant growth. The advantages of LEDs over other light sources are long life, small mass and volume, no infrared radiation, and the solid state nature of the device. For example, an array of 250 LED chips creates an even field of red light with intensities equaling full sunlight yet remains cool to the touch, unlike many conventional photo-biological lamps that require water jacketed cooling.

Other factors

In conventional closed systems, relative humidity is generally in the range of 70–90%. This poses a problem during transplanting to the field where the relative humidity is extremely low. Plantlets *in vitro* normally possess thin cuticle and abnormal stomata and are unable to restrict water loss to the external environment. The high humidity also resembles waterlogged conditions where under such stress, plants may produce more ethylene. The ill-effects of high humidity on *in vitro* plant growth can be easily overcome using a gas permeable vessel or lid.

There are a few reports on the effects of oxygen on plant cultures but none for orchids. Generally, a decrease in oxygen level in culture vessels can be expected with increasing CO_2 concentration especially for CDE. Growth of fungi and bacteria is suppressed with decreasing oxygen concentration and this facilitates working under sterile conditions, especially during large-scale micropropagation. However, important processes like respiration may be inhibited when oxygen concentration is too low. Within limits, the positive effects of low oxygen concentration on growth can be explained by its effect on reducing photorespiratory loss of carbon (see Chap. 4 on Respiration).

9.3. Improving Orchid Cultures

Recent pioneering studies by Kozai and coworkers revealed that the low net photosynthesis of green chlorophyllous shoots/plantlets *in vitro* is largely due to low CO_2 concentration in the container during the periods for carbon fixation. Presence of sugar in conventional closed system discourages photoautotrophy in plantlets. Explants *in vitro* have been thought to have little photosynthetic ability to provide a positive carbon balance and there is a requirement for sugar as a carbon source for their heterotrophic growth in the closed system. Slow growth and high mortality during acclimatisation are some serious consequences of heterotrophic growth (Kozai, 1991). Evidence from recent studies has shown that chlorophyllous explants and plantlets do have photosynthetic ability and they might achieve higher growth rate under photoautotrophic and optimised environmental conditions. The advantages of photoautotrophic micropropagation and the disadvantages of conventional micropropagation are listed in Tables 9.2 and 9.3 respectively.

Gas-permeable culture system

With the understanding that growth of plantlets *in vitro* is highly affected by the gaseous environment in a closed system, the conventional closed system needs to be modified to allow sufficient gaseous exchange. This would result in the removal of headspace ethylene gas and supplying optimal levels of CO_2 (above the CO_2 compensation point) to tissues and plantlets. Gas permeable culture systems adopted must have high light transmittance, low water vapour transmittance and chemically inert properties for normal growth and develop-

Table 9.2. Some advantages of photoautotrophic micropropagation.

1	Loss of plantlets due to contamination is reduced
2	Growth and development of plantlets are promoted under high light (PAR)
3	A larger containers can be used with minimal losses due to contamination
4	Air inside these 'open' containers is not saturated. This allow the plants to transpire and leads to improve stomatal functioning
5	Venting of excess ethylene which is inhibitory to growth. Elevated levels of CO_2 can be added to the plants via diffusion or forced ventilation
6	Application of plant hormones and other organic supplements are minimized
7	Procedures for rooting and acclimatization are simplified
8	The environmental control of growth and development of plants is easier to implement
9	Automation for the micropropagation process using computers and robots is easier to develop

Adapted from Kozai (1991), Cassells & Walsh (1994) and Hew (1994).

Table 9.3. Some disadvantages of conventional closed system of micropropagation.

1	Sugars present in the media may cause biological contamination
2	In the presence of sugar, high light (PAR) is not effective for growth promotion
3	Airtight, small containers must be used to reduce losses due to contamination
4	Air inside these airtight containers is always saturated
5	CO_2 and ethylene levels may change to undesirable levels
6	Plant hormones are often necessary for growth
7	The abnormal environment may induce physiological or morphological disorders, retardation of plantlet growth, somaclonal variation and mutation
8	The disorders may result in high mortality during the acclimatization stage and greater percentages of rejected plantlets
9	Automation for the micropropagation process using computers and robots is difficult to develop

Adapted from Kozai (1991) and Hew (1994).

ment of plantlets. In addition, the gas permeable culture system must be able to prevent the entry of bacteria and fungi.

Several open systems have been developed and possess gas permeable properties that allow sufficient gaseous exchange with the external environment (Tanaka, 1991). When the open system is coupled with CDE, there is a maintenance of high CO_2 level and an outward diffusion of excess ethylene (growth-inhibitory gas). This positive interaction between the open system and CDE may serve to enhance growth of orchid plantlets *in vitro*.

A novel culture system suitable for practical application in micropropagation has been developed (Tanaka, 1991). The culture system makes use of the gas permeability, thermal stability, chemical inertness and electrical reliability of fluorocarbon polymer films for CO_2 enrichment. These culture vessels are generally known as 'Culture Pack.' The set-up of the system is relatively easy to follow. It involves heat-sealing of the film to a desired 'milk-carton' shape and size (Fig. 9.5). A steel metal frame of the appropriate size is added as a

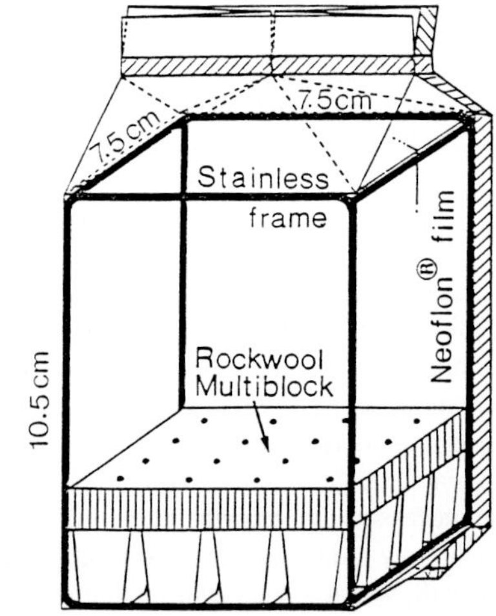

Fig. 9.5. Culture Pack of standard size with a Multiblock rockwool.
Note: During orchid culture, a suitable liquid medium is poured onto the Multiblock rockwool. After inserting the explants, the Culture Pack is closed permanently by 'heat-sealing' the upper portion.

By courtesy of Professor M. Tanaka, Kagawa University, Japan.

support. Sterilised rockwool is used as the artificial substrate where suitable culture medium is absorbed. Neoflon film of 25 μm thickness is found to be suitable for culture as it is autoclavable and can be heat-sealed, and it allows sufficient gaseous exchange without being flimsy. After inoculating of the explants in the culture pack under sterile environment, the opening is heat-sealed to prevent the entry of contaminants.

A viable and economical alternative to obtaining an open culture system is to use a gas-permeable membrane (e.g., Milliseal™, Japan Millipore Ltd.; pore size 0.5 μm) which is an autoclavable air diffusive filter. This procedure requires a slight modification to the many existing culture vessels. Holes of the desired size are drilled into the conventional flask lids (e.g., Magenta™ GA7 lid). The round-shaped Milliseal™, which is self-adhesive, can be pasted over the drilled hole. The modified GA7 lid with the Milliseal™ functions like any ordinary GA7 flask lid, except that it has gas permeable ability without the risk of entry of contaminants (Fig. 9.6). Alternatively, a gas-permeable Magenta™ GA7 lid with a vent (0.22 μm pore size) can be purchased from a suitable supplier. The primary advantage of using Milliseal™ is that the conventional culture system can be easily modified for CDE without the need to revamp the whole system, thereby reducing cost. Both systems offer flexibility in their usage and are extremely useful for *in vitro* CDE of plants.

There are several ways of supplying CO_2 to the culture vessels. CDE can be conducted by gas diffusion method that involves the placing of the gas permeable culture system in a chamber containing elevated levels of CO_2. It is noteworthy that the concentration of CO_2 inside the vessel may not correspond to that in the chamber due to the slow natural diffusion of gases. In addition, the low ventilation rate of natural diffusion usually results in reduced CO_2 concentration around the plantlets. This system permits limited gas exchange between the culture vessels and the chamber containing CO_2-enriched air, and the outward diffusion of ethylene. Alternatively, the process of forced ventilation may be used and it involves the 'pressurised' flushing of elevated CO_2 into the *in vitro* culture system. The build-up of growth inhibitory gas like ethylene is prevented effectively through the constant supply of CO_2. Further, an optimal level of CO_2 is maintained around the plantlets and this may promote photosynthesis. The main drawback of this method is the excessive drying

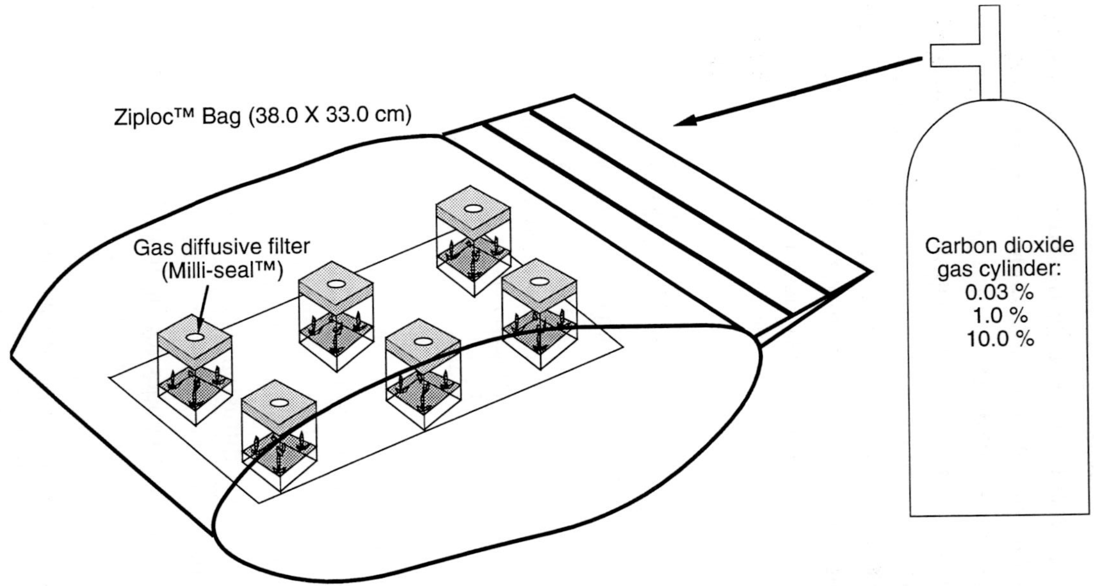

Fig. 9.6. Semi-diffusion forced ventilation (SDFV) system for *in vitro* carbon dioxide enrichment of orchid plantlets.
Note: The SDFV technique involved direct flusing of pressurised non-sterile CO_2 (0.03%, 1% and 10%, balance air) into a transparent plastic bag (38 × 33 cm: Ziploc™, Dow Brand Inc., USA) containing 6 culture vessels (GA7, Magenta™, Sigma Chemical Co., USA) with modified lids until the bag is fully inflated. Each modified GA7 lid has a hole (diameter, 2.5 mm) covered with a self-adhesive gas permeable membrane (Milliseal™, Nihon Millipore Ltd, Japan). Natural diffusion of gases was allowed to take place for two days between the culture vessels and the headspace of the plastic bag before the next flushing session was conducted.

Redrawn from Hew, Hin, Yong, Gouk & Tanaka (1995).

effect of passing gas stream directly onto *in vitro* plants. For example, the use of elevated CO_2 on some plants through forced ventilation gives poor growth results (Buddendorf-Joosten and Woltering, 1994).

It therefore appears that a combination of diffusion and forced ventilation methods will achieve the twofold objectives of supplying adequate CO_2 and maintaining optimal levels of humidity. The recent development of a Semi-Diffusion Forced Ventilation (SDFV) system integrates the advantages from forced ventilation process and natural diffusion (Fig. 9.6). For example, high levels of CO_2 can be maintained in the headspace of a GA7 vessel containing five orchid plantlets using a lid with an air diffusive filter (Fig. 9.7). A more elaborated system of SDFV using an automated CO_2 controlling device is now under development.

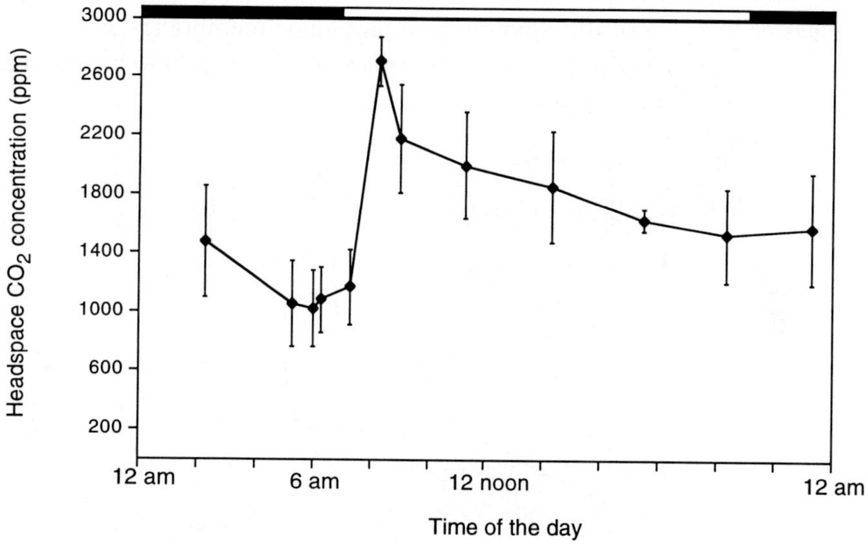

Fig. 9.7. Mean carbon dioxide concentration in the headspace of culture vessels under open system without sucrose after one day of flushing with 1% carbon dioxide.
Note: Each culture vessel contained five three-month old *Mokara* White plantlets.
Light intensity = 80 µmol m^{-2} s^{-1}. External CO_2 = 1% (internal CO_2 of ca. 0.327%) (Mean, ± SE).

Redrawn from Hew, Hin, Yong, Gouk & Tanaka (1995).

Alternative supporting media

Orchid meristem tissues are cultured in defined liquid or solid media where cell proliferation takes place. Studies have shown that this method can be improved by aeration of the medium. Microporous polypropylene membranes have been used as an alternative support of *Cattleya* and *Epidendrum* seedlings, propagation of *Cattleya*, *Cymbidium* and *Dendrobium* protocorms, and plantlet production of *Cattleya*. These membranes are hydrophilic in nature but differ in their pore size, pore density and thickness (Celgard™, Hoechst Celanese Separations Product Div., Charlotte, North Carolina; Membrane Raft™ from Sigma Chemical Co.). The technique of culturing orchid callus tissues on a polypropylene membrane has been evaluated with positive results (Fig. 9.8). The improved growth of orchid callus is probably attributed to greater aeration. This method greatly improved growth and the tedious task of subculturing is made easier. The use of microporous polypropylene membranes as a culture system can be further automated by incorporating the system to a continual-flow system, thus eliminating the need for subculture.

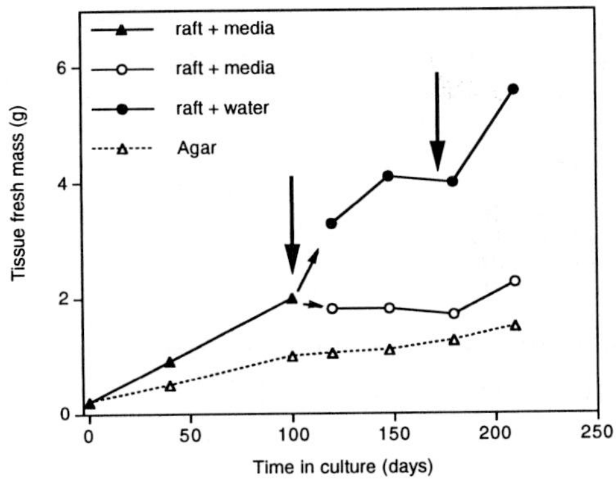

Fig. 9.8. Fresh mass increases of *Laeliocattleya* hybrid tissues growing on Membrane Rafts™ and agar medium.
Note: The arrows indicate the dates of liquid supplementation. The hybrid is a *Laeliocattleya* hybrid (El Cerrito × Spring Fires).

Redrawn from Adelberg, Desamero, Hale & Young (1992).

The older orchid plantlets can also be grown in rockwool instead of using agar as a supporting medium (Fig. 9.9). Better root growth is reported for *Phalaenopsis* growing in rockwool than in agar medium. However, the physiological basis for this observation is still unknown. Growing orchid plantlets in the multi-blocks of rockwool makes transplanting to the smaller

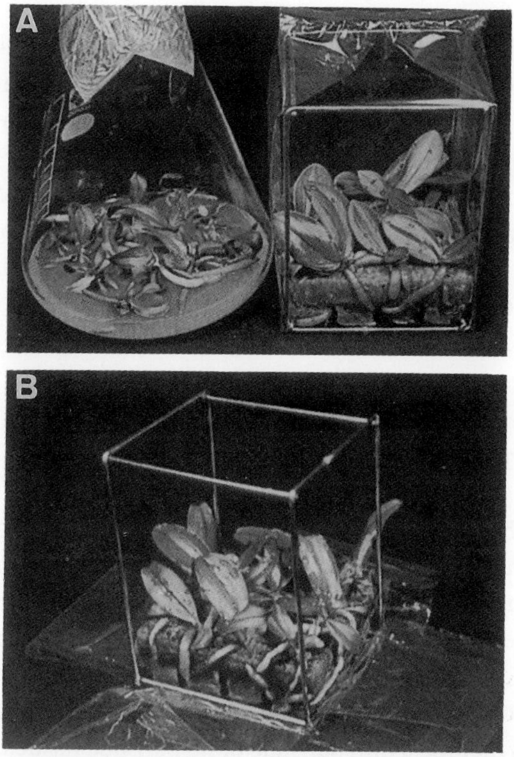

Fig. 9.9. A comparison of the development of regenerated plantlets from *Phalaenopsis* synthetic seeds in agar medium and Culture Pack–Rockwool system.
Note: (A) Left, plantlets are growing on agar medium in a 500 ml flask; Right, plantlets are growing in the standard-size Culture Pack–Rockwool system. The comparison was made after 120 days in culture; (B) Plantlets growing on the Rockwool Multiblock ('*in vitro* community pot') can be easily removed from the Culture Pack by cutting the film using a blade.

By courtesy of Professor M. Tanaka, Kagawa University, Japan.

pots in the greenhouse less stressful as the roots of individual plantlet remain intact. More work is needed to evaluate the suitability of rockwool as an alternative medium for the other orchid genera such as *Dendrobium*, *Aranda* and *Mokara*.

Carbon dioxide enrichment

Millions of orchid plantlets can be mass produced by conventional micropropagation. However, not all plantlets survive when they are transferred from *in vitro* conditions to the field environment. A critical approach to assess success in any micropropagation system is to measure the number of plantlets that survived after being transplanted from i*n vitro* cultures to field conditions. A period of acclimatisation is necessary to ensure that a reasonable number of plantlets survive the transplanting process. CDE has been shown to alleviate these problems although the precise physiological mechanism involved remains unclear.

Reduction in vitrification of in vitro plantlets

Vitrification has been used to describe anatomical, morphological and physical anomalies in tissue culture plantlets. Conditions inducing vitrification include high humidity, supra-optimal supply of minerals and carbohydrate, high levels of plant hormones and low light intensity. Vitreous plantlets are observed to have broad, thick and translucent stems, thick, wrinkled and elongated leaves and poor growth of vitrified roots. Some anatomical and morphological changes in vitreous plants include a thin palisade or no palisade tissues, rich intercellular mesophyll cells, thin deposition of epicuticular waxes, malfunctioning of guard cells and reduced vascular tissues (Ziv, 1991).

Disorders such as reduced chlorophyll content and abnormal organisation of chloroplast are mainly manifested in the leaves. Photosynthesis and transpiration are seriously or badly affected due to the altered leaf morphology. The abnormal plant morphogenesis *in vitro* is found to be highly influenced by the water status and the gaseous phase in culture (Ziv, 1991). High relative humidity in conventional culture system (90% to 100%) has been proposed to be analogous to waterlogging which causes *in vitro* plants to produce more ethylene that eventually accumulates in the culture flasks. In addition, these plants often have poorly formed cuticle and stomata.

A reduction in vitrification for chrysanthemum has been reported for plants grown under lower humidity. Under CDE at suitable light levels, a reduction in relative humidity in the culture vessel is often observed as water is lost from the medium during the enrichment process (Kozai, 1991). Plantlets grown under CDE have thicker cuticular wax layer, proper stomatal function and achieved higher rates of nutrient uptake. This preconditioning process *in vitro* should be beneficial to the transplanting of plantlets to field conditions. Plants need to possess sufficient epicuticular wax and have functional stomata to adapt to the field conditions. For plants grown in conventional culture system, they do not possess the natural morphological characteristics against desiccation. Therefore, a progressive reduction of relative humidity over time in "sweat boxes" is necessary before transfer to the field. In addition, plantlets may be sprayed with anti-transpirants that help to reduce unnecessary water loss during transplanting. However, the frequent growth impairment, phytotoxicity and additional cost associated with the use of anti-transpirants make it unpopular with commercial growers.

In vitro cultured plantlets have been observed to have little or no net photosynthesis immediately after transplanting. Photosynthesis is observed to occur only after two weeks of transfer to the field. Therefore, prolonged acclimatisation is necessary to allow new leaves to be produced (Ziv, 1991). In contrast, plants grown under CDE require little or no sucrose for *in vitro* growth and carbon necessary for growth is produced from its own photosynthesis. These plants would then acquire a certain amount of photosynthetic ability *in vitro* and will not experience 'sudden shock' during transplanting to the field, thus requiring a shorter period of acclimatisation.

Improved rooting of plantlets

When plantlets are removed from conventional closed system, the roots formed *in vitro* frequently die. Attempts to induce *in vitro* rooting are usually expensive and incur between 35% and 75% of the total cost for micropropagation. An experiment using *in vitro* propagated non-rooted grapevines has shown that CDE is highly beneficial to rooting and growth. Secondary root initiation occurs much earlier at 1,200 ppm of CO_2 and the root dry mass is six times higher than that at 350 ppm of CO_2. This experiment demonstrated an increase in the root/shoot ratio of grapevine, implying that more carbon is partitioned to the root organ during CDE. As noted by Kozai (1991), plantlets grown under CDE require no rooting and acclimatisation process. Since CDE aided in the development and growth of roots of plantlets *in vitro*, this would reduce the cost and time for *ex vitro* rooting in the field. At present, the physiological basis of CDE in increasing carbon partitioning to roots is unknown.

Development of a flow system

In vitro tissue culture can be grouped into two categories: Batch and flow cultures (Fig. 9.10). Mass clonal propagation of orchids through batch culture has been the mainstay throughout the world since 1960. With this method, shoot–tip meristems are excised and cultured in a defined liquid or solid medium. Given an appropriate culture medium, the explants proliferate and then differentiate. However, this is essentially a closed system, and the conditions may not be optimal for cell growth. Since the tissues are grown in a fixed volume of medium, depletion of nutrients and accumulation of toxic

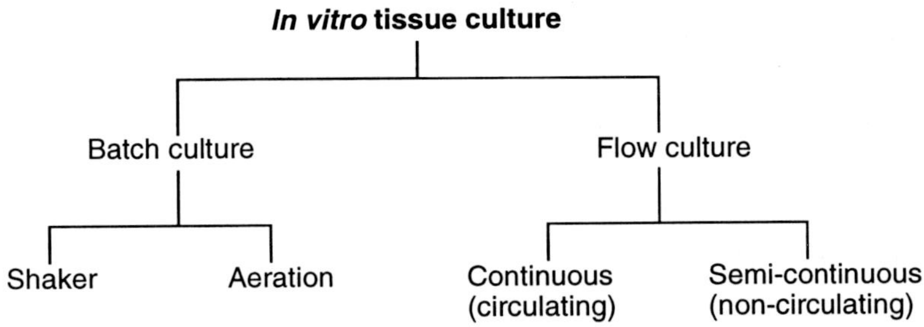

Fig. 9.10. The two categories of *in vitro* orchid tissue culture.

materials are continuous. In addition, the oxygen and carbon dioxide levels, as well as pH of the medium, change considerably with time. To optimise cell and tissue growth, it is important to maintain all important factors at optimal levels. In batch culture, this is only possible by highly frequent subculturing. On the other hand, subculturing involves considerable time and effort, resulting in a major increase of production costs.

Although batch culture is laborious and has many unfavourable aspects, it is still practised widely mainly because of its simplicity. An airflow system for orchid tissue culture was proposed in 1982. This system is similar to the batch culture method, except that the cultures are agitated and aerated with an air flow instead of using shakers. Keeping the tissues constantly agitated and well supplied with oxygen have shown to accelerate growth due to the added aeration. The substantially larger volume of nutrients used in the flow system provides a much more favourable mass-to-volume ratio from the onset. The disadvantages of this system are the same as those of batch culture, except for the better aeration. However, an increase in aeration may cause browning of orchid callus tissues as a result of friction between calli.

A computerised long-term tissue culture system for orchids was devised in 1986. This system reduces labour requirements by minimizing the number of transfers required. The explants remain stationary, and the medium is introduced or removed as needed to achieve optimum growth. Through automated control, explants and tissues in the Automated Plant Culture System (APCS) are aerated and batched intermittently in fresh medium. In this system, the necessity for physical transfer of plant cultures is minimised. APCS provides the essential functions and manipulations associated with traditional plant tissue culture but in different perspectives:

1. Culture medium can be introduced or removed within a single chamber.
2. Labour requirements to culture plant *in vitro* are reduced.
3. Cultures of different plants can be grown in the same chamber.

APCS can be expected to have an impact on the orchid culture industry. However, the system has not been tested on a wide scale or with callus and protocorm-like bodies. There is no information pertaining to the extent of nutrient depletion in the APCS system. For a new system to be implemented and used widely, cost and production output must be better than those of the present batch method. One of the important areas that needs to be addressed is the reduction of labour costs which at present constitute up to 20–30% of production costs. Automation will reduce labour costs. Other factors to be taken into account are productivity and yield. A new system will only be practical when its fundamentals are understood. This can be done by the wide screening of different orchid species for substrate affinity and utilisation.

9.4. *In Vitro* Flowering

The technique of *in vitro* flowering involves the initial explanting of disinfected tissue and the eventual flowering on a defined growth medium in an aseptic environment. Flowering may be induced by subjecting the tissues to various types and concentrations of plant hormones, vitamins, carbohydrate sources, alternate temperatures, light intensities, photoperiodic treatment, and different biologically active organic and inorganic chemicals (Scorza, 1982). The use of *in vitro* flowering is significant as it shortens the breeding period of transgenic orchids, new hybrids and cultivars. For the commercial growers, breeders will see the results of their crosses sooner and this will improve the efficiency of varietal development by shortening the generation interval. A potential application of this approach is that the orchid breeding cycle can be further reduced if seeds can be successfully obtained *in vitro*.

A number of reports are available on the induction of early flowering in orchids using tissue culture procedures (Table 9.4). Investigations of this nature for most higher plants such as tobacco and grapes have focused on the formulation of a suitable media containing cytokinins, auxins, gibberellins, and sugars (Scorza, 1982). Cytokinins promote flowering while auxins are generally inhibitory. GAs affect flower development rather than changing the plant from vegetative to flowering state. Sugars are needed for *in vitro* flowering as an energy source. In orchids, results obtained from several studies indicate that BAP-induced floral bud development requires proper nutritional conditions such as the ratio of carbohydrate and nitrogen. For example, BAP initiates formation of floral buds in *Doriella* Tiny, but prolonged growth in BAP-containing medium will inhibit flower development. In addition, floral bud formation can only take place in media with 10 g l^{-1} sucrose for BAP-induced floral bud initiation.

Table 9.4. Conditions for inducing *in vitro* flowering in some orchids.

Orchid	Duration	Medium
Cymbidium ensifolium	Subcultured plantlets produced terminal flowers in 2–3 months	Murashige and Skoog medium with 1 mg l^{-1} BAP and 0.1 mg l^{-1} NAA
Dendrobium candidum	Induced within 3–6 months from protocorms	Application of polyamines, BAP and NAA are used to achieve a flowering frequency of greater than 36%. Higher flowering frequency (83%) was achieved by growing the cultured shoots in ABA containing medium and transferring them later to Murashige and Skoog medium with BAP.
Doriella Tiny	Induced in 7-month old explants	Initiation of floral buds using BAP (5 mg l^{-1}) in Vacin and Went medium Floral development in a BAP-free Hyponex media.

Adapted from Wang (1988), Wang, Xu, Chia & Chua (1990) and Duang & Yazawa (1994).

9.5. Thin Section Culture

Tissue culture of orchids commonly involved the use of shoot tips and axillary buds to produce protocorm-like bodies (PLBs) which subsequently develop into plantlets under appropriate *in vitro* conditions (Arditti and Ernst, 1993). The production and growth of PLBs from shoot tips and axillary buds by many economically important orchids are generally very slow. For example, the development of plantlets from PLBs of *Aranda* Deborah takes about nine to 12 months. The development of a faster and more productive approach to produce PLBs from shoot tips and axillary buds would be beneficial to the orchid industry.

The concept of thin section culture was earlier proposed by Tran Thanh Van (1981) where thin explants are used to study tissue morphogenesis with minimal physiological influence from nearby tissues. Recently, this idea is used as a means of rapid plant production in orchids. Rapid regeneration of a monopodial orchid *Aranda* Deborah can be obtained using thin section culture. Thin sections (0.6–0.7 mm thick) are obtained by transverse sectioning of a single shoot tip. When cultured in modified Vacin and Went medium with appropriate plant hormones and additives, more than 80 000 plantlets could be produced, compared to nearly 11,000 plantlets produced by the conventional shoot tip culture in a year.

9.6. Synthetic Seeds

The usefulness of synthetic seeds (or encapsulated somatic embryos) has promoted research in this area for many agricultural and horticultural crops. The production of synthetic seeds involves the encapsulation of PLBs in a calcium alginate matrix (Fig. 9.11). Synthetic seeds for several genera such as *Cymbidium*, *Dendrobium*, *Phalaenopsis* and *Spathoglottis* have been obtained using the general encapsulation technology available for other plants with some modifications. As the orchid industry is reliant on micropropagation as a major source of planting material, orchid synthetic seeds may become indispensable as it can be delivered easily like true seeds from commercial tissue culture laboratories to growers (e.g., packed in vials). More research is needed in this area to improve the encapsulation process, ‘germination’ rate, early growth of young plantlets (e.g., using CO_2 enrichment) and to automate the several key processes involved in making the synthetic seeds.

Fig. 9.11 Synthetic orchid seeds.

Note: The protocorm-like-bodies of a *Phalaenopsis* hybrid are encapsulated in calcium alginate beads. The PLBs are 4 mm long. About one thousand synthetic seeds can be produced in 30 min using general encapsulation technology.

By courtesy of Professor M. Tanaka, Kagawa University, Japan.

9.7. Concluding Remarks

The usage of photoautotrophic micropropagation that incorporates gas-permeable culture systems and carbon dioxide enrichment (CDE) appears to be highly feasible for orchid cultures. Sugar should be given at the very early stage of growth since the explants have little chlorophyll and/or little leaf area for photosynthesis. Thus, if sugar is given at the early stage of growth and is subsequently removed (gradually, if necessary), greatest growth rate of plantlets *in vitro* may be obtained. CDE has been proven to be effective in promoting growth at the later stages of plant growth and is able to reduce cost in several areas of traditional micropropagation. However, the additional cost of providing supplemental CO_2 and light, and the usage of new gas-permeable vessels must be considered. For the orchid industry, a logical step to take is to modify the existing conventional closed culture systems or to adopt an entirely new open system in view of the positive research findings.

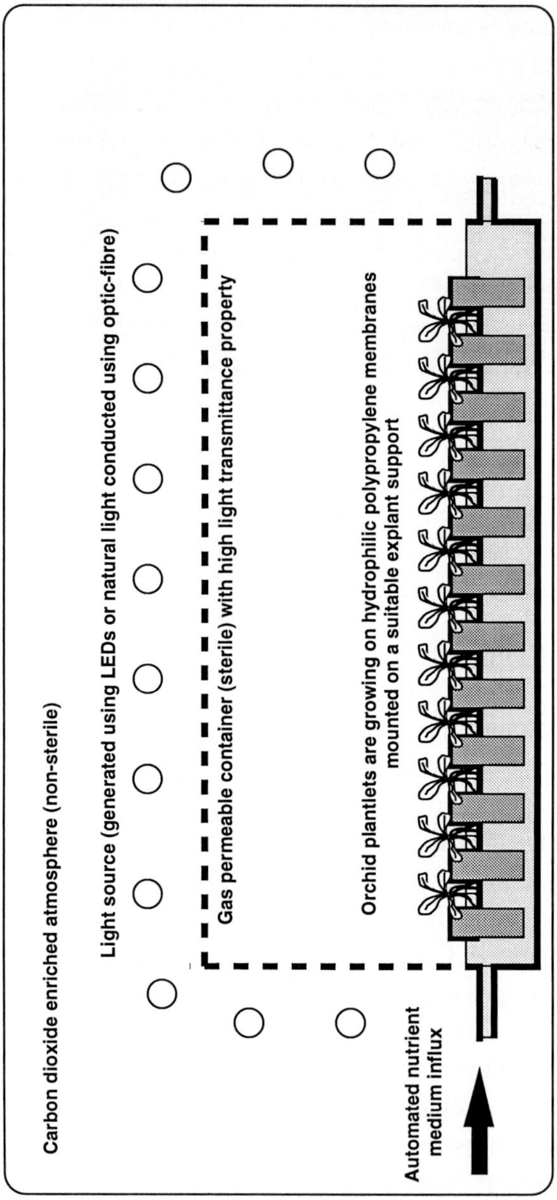

Fig. 9.12. An automated plant culture system for orchids using a liquid flow system under optimized environmental conditions.

In time to come, it is likely that Automated Plant Culture System (APCS) will be used in combination with some of the useful techniques associated with batch cultures such as polypropylene membranes, CO_2 enrichment and gas-permeable cultures. The following is a possible scenario for the orchid industry in the near future: Mass and continual production of uniform callus and protocorm-like bodies are carried out using an appropriate medium whose composition is maintained regularly at optimal levels. Later, the young plantlets are grown photoautotrophically on polypropylene membranes using CO_2 enrichment in the APCS (Fig. 9.12).

9.8. Summary

1. Presence of sugar in the conventional closed system of culture discourages photoautotrophy in orchid plantlets.
2. Both C_3 and CAM orchid tissues and plantlets respond positively to *in vitro* CO_2 enrichment.
3. Ethylene accumulation in the culture system can be alleviated by the use of an open system (e.g., gas-permeable vessels) or to modify existing closed system into an open system using gas-permeable lids or vessels.
4. New cultural technologies such as polypropylene membranes as a supporting medium, LEDs as a light source, gas-permeable vessels and synthetic seeds should be evaluated for large-scale implementation.
5. Flower colour and morphology of new hybrids or transgenic orchids can be evaluated earlier using *in vitro* flowering techniques. More orchid plantlets can be produced using thin-section culture.

General References

Abeles, F. B., Morgan, P. W. and Saltveit, M. E., 1992, *Ethylene in Plant Biology*, 2nd ed. (Academic Press, San Diego), 414 pp.

Aitken-Christie, J., Kozai, T. and Smith, M. A. L., 1995, *Automation and Environmental Control in Plant Tissue Culture* (Kluwer Academic Publishers, Dordrecht), 574 pp.

Arditti, J., 1977, "Clonal propagation of orchids by means of tissue culture — A manual," in *Orchid Biology: Reviews and Perspectives I*, ed. J. Arditti (Cornell University Press, Ithaca, New York), pp. 203–294.

Arditti, J. and Ernst, R., 1993, *Micropropagation of Orchids* (John Wiley and Sons Inc., New York), 640 pp.

Buddendorf-Joosten, J. M. C. and Woltering, E. J., 1994, "Components of the gaseous environment and their effects on plant growth and development *in vitro*," *Plant Growth Regulation* **15**, 1–16.

Debergh, P. C. and Zimmerman, R. H., 1991, *Micropropagation Technology and Application* (Kluwer Academic Publishers, Dordrecht), 484 pp.

Desjardins, Y., 1995, "Factors affecting CO_2 fixation in striving to optimise photoautotrophy in micropropagated plantlets," *Plant Tissue Culture and Biotechnology* **1**: 13–25.

Goh, C. J., 1983, "Asexual mass propagation of orchids and its commercialisation: A review of the present status," in *Plant Cell Culture in Crop Improvement*, eds. S. K. Sen and K. L. Giles (Plenum Press, New York), pp. 319–336.

Hew, C. S., 1994, "Orchid cut-flower production in ASEAN countries," in *Orchid Biology: Reviews and Perspectives, Vol. VI*, ed. J. Arditti (John Wiley and Son Inc., New York), pp. 363–401.

Kozai, T., 1991, "Micropropagation under photoautotrophic conditions," in *Micropropagation Technology and Application*, eds. P. C. Debergh and R. H. Zimmerman (Kluwer Academic Publishers, Dordrecht), pp. 447–469.

Lumsden, P. J., Nicholas, J. R. and Davies, W. J., 1994, *Physiology, Growth and Development of Plants in Culture* (Kluwer Academic Publishers, Dordrecht), 427 pp.

Morel, G. M., 1974, "Clonal propagation of orchids," in *The Orchids: Scientific Studies*, ed. C. L. Withner (Wiley-Interscience, New York), pp. 169–222.

Rao, A. N., 1977, "Tissue culture in the orchid industry," in *Applied and Fundamental Aspects of Plant Cell, Tissue and Organ Culture*, eds. J. Reinert and Y. P. S. Bajaj (Springer-Verlag, Berlin, Heidelberg, New York), pp. 44–69.

Scorza, R., 1982, "*In vitro* flowering," in *Horticultural Reviews, Vol. 4*, eds. D. P. Coyne, D. Durkin and M. W. Williams (Avi Publishing Company Inc., Connecticut), pp. 106–127.

Tanaka, M., 1991, "Disposable film culture vessels," in *Biotechnology in Agriculture and Forestry, Vol. 17, High-Tech and Micropropagation I*, ed. Y. P. S. Bajaj (Springer-Verlag), pp. 212–228.

Tran Thanh Van, K. M., 1981, "Control of morphogenesis in *in vitro* cultures," *Annual Review of Plant Physiology* **32**: 292–311.

Vajrabhaya, T., 1977, "Variations in clonal propagation," in *Orchid Biology: Reviews and Perspectives I*, ed. J. Arditti (Cornell University Press, Ithaca, New York), pp. 177–202.

Ziv, M., 1991, "Vitrification: morphological and physiological disorders of *in vitro* plants," in *Micropropagation Technology and Application*, eds. P. C. Debergh and R. H. Zimmerman (Kluwer Academic Publishers, Dordrecht), pp. 45–69.

References

Adelberg, J., Desamero, N., Hale, A. and Young, R., 1992, "Orchid micropropagation on polypropylene membranes," *American Orchid Society Bulletin* **61**: 688–695.

Adelberg, J. and Darling, J., 1993, "*In vitro* membrane treatment accelerates flowering of *Laeliocattleya* (El Cerrito × Spring Fires)," *American Orchid Society Bulletin* **62**: 920–923.

Burg, S. P. and Burg, E. A., 1967, "Molecular requirements for the biological activity of ethylene," *Plant Physiology* **42**: 144–152.

Cassells, A. C. and Walsch, C., 1994, "The influence of gas permeability of the culture lid on calcium uptake and stomatal function in *Dianthus* microplants," *Plant Cell, Tissue Organ Culture* **37**: 171–178.

Cheng, Y. W. and Chua, S. E., 1982, "The use of air-flow system in plant tissue and organ culture," in *Proc. COSTED Symp. on Tissue Culture of Economically Important Plants*, Singapore (1981), pp. 210–212.

Corrie, S. and Tandon, P., 1993, "Propagation of *Cymbidium giganteum* Wall. through high frequency conversion of encapsulated protocorms under *in vivo* and *in vitro* conditions," *Indian Journal of Experimental Biology* **31**: 61–64.

Duang, J. X. and Yazawa, S., 1994, "*In vitro* floral development in *Doriella* Tiny (*Doritis pulcherrima* × *Kingiella philippinensis*)," *Scientia Horticulturae* **59**: 253–264.

Fonnesbech, M., 1972, "Organic nutrients in the media for propagation of *Cymbidium in vitro*," *Physiologia Plantarum* **26**: 360–364.

Freson, R., 1969, "Action du glucose sur des protocormes de *Cymbidium* Sw. (Orchidaceae) cultives *in vitro*," *Bull. Soc. Roy. Belg.* **102**: 205–209.

Grout, B. W. W. and Aston, H., 1978, "Modified leaf anatomy of cauliflower plantlets regenerated from meristem culture," *Annals of Botany* **42**: 993–995.

Hew, C. S., Chia T. F., Lee, Y. K. and Loh, C. S., 1987, "The need for a flow orchid tissue culture system," *Malayan Orchid Review* **21**: 30–34.

Hew, C. S., Ting, S. K. and Chia, T. F., 1988, "Substrate utilisation by *Dendrobium* tissues," *Botanical Gazette* **149**: 153–157.

Hew, C. S. and Lim, L. Y., 1989, "Mineral uptake by orchid plantlets grown on agar culture medium," *Soilless Culture* **5**: 23–34.

Hew, C. S. and Mah, T. C., 1989, "Sugar uptake and invertase activity in *Dendrobium* tissues," *New Phytologist* **111**: 167–171.

Hew, C. S., Chan, Y. S., Lee, Y. K. and Chia, T. F., 1990, "Culture of orchid tissue on polypropylene membrane," *Malayan Orchid Review* **24**: 78–81.

Hew, C. S., Lim, L. Y. and Low, C. M., 1993, "Nitrogen uptake by tropical orchids," *Environmental and Experimental Botany* **33**: 273–281.

Hew, C. S., Gouk, S. S., Lin, W. S. and Yong, J. W. H., 1995, "Ethylene production by orchid roots," *Lindleyana* **10**: 43–48.

Hew, C. S., Hin, S. E., Yong, J. W. H., Gouk, S. S. and Tanaka, M., 1995, "*In vitro* CO_2 enrichment of CAM orchid plantlets," *Journal of Horticultural Science* **70**: 721–736.

Jones, J. B. and Sluis, C. J., 1991, "Marketing of micropropagated plants," in *Micropropagation Technology and Application*, eds. P. C. Debergh and R. H. Zimmerman (Kluwer Academic Publishers, Dordrecht), pp. 141–154.

Kozai, T., Iwanami, Y. and Fujiwara, K., 1987, "Environmental control for mass-propagation of tissue cultured plantlets. (1) effects of CO_2 enrichment on the plantlet growth during the multiplication stage," *Plant Tissue Culture Letter* **4**: 22–26.

Kozai, T., Oki, H. and Fujiwara, K., 1990, "Photosynthetic characteristics of *Cymbidium* plantlets *in vitro*," *Plant Cell, Tissue and Organ Culture* **22**, 205–211.

Kozai, T., Iwabuchi, K., Watanabe, C. and Watanabe, I., 1991, "Photoautotrophic and photomixotrophic growth of strawberry plantlets *in vitro* and changes in nutrient composition of the medium," *Plant Cell, Tissue and Organ Culture* **25**: 107–115.

Kunisaki, J. T., Kim, K. K. and Sagawa, Y., 1972, "Shoot tip culture of *Vanda*," *American Orchid Society Bulletin* **41**: 435–439.

Lakso, A. N., Reish, B. I., Mortensen, J. and Roberts, M. H., 1986, "Carbon dioxide enrichment for stimulation of growth of *in vitro* propagated grapevines after transfer from culture," *Journal of the American Society of Horticultural Science* **111**: 634–638.

Lakshmanan, P., Loh, C. S. and Goh, C. J., 1995, "An *in vitro* method for rapid regeneration of a monopodial orchid hybrid *Aranda* Deborah using thin section culture," *Plant Cell Reports* **14**: 510–514.

Lim, L. Y., Hew, Y. C., Wong, S. C. and Hew, C. S., 1992, "Effects of light intensity, sugar and CO_2 concentrations on growth and mineral uptake of *Dendrobium* plantlets," *Journal of Horticultural Science* **67**, 601–611.

Preece, J. E. and Sutter, E. G., 1991, "Acclimatisation of micropropagated plants from greenhouse and field," *Micropropagation Technology and Application* (Kluwer Academic Publishers, Dordrecht), pp. 71–93.

Raghavan, V. and Torrey, J. G., 1964, "Inorganic and nitrogen nutrition of the seedlings of the orchid *Cattleya*," *American Journal of Botany* **51**: 264–274.

Singh, F., 1991, "Encapsulation of *Spathoglottis plicata* protocorms," *Lindleyana* **6**: 61–63.

Sisler, E. C. and Wood, E. C., 1988, "Interactions of ethylene and carbon dioxide," *Physiologia Plantarum* **73**: 440–444.

Tanaka, M., Jinno, K., Goi, M. and Higashiru, T., 1988, "The use of disposable fluorocarbon polymer film culture vessel in micropropagation," *Acta Horticulturae* **230**: 73–80.

Tanaka, M., Yoneyama, M., Minami, T. and Noguchi, K., 1993, "Micropropagation of *Phalaenopsis* by using synthetic seeds in film culture vessels," *Proc. of the 14th World Orchid Conference*, Glasgow (1993) (HMSO Publications Centre, UK), pp. 180–187.

Tennessen, D. J., Singsaas, E. L., Sharkey, T. D., 1994, "Light-emitting diodes as a light source for photosynthesis research," *Photosynthesis Research* **39**: 85–92.

Tisserat, B. and Vandercook, C. E., 1986, "Computerised long term tissue culture for orchids," *American Orchid Society Bulletin* **55**: 35–42.

Tripathy, B. C. and Brown, C. S., 1995, "Root-shoot interaction in the greening of wheat seedlings grown under red light," *Plant Physiology* **107**: 407–411.

Wang, G. Y., Xu, Z., Chia, T. F. and Chua, N. H., 1990, "*In vitro* flowering of *Dendrobium candidum*," *Abstracts of the Thirteenth World Orchid Conference*, New Zealand, September 1990. 67 pp.

Wang, X., 1988, "Tissue culture of *Cymbidium*: plant and flower induction *in vitro*," *Lindleyana* **3**: 184–189.

Wilson, G., 1980, "Continuous culture of plant cells using the chemostat principle," in *Advances in Biochemical Engineering* **18**, ed. A. Fiechter (Springer Verlag, Berlin), pp. 101–150.

Ziv, M., 1986, "*In vitro* hardening and acclimatisation of tissue culture plants," in *Plant Tissue Culture and its Agricultural Applications*, eds. L. A. Withers and P. G. Alderson (Butterworths, London), pp. 187–196.

Subject Index

1-aminocyclopropane-1-carboxylic acid (ACC) 258, 261
1-MCP (1-Methyl-cyclopropene) 273
^{14}C-assimilates 205
$^{14}CO_2$ feeding experiments 208
^{32}P uptake 149
6-benzylaminopurine, BAP 178, 184, 185

Abscisic acid (ABA) 188, 253
ACC
    oxidase 258, 263
    synthase 258, 263
acclimatization of plantlets 7, 288
Acetylsalicyclic acid 276
acid phosphatase 253
acidity fluctuation 54, 69, 75
alternative
    oxidase 95
    respiration 95
amino acids 159
aminoethoxyvinylglycine (AVG) 258, 261
Aminooxyacetic acid (AOA) 113, 258
ammonium 143
    nitrate 143
    uptake 155, 157
anther 174
anthesis 248
anthocyanins 247
anti-auxin 178
AOA (aminooxyacetic acid) 113
APCS (Automated Plant Culture System) 311
ascorbic acid oxidase 121
ASEAN 2, 5
aspartate 39
Assimilate
    import 204
    partitioning 198
    transport 198
Astomatal CAM 61
asymbiotic germination 2
ATP 201
Automated Plant Culture System (APCS) 311
auxin 178

β-carboxylation 47, 64
batch culture 310, 311
bioassay 178
bud
    blasting 188
    drop 189
    opening 252
    primordia 170
    removal 189
bud-opening solution 271

C/N ratio 142
C_2 (photorespiratory) cycle 118
C_3
    orchids 77, 293
    pathway 38
    photosynthesis 47, 293
    plants 37

C_4
  pathway 45
  plants 38, 42, 47
CAM (Crassulacean Acid Metabolism) 7
  orchids 75, 77, 293
  pathway 39–41
  plants 47, 50, 118
carbohydrate metabolism 110
catalase 122
chemical regulation 184
chicken manure 138
chlorophyll a/b 45
chloroplast 58
circadian rhythm 103
climacteric rise 113
clonal
  growth 205
  propagation 2, 6
closed system 300
CO_2
  compensation points 46, 118
  concentrating mechanism 40
  diffusion 64
  enrichment 191, 192, 292, 302, 308
Cobalt ions 258, 296
commercial
  fertilisers 138
  holding or bud-opening solutions 275
control of flowering 178
controlled storage 277
cotyledon 96
Culture
  Pack 302
  vessels 289
cuticular transpiration 267
cyanide-resistant
  pathway 95
  respiration 113
Cymbidium Mosaic Virus 84
cytochrome oxidase 95
cytokinin 176, 253, 260
dark respiration 93, 118
day neutral plants 177
day/night temperature 174
decapitation 177, 183
deficiency symptoms 131
deformed flowers 185
dehydrozeatin riboside 176
desiccation 260
development of flowers 168
diffusion of ethylene 303
dormancy 174
drought 75–77
  tolerance 58

elements
  macro- 129
  micro- 129
elevated CO_2 85
emasculation (removal of pollen) 103, 254
Embden Meyerhoff Parnass (EMP) 94
endophytic fungus 121
endosperm 96
epicuticular waxes 308
epiphytic orchids 13
ethylene 254, 293
  action 266
  biosynthesis 258, 262
    inhibitor 256
  sensitivity 264
exodermis 27, 159
explants 7, 290

fertiliser programmes 152
fibres
  optical 299
films
  fluorocarbon polymer 302
floral development 168
flow culture system 292

Flower structure 15–20
  evocation 168
  induction (or flower initiation) 168, 183
    application of 183
  longevity 101
  production peak 189
  senescence 113, 245
    pollination-induced 264
  size 144
flowering 180
  chemical regulation 184
  control 178
  gradient 177
fluorocarbon polymer films 302
foliar application 149
fragrance 105
fragrance production 106
fungal infection 97

GA (see gibberellin)
Gas-permeable
  culture system 300
  membrane 303
GDH (glutamate dehydrogenase) 157
germination 97
gibberellin 174, 176
glucan 99
glucose 99
glutamate dehydrogenase (GDH) 157
glutamine synthetase (GS) 157
glycolic acid oxidase 48, 118, 120
  activity 47
glycolysis 94
glyoxysomes 96, 97, 123
gradient
  flowering 177
'ground' orchids 139
growth habit 11
GS (glutamine synthetase) 157
gynostemia 101

harvestable yield 191, 228
headspace 293
heteroblastic 13
holding solution 271
homoblastic 13
hydroxylquinoline citrate (HQC) 274
Hydroxylquinoline sulphate (HQS) 274
Hypobaric storage 277

import assimilates 204
In vitro
  cultured plantlets 309
  flowering 312
infection
  fungal 97
invertase 99, 176
ion antagonism 145
isoenzyme patterns 122

juvenility 170

keeping quality 268
Kranz anatomy 38
Kreb's cycle 94

lan 1
leaf transpiration 77
leafy orchids 52, 53, 106
life span 101
Light compensation point 68
light emitting diodes (LEDs) 299
light respiration 118
longevity 254
Low-temperature storage 277

Macro-elements 129
malate 39
malonate 111
market potential 2
mass clonal propagation 310

methionine sulfoximine (MSX) 157
metabolism
    carbohydrate 110
micro-elements 129
microporous polypropylene membranes 306
Milliseal™ 303
mineral composition 132
mineral requirements 129
mitochondria 97, 117
Monod relation 291
monopodial 11
mortality rates 7, 288
mulch 142
mulching 139, 161
Münch hypothesis 199
Mycorrhizal fungi 26

N : P : K ratio 136, 137
Neoflon film 303
net photosynthesis 59
nitrate
    activity 297
    reductase (NR) 157
    uptake 152
nitrogen sources 296
nocturnal CO_2 uptake 77

Odontoglassum Ringspot Virus (ORS) 85
ontogeny 74
optical fibres 299
orchid
    cultivation 2
    cut-flowers 117
    industry 2, 6, 198
    production cycle 5
    physiology 5, 9
orchids
    epiphytic 13, 24
    'ground' 139
    leafy 52, 106
    shootless 52, 60
    terrestrial 26
    thick-leaved 49
    thin-leaved 45, 118
organic manure 138
ORSV (Odontoglossum Ringspot Virus) 85
osmotic potential 75
oxaloacetate 39
oxidases 120
oxygenase 39, 120

P/O ratio 113
Paclobutrazol 188
passage cells 159
pentose phosphate shunt 95
PEP (phosphoenolpyruvate) 39
PEPC 39
peroxidase 121
peroxisomes 118
phloem loading 199, 200
phosphoenolpyruvate (PEP) 39
Phosphoglycerate (PGA) 47
photoassimilates 168
photoautotrophic micropropagation 300
photoperiodism 170
photorespiration 118
photosynthesis
    net 59
    pathways 37
    regenerative 54, 66
photosynthetically active radiation (PAR) 289
phyllotaxy 203
Physan 274
pig manure 139
plant morphogenesis 308
plasticity of CAM 77
pollination 101, 254, 255
pollination-induced flower senescence 264
pollutants 82

polyethylene glycol 75
polyphenol oxidase 121
post-harvest storage 117
post-illumination outburst 45
Post-pollinated phenomena 101, 254
postharvest handling 269
preharvest conditions 269
pseudobulb 13, 62, 170, 208, 228
pulsing 271
pyruvate dehydrogenase 97

quantum yield 69

ratio
    C/N 142
    N : P : K 136
    P/O 113
    root/shoot 310
RC (Respiratory Control) 113
relative growth rates 297
Respiration
    alternative 95
    cyanide-resistant 95
    dark 93, 118
    light 118
respiratory
    control 113
    drift 109
    pathways 95
    quotient 110
    rhythm 103
    substrate 110, 267
resupination 17
rockwool 307
root
    Aerial root 23, 54
    feeding 149
    respiration 106
    structure 27
    terrestrial 26, 55
root/shoot ratio 310
RQ 103, 110
RUBISCO 118–120
RUBPC 37, 40
RUBPC/PEPC ratio 68

SAM 258
seasonality of flowering 179
seeds
    germination 97, 98
    synthetic 314
Semi-diffusion forced ventilation (SDFV) system 305
senescence 245
sensitivity to ethylene 264
shootless orchids 52, 60, 199
sidewide lighting 297
silver
    nitrate 271
    thiosulphate 272
sink 198
    activity 160, 209
    demand 78, 87
sink to source transition 208
sink-limited 228
Sodium
    fluoride 111
    thiosulphate 271
somaclone 236
source-limited 228
staggering
    of flower production 189, 191
starch 15, 40
stomata 13, 21, 61, 267
stomatal transpiration 267
storage
    controlled 277
    hypobaric 277
    low-temperature 277
    post-harvest 117

technology 276
subculturing 7
sucrose 99
sucrose
synthase 176
sugars on orchid tissue cultures 290
symplastic or apoplastic pathway 200
sympodial 11
synthetic seeds 314
systems
flow culture 292
gas permeable culture 300
semi-diffusion forced ventilation (SDFV) 305

teliosome 160
terrestrial roots 26, 55
thick-leaved orchids 49, 215, 220
thin section culture 313
thin-leaved orchids 45, 118
tissue
analysis 130
culture 288
titratable acidity 50, 69, 73
Tobacco Mosaic Virus 84
Tolerance
of drought 59
Transpiration
cutricular 267
leaf 77
stomatal 267

Vacin and Went media 153
Values
$\delta^{13}C$ 41, 66
vascular geometry 203
vase-life
extension 270
seasonal variation 269
velamen 27, 159
vernalization 170
virus infection 87
vitrification 308

water
use efficiency 54
loss 267
stress 75
waxes
epicuticular 308

yield
harvestable 191, 232

zeatin riboside 176

Plant Index

Aeridachnis Bogor 106
Agave 75
Angraecum giryamae 20
Arachnis 33, 64–69, 120, 121, 139, 260
Arachnis Maggie Oei 51, 106, 120, 139, 177, 191, 253
Aranda 13, 23, 24, 26, 49, 69, 75, 77, 99, 101, 104, 106, 109–111, 113–115, 123, 132, 137–139, 145, 149, 151, 161, 177, 180, 184, 185, 192, 193, 204, 226, 232, 253, 264, 269–272, 291, 308
Aranda Christine 109, 180, 184, 185, 189, 191, 269
Aranda Deborah 139, 177, 184, 314
Aranda Hilda Galstan 177
Aranda Kooi Choo 134, 184
Aranda Lucy Laycock 177
Aranda Meiling 177
Aranda Nancy 139, 177
Aranda Noorah Alsagoff 134, 137, 149, 178, 220, 226
Aranda Peter Ewart 184
Aranda Tay Swee Eng 153, 220
Aranda Wendy Scott 51, 101, 106, 137, 142, 177
Aranthera Beatrice Eng 184, 188
Aranthera James Storie 51, 106, 139, 178, 184
Arundina 85, 101, 252, 253
Arundina graminifolia 45, 47, 51, 68, 73, 122, 252
Ascocentrum 33
Boronia 176
Brassavola nodosa 105
Bromheadia 159
Bromheadia finlaysoniana 155

C. canaliculatum (see *Cymbidium canaliculatum*)
C. madidum (see *Cymbidium madidum*)
Cattleya 59, 61, 99, 120, 131, 132, 134, 137, 143, 145, 149, 169, 178, 260, 297, 306
Cattleya aurantica 96
Cattleya Bow Bell 51
Cattleya mossiae 101
Cattleya skinneri 101
Chiloschista phyllorhiza 61
Cicer arietinum 204
Citrus sinensis 228
Coelogyne mayerianna 51
Coelogyne rochussenii 51
Cymbidium 2, 8, 26, 47, 49, 64, 68, 101, 121, 132, 134, 137, 159, 169, 173, 174, 178, 188, 192, 193, 234, 253, 260, 264, 291, 306, 314
Cymbidium canaliculatum 45, 49
Cymbidium ensifolium 188
Cymbidium faberi 253
Cymbidium lowianum 101, 102, 122
Cymbidium madidum 45, 47, 49, 52
Cymbidium Reporsa 293

Cymbidium roseum 174
Cymbidium sinense 23, 68, 77, 122, 143

Dactylorhiza 121
Dactylorhiza purpurella 121
Dendrobium 2, 17, 24, 30, 64, 69, 73, 75, 99, 138, 139, 151, 159, 161, 169, 177, 178, 183–185, 188, 189, 192, 204, 215, 220, 226, 234, 249, 264, 269, 270, 272, 274, 278, 290–292, 306, 308, 314
Dendrobium bigibbum 188
Dendrobium crumenatum 13, 174
Dendrobium Jacquelin Concert × Jester 185
Dendrobium Jacquelyn Thomas 181
Dendrobium Jashika Pink 220, 226
Dendrobium Lousiae Dark 184
Dendrobium Madam Uraiwan 185
Dendrobium Mary Mak 185
Dendrobium Multico White 153
Dendrobium nobile 122, 170, 174
Dendrobium phalaenopsis 131, 170
Dendrobium Pompadour 248, 249, 269–271, 274–277
Dendrobium Rong Rong 215, 220, 226
Dendrobium Sri Siam 188
Dendrobium superbum 20
Dendrobium taurinum 51, 73
Dendrobium Youppadeewan 274
Disa polygonoides 97
Disperis fanniniae 97
Doriella Tiny 313

Encyclia tampensis 64, 82
Epidendrum elongatum 84, 85
Epidendrum regidum 82
Eulophia graminea 30

Galeola septentrionalis 98

Habenaria 26

Holttumara Loke Tack Tip 184
Holttumara Maggie Mason 178
Kalanchoe daigremontiana 49

Laeliocattleya 132, 134
Laeliocattleya Aconcagua 134

Mokara 13, 23, 69, 85, 138, 142, 192, 193, 264, 308
Mokara Chark Kuan 184
Mokara White 33, 85, 293

Oncidium 2, 24, 63, 68, 138, 139, 161, 184, 192, 193, 204, 226, 232, 234–236, 276,
Oncidium flexuosum 51
Oncidium Goldiana 13, 23, 33, 45, 62, 63, 66, 68, 69, 85, 101, 104, 138, 181, 189, 191, 205, 208, 215, 226, 228, 235, 236, 238, 248, 253, 256, 271, 274
Oncidium Gower Ramsey 153, 185, 205
Oncidium hybrids 205
Oncidium Norman Gaunt 20
Oncidium Taka 205
Ophyrs 121

Paphiopedilum 169
Paphiopedilum barbatum 174
Phalaenopsis 2, 8, 62, 64, 69, 75, 77, 81, 132, 134, 137, 144, 149, 174, 176, 188, 192, 193, 235, 260, 264, 307, 314
Phalaenopsis amabilis 122, 174, 176, 192
Phalaenopsis aphrodite 253
Phalaenopsis Dos Pueblos 134
Phalaenopsis schilleriana 173, 192
Polystachya culiviformis 173

Rhizoctonia 121

Sophrolaeliocattleya 84, 85

Spathoglottis 26, 314
Spathoglottis plicata 23, 28, 33, 51, 68, 73

Taeniophyllum malianum 61

Vanda 13, 24, 28, 33, 64, 69, 104, 178, 189, 191, 193, 277, 278
Vanda hookerana 33
Vanda Miss Joaquim 20, 33, 81, 144, 177, 178, 189, 191, 277
Vanda Petamboeran 258
Vanda Rose Marie 258
Vanda suavis 20
Vanda teres 33
Vanilla 131